Integrated Systems, Design and Technology 2010

Madjid Fathi (Ed.), Alexander Holland,
Fazel Ansari, and Christian Weber (Co-Eds.)

Integrated Systems, Design and Technology 2010

Knowledge Transfer in New Technologies

Prof. Madjid Fathi (Editor)
Institute of Knowledge Based Systems
and Knowledge Management
University of Siegen
Hölderlinstrasse 3, 57068, Siegen,
Germany
E-mail: fathi@informatik.uni-siegen.de

Fazel Ansari (Co-Editor)
Institute of Knowledge Based Systems
and Knowledge Management
University of Siegen
Hölderlinstrasse 3, 57068, Siegen,
Germany
E-mail: ansari@kmis.uni-siegen.de

Alexander Holland (Co-Editor)
Institute of Knowledge Based Systems
and Knowledge Management
University of Siegen
Hölderlinstrasse 3, 57068, Siegen,
Germany
E-mail: alex@informatik.uni-siegen.de

Christian Weber (Co-Editor)
Institute of Knowledge Based Systems
and Knowledge Management
University of Siegen
Hölderlinstrasse 3, 57068, Siegen,
Germany
E-mail: weber.isdt@uni-siegen.de

ISBN 978-3-642-17383-7

e-ISBN 978-3-642-17384-4

DOI 10.1007/978-3-642-17384-4

© Springer Science + Business Media B.V. 2011

This work is subject to copyright. All rights are reserved, whether the whole or part of the material is concerned, specifically the rights of translation, reprinting, reuse of illustrations, recitation, broadcasting, reproduction on microfilm or in any other way, and storage in data banks. Duplication of this publication or parts thereof is permitted only under the provisions of the German Copyright Law of September 9, 1965, in its current version, and permission for use must always be obtained from Springer. Violations are liable to prosecution under the German Copyright Law.

The use of general descriptive names, registered names, trademarks, etc. in this publication does not imply, even in the absence of a specific statement, that such names are exempt from the relevant protective laws and regulations and therefore free for general use.

Typesetting: Data supplied by the authors

Cover Design: Scientific Publishing Services Pvt. Ltd., Chennai, India

Printed on acid-free paper

9 8 7 6 5 4 3 2 1

springer.com

This book is dedicated to the memory of
Prof. Caro Lucas

Preface

Knowledge creation and technological experiences resulting from modern production life cycles are definitely the most economical and important intellectual capitals in the current manufacturing endeavors. These are together also the basis for enabling industrial competition through managing and identifying organizational and product related needs and opportunities; e.g. health care systems society needs clean environment, sustainable production life cycles needs flexible approachable design and engineering of materials whilst valuable materials are needed for renewable energies and the production of fuel-cells. These efforts requires conceptual modeling techniques, integrating tools and the right managing knowledge, which can result from human expertise based on experiences and human creativities. Integration of components, design of structures and managing knowledge inherent in engineering is a difficult and complex endeavor. A wide range of advanced technologies such as smart materials and their approaches in alternative energy have to be invoked in providing assistance for knowledge requirements ranging from acquisition, modeling, (re)using, retrieving, sharing, publishing and maintaining of knowledge. Integration, design and management with regards to Knowledge Management originate at least on three roots.

Firstly, suppliers of new and existing technology and academics in this field have developed and created opportunities of supporting knowledge tasks by system solutions. Secondly, organization and human relations professionals have recognized the need for using the opportunities of a highly increasing educated work force in modern societies. Finally, strategic management has recognized that the optimal use of intellectual capabilities may be the best source for sustaining competitiveness in the global economy.

Definitely it is clear that we will approach knowledge technologies for industrial organization. However we need novel and sustainable methodologies to integrate acquired human resource values and experiences beyond existing IT-tools and most importantly we have to prepare new ways of sharing and transferring of knowledge and technology. For this reason, we need interdisciplinary understanding, which helps us to reach out for the next levels of innovation.

The annual International Joint Conference on Integrated Systems, Design and Technology (ISDT), born from this vision, is a meeting with this conceptual ideology based on integration of strategies for developing complex systems and characterized by its multidimensional efforts, ideas and concepts to observe and handle high complexities in different but interdisciplinary connected fields of application.

We intend to bring different researchers to share their knowledge and lab-experiences to accomplish a technology matching for future societies to fill available "know-do" gaps. For this sake we have invited, met and gathered through the years; distinguished researchers with the special intention to know about the close requirements of technologies to expand together our horizons in the understanding of the needs for knowledge transfer among our colleagues in the fields of Alternate Energy, Composite Engineering, Micro-Electronics and Computational Intelligence.

To strive towards these lines of vision this volume is composed of three sections building an interdisciplinary triangle of integrated systems, design and technology.

- Advanced materials and design
- Advanced energy utilization
- Computational and intelligent systems engineering

It is my sincere hope that this publication is a new step towards interdisciplinary thinking, resulting in new works fueling the knowledge transfer in new technologies.

July 2010

Madjid Fathi

Contents

Advanced Materials and Design

Using Homogenization by Asymptotic Expansion for Micro – Macro Behavior Modeling of Traumatic Brain Injury 3
Y. Remond, R. Abdel Rahman, D. Baumgartner, S. Ahzi

Diamond Coating of Hip Joint Replacement: Improvement of Durability 13
Z. Nibennaoune, D. George, S. Ahzi, Y. Remond, J. Gracio, D. Ruch

CAPAAL and CAPET – New Materials of High-Strength, High-Stiff Hybrid Laminates 23
Bernhard Wielage, Daisy Nestler, Heike Steger, Lothar Kroll, Jürgen Tröltzsch, Sebastian Nendel

Challenges in the Assembly of Large Aerospace Components 37
Mozafar Saadat

Improvement of Thermal Stability and Fire Behaviour of pmma by a (Metal Oxide Nanoparticles/Ammonium Polyphosphate/Melamine Polyphosphate) Ternary System ... 47
B. Friederich, A. Laachachi, M. Ferriol, D. Ruch, M. Cochez, V. Toniazzo

Investigation of Mechanical Properties and Failure Behaviour of CFRP, C/C and C/C-SiC Materials Fabricated by the Liquid-Silicon Infiltration Process in Dependence on the Matrix Chemistry 59
Bernhard Wielage, Daisy Nestler, Kristina Roder

Investigation of Polymer Melt Impregnated Fibre Tapes in
Injection Moulding Process 67
Jürgen Tröltzsch, Lothar Kroll

Numerical Analysis of Relaxation Test Based on Prony
Series Material Model 79
*Mahdi Mottahedi, Alexandru Dadalau, Alexander Hafla,
Alexander Verl*

Powder Metallurgy of Particle-Reinforced Aluminium
Matrix Composites (AMC) by Means of High-Energy Ball
Milling 93
*Daisy Nestler, Steve Siebeck, Harry Podlesak, Swetlana Wagner,
Matthias Hockauf, Bernhard Wielage*

Initial Stress Behaviour of Micro Injection-Moulded
Devices with Integrated Piezo-Fibre Composites 109
*Lothar Kroll, Marco Walther, Wolfgang Nendel, Michael Heinrich,
Matthias Klärner, Jürgen Tröltzsch*

Development of Particle-Reinforced Silver-Based Contact
Materials by Means of Mechanical Alloying 121
Bernhard Wielage, Thomas Lampke, Daisy Nestler, Heike Steger

Design of Sports Equipment Made from Anisotropic
Multilayer Composites with Stiffness Related Coupling
Effect 133
J. Kaufmann, L. Kroll, E. Paessler, S. Odenwald

Structural Optimization of Fibre-Reinforced Composites
for Ultra-Lightweight Vehicles 143
Bernhard Wielage, Tobias Müller, Daisy Nestler, Thomas Mäder

Optimisation and Characterisation of Magnetoelastic
Microsensors 155
Bernhard Wielage, Thomas Mäder, Daisy Nestler

Advanced Energy Utilization – Fuel Cell

Challenges and Opportunities in PEM Fuel Cell Systems 171
Amir M. Niroumand, Mehrdad Saif

An Overview of Current Research Activities on PEM Fuel
Cell Technology at Florida Atlantic University 185
Amir Abtahi, Ali Zilouchian

Computational and Intelligent Systems Engineering

BELBIC and Its Industrial Applications: Towards
Embedded Neuroemotional Control Codesign 203
Caro Lucas

Intelligent 3D Programmable Surface 215
Michael Fielding, Samer Hanoun, Saeid Nahavandi

Manage Competence Not Knowledge 227
A.G. Hessami, M. Moore

Entropy Measure and Energy Map in Machine Fault
Diagnosis 243
R. Tafreshi, F. Sassani, H. Ahmadi, G. Dumont

Enterprise Information Management for the Production of
Micro- and Nanolayered Devices 267
Rainer Brück

An Interdisciplinary Study to Observe Users' Online
Search Tasks and Behavior for the Design of an IS&R
Framework 279
I-Chin Wu

A Generic Knowledge Integration Approach for Complex
Process Control 293
Stefan Berlik

A Novel Hybrid Adaptive Nonlinear Controller Using
Gaussian Process Prior and Fuzzy Control 301
*H.R. Jamalabadi, F. Boroomand, C. Lucas, A. Fereidunian,
M.A. Zamani, H. Lesani*

Image Based Analysis of Microstructures Using Texture
Detection and Fuzzy Classifiers 313
Lars Hildebrand, Thomas Ploch

Fulfilling the Quality Demands of the Automotive Supply
Chain through MES and Data Mining 321
Ralf Montino, Peter Czerner

Evaluation of Professional Skill and *Kansei* Based on
Physiological Index: Toward Practical Education Using
Professional *Kansei* 331
Koji Murai, Yuji Hayashi

WLoLiMoT: A Wavelet and LoLiMoT Based Algorithm for
Time Series Prediction 345
Elahe Arani, Caro Lucas, Babak N. Araabi

Author Index.. 357

Advanced Materials and Design

Using Homogenization by Asymptotic Expansion for Micro – Macro Behavior Modeling of Traumatic Brain Injury

Y. Remond, R. Abdel Rahman, D. Baumgartner, and S. Ahzi

University of Strasbourg, CNRS, 2 rue Boussingault., 67000 Strasbourg, France
The French University in Egypt, El Chourouq City, Cairo, Egypt
remond@unistra.fr, rania.elanwar@gmail.com,
baumgartner@unistra.fr, ahzi@unistra.fr

Abstract. Bridging veins are frequently damaged in traumatic brain injury. These veins are prone to rupture in their subdural portion upon head impact, due to brain-skull relative motion, giving rise to an acute subdural hematoma. To understand the biomechanical characteristics of this hematoma, we take into account the periodical distribution of bridging veins in the sagittal plane. This allowed the use of the method of homogenization by asymptotic expansion to calculate the homogenized elastic properties of the brain-skull interface region. The geometry of this zone was simplified and a representative volume element was developed comprising: sinus, bridging vein, blood circulating inside them, surrounding cerebrospinal fluid and tissues. The heterogeneous elementary cell is transformed to an anisotropic homogenous equivalent medium and the homogenized elastic properties were calculated. The macroscopic homogenized properties resulted from the current study can be incorporated into a finite element model of the human head at macroscopic scale. The main results of this homogenization theory are the calculation of the local stress field into the elementary cell, as well as its homogeneous anisotropic properties at the macroscopic scale.

Keywords: Subdural hematoma, homogenization, mechanical behavior, brain – skull interface, bridging veins.

1 Introduction

ASDH is a potentially devastating, yet curable extra axial fluid collection within the potential subdural space. It is classically associated with head trauma including rapid acceleration and deceleration that produce rupture of cortical arteries but most often tearing of BVs as they cross from the brain surface to dural sinus as mentioned by [1], [2] and [3]. These veins are prone to rupture at the point of their entry into the SSS according to [4] and [5]. Moreover, due to the histological composition of the BVs, the subdural portion of a BV is more fragile than the subarachnoidal portion as explained by [6] and [7]. The rupture of BVs is induced by the brain-skull relative motion following a head impact as described by neuropathologists and by experimental work conducted in vivo by [8], [9] and [10]. In addition, [11] had the same observations as they studied the brain-skull relative motion in vitro.

Owing the previous facts, motion between the brain and the skull has been considered potentially important to head injury. That is why a primary concern in finite element modeling of the human head has been the way of modeling the constituents of the brain-skull interface. Another important issue was the selection of materials properties as pointed by [12]. The use of normal finite element methods to find the local behavior at each point of a complicated biological structure such as brain-skull interface region being very difficult, the use of homogenization methods on mesoscopic scale is becoming indispensable. The aim of the present work is to find the homogenized mechanical properties of the segment representing the junction between the SSS and the BV and the surrounding constituents. This zone will be considered as a heterogeneous medium which will be transformed to another one mechanically equivalent anisotropic and homogeneous. The arrangement of BVs being almost periodic along the sinus, the method of homogenization by asymptotic expansion will be used. The calculations are effectuated on a representative volume element (RVE) with a simplified geometry. To achieve this goal, the anatomy, the geometrical arrangement and the mechanical properties of the tissues forming the segment in question must be well defined.

2 Material Description

The investigation zone represents the brain – skull interface region in the location of the superior sagittal sinus. This zone comprises the sinus, bridging veins, blood circulating inside them, and the surrounding cerebrospinal fluid and tissues. The tissues dimensions used in the current study were extracted from the work [6], [12], [13], [14] and [15]. Also, the mechanical properties were found in the literature in the work of [16], [17], [18] and [19]. The chosen geometry and mechanical properties are illustrated in table (1).

Table 1. Geometry and mechanical properties of the RVE constituents

Tissues	Geometry (mm)	Mechanical properties	
		Elastic modulus (E) (Mpa)	Poisson's ratio (υ)
SSS	Thickness = 0.5	31.5	0.45
BVs	Diameter = 1.5	$E_l = 6.43$	$\upsilon_{lt} = 0.385$
	Wall thickness = 0.25	$E_t = 2.4$	$\upsilon_{tt} = 0.49$
Subarachnoid space (CSF)	Thickness = 1.5	0.012	0.49
Blood		0.012	0.49

Using Homogenization by Asymptotic Expansion 5

All the materials in the current study were considered as linear elastic. The blood, in the current study, being an incompressible fluid enclosed in a membrane, it was suitable to choose an elastic modulus to describe its behavior. All the constituents were considered as isotropic materials while the BV was treated as orthotropic material composed of collagen fibers surrounded by a matrix of elastin.

3 Periodicity Problem

BVs can be considered as symmetric on both sides of the SSS and evenly distributed along the axe of the sinus in the sagittal plane as shown in figure 1(a). So, we consider the distribution of BVs to be almost periodic in sagittal direction. Thus we can apply the homogenization method by asymptotic expansion in our current study.

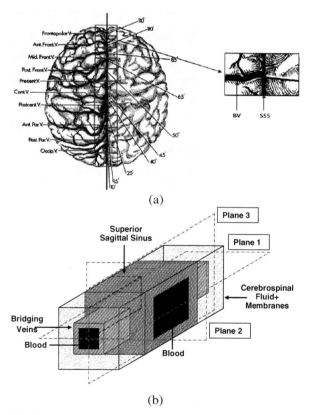

Fig. 1. (a) Distribution of BVs along the SSS as shown by [20] with a zoom on the SSS – BV junction (b) The Representative Volume Element

The investigation zone will be reduced to the simple geometrical 3D unit cell shown in figure 1 (b). This unit cell is chosen to be the representative volume element (RVE) used in homogenization calculations.

4 Homogenization by Asymptotic Expansion

If L is the characteristic dimension of the structure, then ηL will define the size of the RVE. If X is the slow variable reflecting the variation of properties magnitude on the structure scale, Y is the local variable describing the rapid variation on the RVE, then X and Y are related together through the relation:

$$Y = \frac{X}{\eta} \tag{1}$$

If $\vec{U}(X,Y)$ is the displacement field, thus it may be expressed in term of infinite series expansion as done by Sanchez et al. (1974), Ladeveze et al. (1985), Dumont et al. (1987), and Devries et al. (1989):

$$\vec{U}(X,Y) = \vec{U}_0(X,Y) + \eta \vec{U}_1(X,Y) + \eta^2 \vec{U}_2(X,Y) + \dots \tag{2}$$

Where $\vec{U}_1$ is the solution of a problem which the variational formulation is $\vec{U}_1 \in W(M)$, $W(M)$ is the displacement space. By minimizing over $W(M)$:

$$\vec{U} \rightarrow \frac{1}{2} \int_M K_{ijkl} \, \varepsilon_{y_{ij}}(\vec{U}) \, \varepsilon_{y_{kl}}(\vec{U}) dM + \int_M K_{ijkl} \, \varepsilon_{x_{ij}}(\vec{U}_0) \, \varepsilon_{y_{kl}}(\vec{U}) dM \tag{3}$$

M is the unit cell (RVE), K represents the local characteristics
$\vec{U}/U^+ = U^-$, U^+ and U^- represent the values of $\vec{U}$ on 2 opposites faces of the RVE.

On the other hand, it is possible to find $\vec{U}_1$ by:

$$K_{ijkl} \, \varepsilon_{y_{kl}}(\vec{U}_1) = H_{ijkl} \, \varepsilon_{x_{kl}}(\vec{U}_0) \tag{4}$$

Where H is a symmetric linear operator similar to K.

The homogenized behavior of the periodic medium can be expressed in term of K^*, where,

$$K^*_{ijkl} = \frac{1}{vol(M)} \int_M (H_{ijkl} + K_{ijkl}) dM \tag{5}$$

Knowing that $\varepsilon_y(\vec{U}_1)$ and $\varepsilon_x(\vec{U}_0)$ are linearly dependant as shown in relation (3), $\vec{U}_1 = A_{rs}\varepsilon_{rs}(\vec{U}_0)$, $A_{rs} \in W(M)$. Thus the variational formulation (3) becomes: Find $A_{rs} \in W(M)$ by minimizing the following relations:

$$\vec{U} \to \frac{1}{2} \int_M K_{ijkl} \, \underset{y\,ij}{\varepsilon}(\vec{U}) \, \underset{y\,kl}{\varepsilon}(\vec{U}) dM + \int_M K_{rskl} \, \underset{y\,kl}{\varepsilon}(\vec{U}) dM \tag{6}$$

$$\underset{ijrs}{K^*} = \frac{1}{vol(M)} \int_M (K_{ijrs} + K_{ijkl}\varepsilon_{kl}(\vec{U})) dM \tag{7}$$

The charging cases are separately treated and a double formulation is applied on the unit cell as follows:

- A displacement formulation is used in the case of normal loading.
- A stress formulation is applied to find the values of the shear coefficients.

First we treated the case of displacement field affine per part and continue in normal loading situations on the space $W_0(M)$, where $W_0(M) = \{\vec{U}/U_1 = 0|_{0,L_1}; U_2 = 0|_{0,L_2}; U_3 = 0|_{0,L_3}\}$ for $r,s = (1,1), (2,2), (3,3)$ and we will get the values of $\underset{ijkl}{K^*}$ where $i = j$. Second, for the shear coefficients, we used the duality on the displacement spaces, the stress field is assumed to be constant per part and continue on the interfaces between parts. The dual formulation in stress is: find $\bar{\sigma} \in W^*(M)$, where $W^*(M)$ are the spaces for statically admissible fields, by minimizing the form:

$$\frac{1}{2} \int_M Tr\left[\bar{\sigma} K^{-1} \bar{\sigma}\right] dM - \int_M Tr\left[\varepsilon(\vec{U}_0)\bar{\sigma}\right] dM \tag{8}$$

5 Local Stresses Calculation

The local stresses are given in the following form:

$$\sigma_{loc} = K\left\{\underset{x}{\varepsilon}(\vec{U}_0) + \underset{y}{\varepsilon}(\vec{U}_1).\frac{1}{\eta}\right\} \tag{9}$$

By taking the first two terms only of the development of the displacement field in (2) and by using the linear relation between $\vec{U}_0$ and $\vec{U}_1$ in (4), the form of the local stress becomes:

$$\sigma_{loc} = (H+K) \underset{x}{\varepsilon}(\vec{U}_0) \tag{10}$$

This is equivalent to:

$$\sigma_{loc} = K \underset{x}{\varepsilon}(\vec{U}_0) + K \underset{y}{\varepsilon}(\vec{U}_1) \tag{11}$$

Hence, the final forms of local stresses are:

$$\sigma_{11}^{loc} = K_{11}(\varepsilon_{11}^0 + \varepsilon_{11}') + K_{12}(\varepsilon_{22}^0 + \varepsilon_{22}') + K_{13}(\varepsilon_{33}^0 + \varepsilon_{33}') \tag{12}$$

$$\sigma_{22}^{loc} = K_{12}(\varepsilon_{11}^0 + \varepsilon_{11}') + K_{22}(\varepsilon_{22}^0 + \varepsilon_{22}') + K_{23}(\varepsilon_{33}^0 + \varepsilon_{33}') \tag{13}$$

$$\sigma_{33}^{loc} = K_{13}(\varepsilon_{11}^0 + \varepsilon_{11}') + K_{23}(\varepsilon_{22}^0 + \varepsilon_{22}') + K_{33}(\varepsilon_{33}^0 + \varepsilon_{33}') \tag{14}$$

$$\sigma_{23}^{loc} = 2K_{44}(\varepsilon_{23}^0 + \varepsilon_{23}') \tag{15}$$

$$\sigma_{13}^{loc} = 2K_{55}(\varepsilon_{13}^0 + \varepsilon_{13}') \tag{16}$$

$$\sigma_{12}^{loc} = 2K_{66}(\varepsilon_{12}^0 + \varepsilon_{12}') \tag{17}$$

ε^0 is the global strain in the REV

ε' is the local strain in tissues

$K_{11}, K_{22}, \cdots$ are the rigidity coefficients of different tissues

It must be noted that the values of $\sigma_{23}^{loc}, \sigma_{13}^{loc}, \sigma_{12}^{loc}$ correspond directly to $K_{44}^*, K_{55}^*, K_{66}^*$ respectively.

To conclude we can state that the local and global (homogenized) behavior of a given periodic medium may be found as follows in table (2):

Table 2. The local and global behavior of a given periodic medium

	Local behavior	Global behavior
Displacement	$\overrightarrow{U_0} + \eta\overrightarrow{U_1}$	$\overrightarrow{U_0} = \overrightarrow{U_0}(X)$
Deformation	$\underset{x}{\varepsilon}(\overrightarrow{U_0}) + \dfrac{1}{\eta}\underset{y}{\varepsilon}(\overrightarrow{U_1})$	$\underset{0}{\vec{\varepsilon}} = \underset{x}{\varepsilon}(\overrightarrow{U_0})$
Stress	$\sigma_{loc} = (H+K)\underset{x}{\varepsilon}(\overrightarrow{U_0})$	$K^* \underset{x}{\varepsilon}(\overrightarrow{U_0})$
Elastic coefficient	$K_{loc} = K(Y)$	$K^* = \dfrac{1}{V^*}\underset{M^*}{\int}(H + K)\, d\Omega$

6 Results

The numerical values of tissues elastic properties were substituted in the previous relations to get the values of the homogenized properties of the periodic structure. Then the following rigidity matrix, for orthotropic materials, was filled:

$$\begin{pmatrix} 2.56 & 1.46 & 1.38 & 0 & 0 & 0 \\ 1.46 & 9.15 & 3.9 & 0 & 0 & 0 \\ 1.38 & 3.9 & 7.6 & 0 & 0 & 0 \\ 0 & 0 & 0 & 1.2 & 0 & 0 \\ 0 & 0 & 0 & 0 & 1.2 & 0 \\ 0 & 0 & 0 & 0 & 0 & 1.2 \end{pmatrix}$$

An example of the resulting elastic modulii and Poisson's ratios are illustrated in table (3).

Table 3. The homogenized elastic modulii and Poisson's ratios

Homogenized Elastic and Shear Modulii (MPa)					Homognized Poisson's Ratios		
E_1	E_2	E_3	$G_{12} = G_{13}$	G_{23}	υ_{12}	υ_{13}	υ_{23}
1.54	6.9	5.7	0.6	0.6	0.1	0.13	0.1

It can be noticed that the studied RVE was highly anisotropic. The anisotropy governing the tensile behavior of our structure is deduced from the large difference existing between the values of the elastic modulii. On the other hand, the shear modulii in the three planes were always equal.

7 Conclusion

The brain-skull interface region in general and the SSS-BV complex in particular should be deeply investigated using modern computerized methods. These investigations will allow an accurate description of the mechanism of ASDH arising from BVs rupture. Homogenization methods should be attempted to predict the global and local responses of the complex biological structures through the construction of a more simplified models. The geometry and mechanical properties of the models constituents must be well defined to ensure obtaining results with an acceptable degree of certainty.

References

1. Maxeiner, H.: Detection of ruptured cerebral bridging veins at autopsy. Forensic Science International 89, 103–110 (1997)
2. Maxeiner, H., Wolff, M.: Pure subdural hematomas: a postmortem analysis of their form and bleeding points. Neurosurgery 50, 503–509 (2002)

3. Ly, J.Q., Sanders, T.G., Smirniotopoulos, J.G., Folio, L.: Subdural Hematoma. Military Medicine 171, 1–5 (2006)
4. Arslan, O.: Neuroanatomical basis of clinical neurology. Parthenon Publishing (2001)
5. Fenderson, B.A., Rubin, R., Rubin, E.: Linppincott's review of pathology illustrated interactive Q and A. Lippincott Williams and Wilkins (2006)
6. Yamashima, T., Friede, R.L.: Why do bridging veins rupture in the virtual subdural space? Journal of Neurology, Neurosurgery, and Psychiatry 47, 121–127 (1984)
7. Vignes, J.R., Dagain, A., Liguoro, D., Guérin, J.: A hypothesis of cerebral venous system regulation based on a study of the junction between the cortical bridging veins and the superior sagittal sinus. Journal of Neurosurgery 107, 1205–1210 (2007)
8. Willinger, R., Taleb, L., Kopp, C.M.: Modal and temporal analysis of head mathematical models. Journal of Neurotrauma 12(4), 743–754 (1995)
9. Bayly, P.V., Cohen, T.S., Leister, E.P., Ajo, D., Leuthardt, E.C., Genin, G.M.: Deformation of the human brain induced by mild acceleration. Journal of Neurotrauma 22(8), 845–856 (2005)
10. Sabet, A.A., Christoforou, E., Zatlin, B., Genin, G.M., Bayly, P.V.: Deformation of the human brain induced by mild angular head acceleration. Journal of Biomechanics 41, 307–315 (2008)
11. Hardy, W.N., Mason, M.J., Foster, C.D., Shah, C.S., Kopacz, J.M., Yang, K.H., King, A.I., Bishop, J., Bey, M., Anders, W., Tashman, S.: A study of the response of the human cadaver head to impact. Stapp Car Crash journal 51, 17–80 (2007)
12. Kleiven, S., Hardy, W.N.: Correlation of a finite element model of human head with local brain motion – consequences for injury prediction. Stapp Car Crash Journal 46, 123–144 (2002)
13. Lee, M.C., Haut, R.C.: Insensitivity of tensile failure properties of human bridging veins to strain rate: Implication in biomechanics of subdural hematoma. Journal of Biomechanics 22(6), 537–542 (1989)
14. Henrikson, R.C., Kaye, G.I., Mazurkiewicz, J.E.: Histology, National Medical Series for independent Study. Lippincott Williams and Wilkins (1997)
15. Bashkatov, A.N., Genina, E.A., Sinichkin, Y.P., Kochubey, V.I., Lakodina, N.A., Tuchin, V.V.: Glucose and mannitol diffusion in human dura mater. Biophysical Journal 85, 3310–3318 (2003)
16. Melvin, J.W., McElhaney, J.H., Roberts, V.L.: Development of a mechanical model of the human Head – Determination of tissue properties and synthetic substitute materials. SAE Paper, 700–903 (1970)
17. Ruan, J.S., Khalil, T.B., King, A.I.: Finite element modeling of direct head impact. SAE, 933–1114 (1993)
18. Huang, Y., Lee, M.C., Chiu, W.T., Chen, C.T., Lee, S.Y.: Three-dimensional finite element analysis of subdural hematoma. Journal of Trauma 47, 538–544 (1999)
19. Monson, K.L., Goldsmith, W., Barbaro, N.M., Manley, G.T.: Significant of source and size in the mechanical response of human cerebral blood vessels. Journal of Biomechanics 38, 737–744 (2005)
20. Rhoton, A.L.: The cerebral veins. Neurosurgery 51, 159–205 (2002)
21. Sanchez-Palencia, E.: Comportement local et macroscopique d'un type de milieux physiques hétérogènes. International Journal of Engineering Science 12, 331–351 (1974)

22. Ladeveze, P., Proslier, L., Remond, Y.: Reconstruction of a 3-D composite behaviour from a local approach. In: Conference paper. Metallurgical Society Inc., Warrendale (1985)
23. Dumont, J.P., Ladeveze, P., Poss, M., Remond, Y.: Damage mechanics for 3-D composites. Composites Structures 8(2), 119–141 (1987)
24. Devries, F., Dumontet, H., Duvaut, G., Lene, F.: Homogenization and damage for composite structures. International Journal for Numerical Methods in Engineering 7(2), 285–298 (1989)

Diamond Coating of Hip Joint Replacement: Improvement of Durability

Z. Nibennaoune, D. George, S. Ahzi, Y. Remond, J. Gracio, and D. Ruch

University of Strasbourg, CNRS, 2 rue Boussingault, 67000 Strasbourg, France
University of Aveiro, TEMA, 3810-193 Aveiro, Portugal
CRP Henri Tudor, DAMS, 4002 Esch-sur-Alzette, Luxembourg
nibzhor@yahoo.fr, george@unistra.fr, ahzi@unistra.fr,
remond@unistra.fr, jgracio@ua.pt, david.ruch@tudor.lu

Abstract. Despite the success of surgical implants such as artificial hip, materials used in these procedures still do not satisfy the demands of human life time functioning. Currently used materials such as titanium alloys, ceramics and polymers are degraded after about a dozen years of use. Diamond coating technology has proven to be efficient in the performance of human joints. In this study, we have investigated the deposition of diamond thin films on Ti6Al4V using a new developed time-modulated Chemical Vapour Deposition (TMCVD) method. Finite element simulations were used to analyse the development of residual stresses. Micro Raman spectroscopy was also used to evaluate the residual stresses and compare with numerical models.

Keywords: Hip joint replacement, Prosthesis, Durability, Diamond coating, Ti6Al4V, TMCVD, Residual stresses.

1 Introduction

As advances have been made in medical sciences, the aging population has increased significantly. More organs joints and other critical body parts will wear out and must be replaced to maintain a good quality of life for elderly people [1]. Some implants play a major role in replacing or improving the function of major body system (skeletal, circulatory, nervous...). Materials that are used in these replacements are by definition biomaterials. Some common implants include orthopaedic devices such as total knee, hip, shoulder, and ankle joint replacements; cardiac implants such as artificial heart valves; soft tissue implants such as breast implants; and dental implants to replace teeth/root systems and bony tissue in the oral cavity. In this work, we focus on hip joint replacements.

The human hip has to carry about 2.5 times the body weight of most full-grown humans and this for many decades. The hip joint has been optimized by nature for such a permanent stress with about 160 million cycles of alternating loads [2]. These forces and strains consist of the body weight and the stabilizing muscles.

Arthritis, obesity, osteoporosis, as well as extreme sport activities can have considerably negative consequences for this high-performance design.

Hip Joint Replacement or Total Hip Replacement is a surgery to replace all or part of the hip joint with an artificial device to restore joint movement (prosthesis). Hip joint replacement is mostly done in older people. The operation is usually not recommended for younger people because of the strain they can put on the artificial hip while ageing. The indications for the replacement of the hip joint include [3]:

- hip pain that has failed to respond to conservative therapy (medication for 6 months or more)
- hip osteoarthritis or arthritis confirmed by X-ray
- inability to work, sleep, or move because of hip pain
- loose hip prosthesis
- some hip fractures
- hip joint tumors

Titanium alloys are known for their light weight and high strength properties which make them suitable for applications ranging from aircraft structures to artificial human body parts [4]. Titanium alloys (Ti6Al4V) is used extensively in femoral prosthesis because of its high mechanical strength, low density, good corrosion resistance and relatively low elastic modulus [5]. However, their tribological behaviour has frequently been reported to be poor, due to a high coefficient of friction and poor performance under abrasive and adhesive wear [6]. The coating of this alloy with diamond [4] is of tremendous interest because of the coatings hardness, inertness, wear and corrosion resistance, low coefficient of friction, and biocompatibility. In all the coatings, the adhesion of the diamond film is a key to determine its usefulness.

For diamond coating, residual stresses are mainly a sum of the thermal and intrinsic stresses [7, 8]. Thermal stresses form during the cooling process after deposition and are due to the large difference of thermal expansion coefficients between the coating and the substrate. At room temperature, α diamond ~ 1.3×10^{-6} /K [9] and α Ti6Al4V ~ 8.6×10^{-6} /K [10]. The intrinsic stresses are due to structural mismatch between the film and the substrate and also the presence of impurities and structural defects in diamond coatings. The finite element method was used to simulate the cooling process of diamond thin films, between 0,5 μm and 8 μm, deposited at temperatures ranging from 600 °C to 900 °C. The film debonding and cracking is discussed and numerical results are compared with data available in the literature.

In this study, we have investigated the deposition of diamond films on Ti6Al4V using a new developed time-modulated Chemical Vapour Deposition (TMCVD) [11]. The methane content in the gas phase was time-modulated by changing its flow rate during the entire diamond deposition. The repetition of this procedure gives the name 'time-modulated. With the conventional hot filament (HFCVD) process, the methane concentration was kept constant during deposition). Generally, diamond deposition consists of two stages : the nucleation stage and the film growth stage. It is known that diamond nucleation density increases with CH_4 concentration. However, depositions at higher CH_4 concentrations deteriorate the crystalline morphology of the film [12]. The TMCVD process enables to modulate the CH_4 concentration in order to optimize both stages of the conventional CVD process.

In a second part of this work, we have studied the residual stresses that develop in the diamond coating of titanium. We have particularly used the finite element method to simulate the development of the residual stresses during cooling of the diamond coating.

2 Experimental

2.1 Substrate Pre-treatment and TMCVD Processes

Diamond films were deposited onto Ti6Al4V substrates (5 x 5 x 0.5 mm) using the TMCVD technique. The Ti6Al4V substrates were polished with Silicon carbide P # 4000 and then with diamond paste (15.0, 7.0 and 0.25 μm) before the film deposition. This was made in order to enhance the nucleation density. After each polishing step, the substrates were ultrasonically cleaned in acetone and ethanol for one minute in order to remove any loose abrasive particles. Prior to the diamond deposition, the samples were ultrasonically abraded for 40 minutes in diamond suspension in order to provide a uniform distribution of nucleation sites. The suspensions were prepared using 0.25 μm grain size of commercial diamond. One carat of diamond powder was mixed with 20 ml of pure methanol. In addition, the tantalum filaments were pre-carburized before the deposition process.

The reactor used for the deposition was a conventional stainless steel hot filament CVD (HFCVD) with a 0.5 mm diameter tantalum wire at a distance of 5 mm from the substrate after carburization. The pressure within the deposition chamber was 30 Torr. The gas mixture was 0.5 to 2.5 % methane in hydrogen. The hydrogen had a flow rate of 200 sccm. The filament temperature was kept above 1800 °C. The substrate temperature was measured by a thermocouple between 700 °C and 900 °C and mounted right below the molybdenum (Mo) substrate holder during the deposition process. The deposition time was 3 hours.

The CH_4 flow rate was modulated during deposition. At first, the methane is introduced into the CVD reactor at a high flow rate of 2.5 % for a 2.5 min of time. The methane flow rate is then lowered to 0.5 %. This second flow rate is maintained for a 15 min of time longer than the high one. This procedure is repeated several times during the CVD process. The repetition of this procedure gives the name 'time-modulated' (see Figure 1).

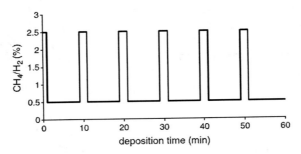

Fig. 1. Characterization of the time modulation of CH_4/H_2 ratio during the TMCVD process

2.2 Experimental Results

A number of specimens were made to evaluate the effect of the modulated times in the TMCVD process. For all specimens, it was observed that the shape and position of the Raman spectra depended on the gas flow modulation. The characteristic diamond line was always observed around 1332 cm^{-1} as shown on Figure 2 and another broader band between 1540 cm^{-1} and 1590 cm^{-1} corresponds to hybridised sp^2 amorphous carbon atoms. This larger band is not graphite since no second order signals between 2400 cm^{-1} and 2800 cm^{-1} is observed. The variation of this amorphous carbon phase is very different from sample to sample and leads to different residual stresses and diamond growth for each case. For more details see reference [11].

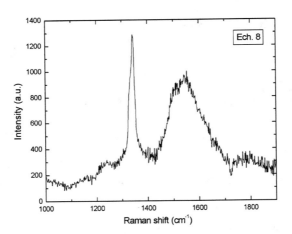

Fig. 2. Example of Raman experimental measurement.

The intensity and quality of the diamond peak requires adequate nucleation and diamond growth, both being dependant on the experimental parameters of the TMCVD method used. Various quality of deposited diamond crystals were observed depending on the variation of the experimental TMCVD parameters. Figure 3 shows the obtained diamond film of sample 8 above. In this case, the surface finish (diamond surface quality) was good but high residual stresses, up to -5.95 GPa, were measured.

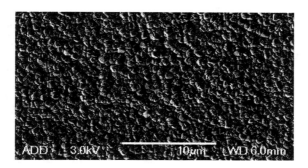

Fig. 3. MEB picture of the deposited diamond film on sample 8

Table 1 shows the evaluated residual stresses corresponding to the different Raman spectra. The Raman peak shape not being always a single peak (as shown on Figure 2 for sample 8), the theory developed by Ager III and Drory [13] was used to compute the split Raman shape residual stresses.

Table 1. Shape and position of the Raman spectra and evaluated residual stresses

Sample No	Raman peak shape	Raman peak position (cm^{-1})	Biaxial stress according to Raman shift (GPa)
3	split	1345.7	-5.43
		1332.0	-0.11
5	single	1339.9	-4.45
7	split	1342.0	-3.84
		1331.9	-0.08
		1326.9	+5.48
8	single	1342.5	-5.95

The calculated residual stresses showed various intensities and all being dependent on the experimental parameters. Similar stress intensities were found by Rats et al. [9] with compressive residual stresses between -4 GPa and -8 GPa for film thickness varying between 1 μm and 10 μm and for deposition temperatures between 500 °C and 900 °C.

3 FEM Simulations of Residual Stresses

The development of residual stresses in the TMCVD process is mainly coming from the cooling process after the film deposition. 2D and 3D elastic finite element models were developed. Each material was assumed isotropic and with a thermo-elastic behaviour. The diamond-substrate interface is supposed to be perfect. A detailed description of the analysis is provided in Nibennanoune et al. [14]. The analysis and residual stresses are assumed purely coming from the thermo-elastic material cooling.

Different temperatures, film thickness and geometries were used. In all cases, compressive residual stresses in the films were observed at the centre line of the geometry. Figure 4 shows the evolution of compressive residual stress as a function of the film thickness. As expected, the residual stress decreases with the increasing film thickness. The numerical model being purely elastic with perfect bonding, the residual stresses are in equilibrium and therefore the stresses present in the substrate will increase with the film thickness.

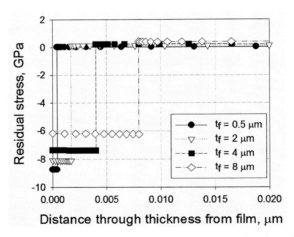

Fig. 4. Numerical tangential residual stresses at the centre line as a function of film thickness.

The film thickness being only between 0.5 μm and 8 μm, the stress intensities in the film vary between -6 GPa and -9 GPa. The sum of positive and negative residual stresses in the film and substrate is in full equilibrium.

In these analyses, we also observed important edge effect since there is a sharp stress gradient between the film and the substrate. Therefore, shear stresses will develop at the interface edges. These are presented in Figure 5 below.

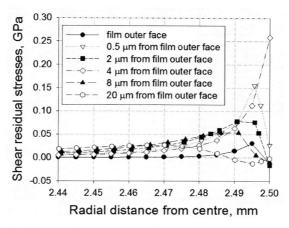

Fig. 5. Numerical shear residual stresses on the edges as a function of film thickness.

The maximum shear stress is observed near the film-substrate interface at about 250 MPa and decreases both towards the centre of the specimen and away from the film-substrate interface. This can, also explain why some fracture or delamination appears on the specimen edges when no damage is observed in the centre part.

The calculated intensity and distribution of shear residual stresses vary also with the film thickness and deposition temperature.

Finally, numerical results were compared with literature experimental data. This is presented in Figure 6.

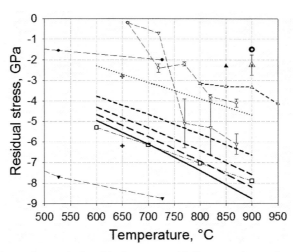

Fig. 6. Comparison of numerical residual stresses with literature data and associated legend.

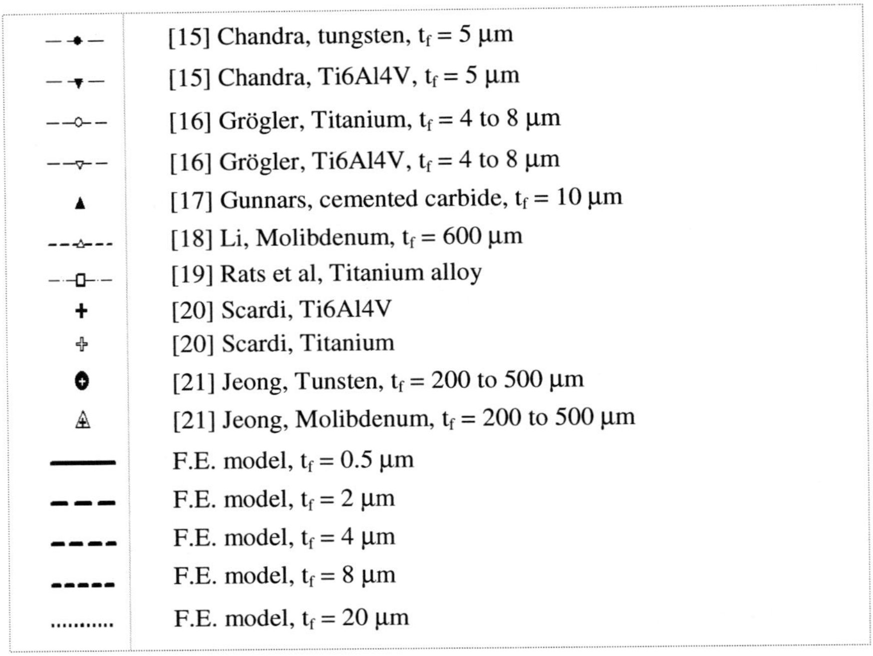

— ◆ —	[15] Chandra, tungsten, $t_f = 5$ μm
— ▼ —	[15] Chandra, Ti6Al4V, $t_f = 5$ μm
— ◇ —	[16] Grögler, Titanium, $t_f = 4$ to 8 μm
— ▽ —	[16] Grögler, Ti6Al4V, $t_f = 4$ to 8 μm
▲	[17] Gunnars, cemented carbide, $t_f = 10$ μm
— ◇ —	[18] Li, Molibdenum, $t_f = 600$ μm
— □ —	[19] Rats et al, Titanium alloy
✦	[20] Scardi, Ti6Al4V
✦	[20] Scardi, Titanium
◉	[21] Jeong, Tunsten, $t_f = 200$ to 500 μm
⬚	[21] Jeong, Molibdenum, $t_f = 200$ to 500 μm
——	F.E. model, $t_f = 0.5$ μm
− − −	F.E. model, $t_f = 2$ μm
− − − −	F.E. model, $t_f = 4$ μm
- - - - -	F.E. model, $t_f = 8$ μm
··········	F.E. model, $t_f = 20$ μm

Fig. 6. (*continued*)

The numerical results are in fair agreement with experimental data although only a linear thermo-elastic analysis was carried out. Similar results are observed with increasing the film thickness and temperatures. Similar stress intensities are observed when similar material properties are used. Comparable stress intensities and distributions were observed but the experimental procedures and material microstructures need to be determined precisely to assess the exact correlations between numerical and experimental data. Care should be taken in the comparisons of these results since most published data are on different substrates than our Ti6Al4V.

4 Conclusion

This work reports on the improvement of diamond film deposition on Ti6Al4V substrate using the newly developed TMCVD technique. The deposited films were characterised using the Raman spectroscopy and residual stresses evaluated using the peak shift method. In addition, a finite element analysis was carried out and showed good agreement with the measured experimental residual stresses and literature experimental data.

References

1. Liang, H., Shi, B., Fairchild, A., Cale, T.: Applications of plasma coating in Artificial joints: an overview. Vacuum 73, 317–326 (2004)
2. http://www.usbjd.org/media/ index.cmf?pg=press_full.cmf&prID=15 (visited 03.30.2006)
3. Anract, P., Rosencher, N., Eyrolle, L., Tomeno, B.: L'environnement médical de la prothèse totale de hanche. Presse Méd 25, 1069–1075 (1996)
4. Fan, W.D., Jagannadham, K., Narayan, J.: Adhesion of diamond films on Ti-6Al-4V alloys. Surface and Coatings Technology 91, 32–36 (1997)
5. De Barros, M.I., Vandenbulcke, L.: Plasma-assisted chemical vapor deposition process for depositing smooth diamond coatings on titanium alloys at moderate temperature. Diamond and related materials 9, 1862–1866 (2000)
6. Gutmanas, E.Y., Gotman, I.: Pirac Ti nitride coated Ti6AL4V head against UHMWPE acetabular cup-hip wear simulator study. Journal of material science 15, 327–330 (2004)
7. Hollmann, P., Alahelisten, A., Olsson, M., Hogmark, S.: Residual stress, Young's Modulus and fracture stress of hot flame deposited diamond. Thin Solid Films 270, 137–142 (1995)
8. Kuo, C.T., Lin, C.R., Lien, H.M.: Origins of the residual stress in CVD diamond films. Thin Solid Films 290, 291, 254–259 (1996)
9. Peng, X.L., Tsui, Y.C., Clyne, T.W.: Stiffness, residual stresses and interfacial fracture energy of diamond films on titanium. Diamond and Related materials 6, 1612–1621 (1997)
10. Chandra, L., Chhowalla, M., Amaratunga, G.A.J., Clyne, T.W.: Residual stresses and debonding of diamond films on titanium alloy substrates. Diamond and Related Materials 5, 674–681 (1996)
11. Nibennaoune, Z., George, D., Antoni, F., Santos, J., Cabral, G., Ahzi, S., Ruch, D., Gracio, J., Remond, Y.: Towards Optimization of Time Modulated Chemical Vapour Deposition for Nanostructured Diamond Films on Ti6Al4V. Journal of Nanoscience and Nanotechnology 10, 2838–2843 (2010)
12. Cabral, G., Reis, P., Polini, R., Titus, E., Ali, N., Davim, J.P., Grácio, J.: Cutting performance of time-modulated chemical vapour deposited diamond coated tool inserts during machining graphite. Diamond & Related Materials 15, 1753–1758 (2006)
13. Ager III, J.W., Drory, M.D.: Quantitative measurement of residual biaxial stress by Raman spectroscopy in diamond grown on a Ti alloy by chemical vapor deposition. Physical Review B 48, 2601–2607 (1993)
14. Nibennaoune, Z., George, D., Ahzi, S., Ruch, D., Remond, Y., Gracio, J.: Numerical Simulation of Residual Stresses in Diamond Coating on Ti-6Al-4V Substrate. Thin Solid Films 518, 3260–3266 (2010)
15. Chandra, L., Chhowalla, M., Amaratunga, G.A.J., Clyne, T.W.: Residual stresses and debonding of diamond films on titanium alloy substrates. Diamond and Related Materials 5, 674–681 (1996)
16. Grögler, T., Zeiler, E., Horner, A., Rosiwal, S.M., Singer, R.F.: Microwave-plasma-CVD of diamond coatings onto titanium and titanium alloys. Surface and Coating Technology 98, 1079–1091 (1998)
17. Gunnars, J., Alahelisten, A.: Thermal stresses in diamond coatings and their influence on coating wear and failure. Surface and Coating Technology 80, 303–312 (1996)

18. Li, C., Li, H., Niu, D., Lu, F., Tang, W., Chen, G., Zhou, H., Chen, F.: Effects of residual stress distribution on the cracking of thick freestanding diamond films produced by DC arc jet plasma chemical vapor deposition operated at gas recycling mode. Surface and Coating Technology 201, 6553–6556 (2007)
19. Rats, D., Vanddenbulcke, L., Herbin, R., Benoit, R., Erre, R., Serin, V., Sevely, J.: Characterisation of diamond films deposited on titanium and its alloys. Thin Solid Films 270, 177–183 (1995)
20. Scardi, P., Leoni, M., Cappuccio, G., Sessa, V., Terranova, M.L.: Residual stress in polycristalline diamond/Ti-6AP-4V systems. Diamond and Related Materials 6, 807–811 (1997)
21. Jeong, J.H., Lee, S.Y., Lee, W.S., Baik, Y.J., Kwon, D.: Mechanical analysis for crack-free release of chemical-vapour-deposited diamond wafers. Diamond and Related Materials 11, 1597–1605 (2002)

CAPAAL and CAPET – New Materials of High-Strength, High-Stiff Hybrid Laminates

Bernhard Wielage, Daisy Nestler, Heike Steger, Lothar Kroll,
Jürgen Tröltzsch, and Sebastian Nendel

Chemnitz University of Technology, Institute of Materials Science and Engineering,
Erfenschlager Str. 73, 09125 Chemnitz, Germany
Chemnitz University of Technology, Professorship of Light-Weight Construction
and Plastics Processing, Reichenhainer Str. 70, 09126, Chemnitz, Germany
bernhard.wielage@mb.tu-chemnitz.de,
daisy.nestler@mb.tu-chemnitz.de,
heike.steger@mb.tu-chemnitz.de,
lothar.kroll@mb.tu-chemnitz.de,
juergen.troeltzsch@mb.tu-chemnitz.de,
sebastian.nendel@mb.tu-chemnitz.de

Abstract. The natural resources available for national and international economic development are limited. A gentler and more efficient use of available energy and materials in all sectors is essential. In mobile applications in particular, in which large masses are moved and accelerated (e.g. automotive, railway, aircraft and in machinery and equipment), a consequent lightweight construction is necessary for a significant saving of energy.

The highest potential of lightweight constructions is in the field of fibre- and textile-fibre-reinforced composites. In some applications, a defined ductility is necessary. To achieve these requirements, high-strength and high-stiff carbon-fibre-reinforced plastics (CFRP) were combined with metal foils by means of hybrid laminates.

The focus of the current investigations is based on the development of hybrid laminates (CAPET and CAPAAL) which consist of metal foils (Ti- and Al-alloy foils) and layers of CFRP. By means of thermoplastic CFRP matrices (e.g. PA and PEEK), disadvantages of the thermosetting plastic matrix can be avoided. The excellent formability of the thermoplastic matrices should be named in particular. Within this contribution first results of the joint project will be presented.

Keywords: hybrid laminates, high-strength and high-stiff, thermoplastic CFRP, CAPAAL, CAPET.

1 State of the Art

1.1 Introduction

Because of its specific properties and its large structural diversity, a combination of semi-finished fibre and textile products with thermosetting and thermoplastic matrices is indispensable.

Continuous filament-reinforced plastics are characterised by a flexible adaption of the material structure and so with a specific adjustment of the material properties regarding their applied forces. Besides these excellent light-weight construction properties, often a defined ductility is required from these composite materials. Such ductility can only be achieved by means of metals, especially light-metals.

To meet the above-mentioned contrary requirements, a suitable combination of the high-strength and high-stiff carbon-fibre-reinforced plastics with metal foils by means of hybrid laminates is necessary. So, metal foils were combined with CFRP layers to generate hybrid laminates. A wide variation of light-weight construction properties can be achieved with this combination.

1.2 Hybrid Laminates

Commercially available hybrid materials consist of steel sheets and plastic layers (registered trade name BONDAL®). They are used to reduce the structure-borne sound [1]. Furthermore, aluminium polymer laminates (e.g. ALUCOBOND) are used [2].

Hybrid laminates with continuous fibre-reinforced thermosetting plastic matrix layers have been developed, especially to improve the light-weight construction properties as well as the damage tolerance. In this regard, the following materials

- GLARE (glass-fibre-reinforced plastics/
aluminium foil laminate)
- ARALL (aramid-fibre-reinforced plastics/
aluminium foil laminate)
- CARALL (carbon-fibre-reinforced plastics/
aluminium foil laminate) and
- TiGr (titanium-graphite laminate)
are named. Different research activities have operated with these materials [3-5].

GLARE for example is used in aircraft constructions, especially within the Airbus A380 [3]. Unfortunately, the stiffness of these hybrid materials is much lower than the stiffness of aircraft-typical aluminium alloys. Furthermore, the strength is lower than with unidirectional fibre-reinforced materials so that such hybrid systems are not suitable for mass production in case of the very complex and so expensive production processes [6-10].

2 Research Programme

A combination of thin metal foils with carbon-fibre-reinforced thermoplastics is an appropriate possibility to produce high-stiff and high-strength multilayer composites by means of mass production technologies. This field has opened a very interesting research potential. For the network of scientists with different specialisations in lightweight construction, plastics processing, composite materials, forming, modelling and simulation, this research potential is the challenge. The objective should be achieved within the joint DFG research project "High-strength and high-stiffness hybrid layer composites for mass production of lightweight constructions".

Thermoplastic matrix systems are appropriate for short cycle times of a prospective mass production process. To obtain very high specific strength and stiffness, carbon fibre reinforcements were used. Due to their high lightweight construction potential in combination with the carbon-fibre-reinforced plastics (CFRP), titanium and aluminium alloys were selected from a wide range of metals. So two different hybrid laminate systems are part of the investigation area:

- Ti/CF-PEEK (ultra-high-strength, ultra-high-stiff)
- Al/CF-PA (high-strength, high-stiff).

Thus the hybrid laminates consist of titanium or aluminium alloys with layers of carbon-fibre-reinforced thermoplastics.

They were registered as

- CAPET (carbon-fibre-reinforced polyetheretherketone/titanium foil laminate).
- CAPAAL (carbon-fibre-reinforced polyamide/aluminium foil laminate) and

Figure 1 presents the structure with three metallic layers and two CFRP layers (5 multidirectional foils).

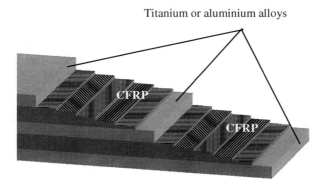

Fig. 1. Scheme of the hybrid laminates

For the combination of the titanium alloys, a high-quality thermoplastic matrix material will be preferred. A conventional thermoplastic matrix material will be used for the combination of the aluminium alloy foils and the CFRP layers. The used materials are systematised in Table 1.

Some advantages argue for the use of the thermoplastic matrices. For example the attenuation of local stress peaks, the possibility of the production with short cycle times, appropriate forming possibilities as well as small radiants of curvature will be named.

A further objective of the joint project is the determination of limitations for the use of the surface finishing of the two selected material systems. The metallic surfaces of the applied hybrid laminates provide the opportunity of conventional surface finishing technologies.

Different basic questions regarding the material development as well as technological or material / mechanical questions will be solved. Therefore, an interdisciplinary cooperation of different parts like development, production, characterisation, modelling and simulation is necessary.

By means of comprehensive numerical observations of the damaging mechanisms as well as the degradation of the hybrid laminates, the comprehension of the material can be enlarged. This characteristics can be the basis of further calculations by means of simulation models.

Very different material properties of the metal foils as well as the CFRP layers within the hybrid laminates complicate the adaptation of the laminates. Therefore, an appropriate interface-engineering of the polymer-matrix/metal interface is the objective. The main focus is on ecological surface treatment of the components as well as on compatible consolidation technology.

The objective of the project is an optimal bonding between the metal foils and the CFRP layers. The disadvantages of the single components should be avoided and the advantages invigorated. So another focus of the scientific work within the joint project is on the continuous and integrated investigation of the whole development process of the hybrid laminates. The following points will be answered:

- Development and production of thermoplastic pre-impregnated CFRP semi-finished materials
- Selection of appropriate titanium and aluminium foils (alloys; dimensioning)
- Development and characterisation of an appropriate polymer-matrix/metal interface (interface engineering)
- Characterisation of the tribological and corrosive properties of the hybrid laminates as well as the aging properties
- Development of appropriate manufacturing technologies and concepts
- Development of appropriate further processing and forming technologies
- Generation of material models for the description of the materials as well as the implementation into FEM software
- Continuous simulation of deformation and failure
- Investigation of failure mechanisms as well as numerical simulation of the failure

Table 1. Specification of the hybrid laminates

Layer composition 1 (CAPET):	
Titanium alloy:	TiAl3V2.5
CFRP:	- Carbon fibres in unidirectional and multidirectional direction - PEEK (poly-ether-ether-ketone)
Forming temperature	< 673 K
Layer composition 2 (CAPAAL):	
Aluminium alloy:	AlSi1.2Mg0.4
CFRP	- Carbon fibres in unidirectional and multidirectional direction - PA (polyamide)
Forming temperature	< 393 K
Further information:	
Laminate structure	Metal/CFRP/metal/CFRP/metal
Thickness of CFRP and metal foils	< 500 µm
CFRP	Four single layers with 60 % fibre volume content

In a second step after the basic investigations, the results will be assigned to a demonstrator (Figure 2).

Fig. 2. Demonstrator

3 Current Investigations

The investigations within the joint project are in an early state. This first phase is characterised by

- analyses of the titanium and aluminium alloys,
- the development of the semi-finished thermoplastic CFRP materials,
- the development of an appropriate interface between the metal foils and the CFRP layers as well as
- the first studies concerning the material models.

This contribution presents some first results as well as the planned strategies within the joint project concerning the semi-finished CFRP materials as well as the interface engineering between these layers.

3.1 Thermoplastic CFRP

To produce the semi-finished thermoplastic CFRP materials, special equipment was established.

This special equipment, which is named unidirectional fibre foil unit, represents a combination of a plastic-foil-manufacturing unit as well as a textile manufacturing unit (Figure 3).

By means of the continuous work principle, this equipment is suitable for mass production. The principle for the thermoplastic CFRP production is the layered arrangement of the carbon fibre layers and the thermoplastic foils (Figure 3).

The impregnation of the fibre layers occurs by means of pressure and temperature. So the molten thermoplastic matrix encloses the fibres (Figure 5).

An appropriate expanding of the carbon fibre rovings is also necessary to obtain a suitable impregnation with the matrix material. In Figure 4, the effect of an adequate and an inadequate impregnation is displayed [11].

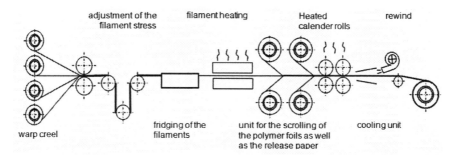

Fig. 3. Principle of the unidirectional fibre foil unit [11]

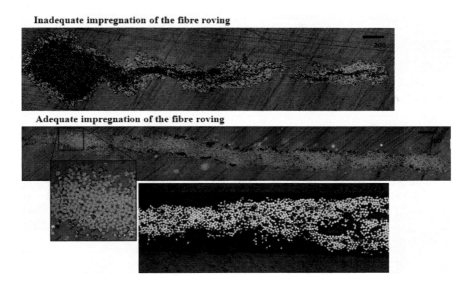

Fig. 4. Adequate and inadequate impregnation of the fibre rovings [11]

The current investigations deal with optimal parameters to obtain an appropriate semi-finished CFRP material for the further joining with the metal foils to hybrid laminates. Another objective within the sub-project is the production of CFRP layers with few unidirectional layers which are positioned in multidirectional arrangement at different angels (e.g. 0°/45°/90°). Figure 6 displays the principle of a cuttling and wrapping unit. By means of this unit, the multidirectional layers will be generated directly after the production of the unidirectional layers within the unidirectional fibre-foil unit.

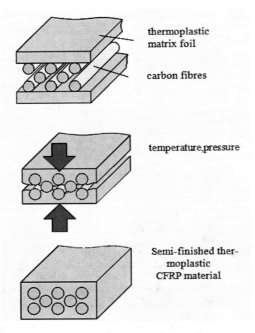

Fig. 5. Impregnation of the fibre filaments with the thermoplastic matrix material [11]

Furthermore, the sub-project is going to consolidate the semi-finished CFRP materials with the metal foils. Additionally, these CFRP materials are important to achieve the mechanical properties for the material models within the "modelling" sub-project.

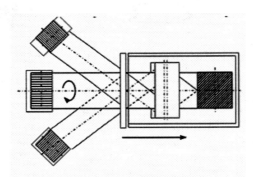

Fig. 6. Principle of the cuttling and wrapping unit

3.2 Interface Engineering

The structure and the mechanical properties of the interface are essential for the mechanical properties of composite materials. Besides the properties of the single materials, the most important property of the hybrid laminate is the optimal interface between the thermoplastic matrix of the CFRP and the metal foils. The combination of different material systems, e.g. metal and polymer, is extremely demanding.

The interface requires a defined transition between the titanium and aluminium layers and the polymer layers. Surfaces with defined structures cause reproducible adhesion systems as well as defined properties.

The main objective of the sub-project "Interface-Engineering" is the development of an optimal interface between the CFRP layers and the metal foils of the CAPAAL and CAPET hybrid materials by means of different surface treatments of the metal foils.

The systematic modification of the surfaces by mechanical treatment, in particular by mechanical blasting, is well-known and has been applied in various areas for decades. The chemical treatment, especially the etching and pickling processes, are well-known to manipulate the surfaces [12, 13]. Metals such as aluminium and titanium allow the generation of defined aluminium oxide and titanium oxide layers on the surfaces. Apart from the anodic oxidation, which is also known as eloxation, a relatively new method has become possible. Using the plasma-electrolytic anodic oxidation, a defined oxide coating on titanium and aluminium can be obtained [14, 15].

Within the project, three different surface treatments are under examination:

- mechanical treatment
- chemical treatment
- plasma-electrolytic anodic oxidation

It is planned to investigate the influence of these treatments in single as well as in appropriate combinations. The focus of the investigations in the first phase of the project was on the selection of different surface treatment methods and the determined microstructuring of the metal surfaces. Different methods, e.g. scanning electron microscopy, 3D-profilometry and roughness measurements, were used to characterize the treated metal foils in comparison with the initial conditions. The current investigations are dealing with the joining of the treated materials. So, a first selection of appropriate treatments is possible.

Mechanical Treatment
The first investigations of the mechanical treatment, in particular the mechanical blasting, have shown that it is possible to obtain the roughness of the surface of the metal foils in a wide range. Therefore, different blasting media like regular aluminium oxide (brown aluminium oxide), specially fused aluminium oxide (white aluminium oxide), silicon carbide as well as glass pearls were tested for the treatment of the aluminium and titanium foils. Figure 7 illustrates the development of the surface roughness of the aluminium foil (initial state roughness $R_z = 5.9$ μm, $S = \pm 1$ μm) with different blasting pressures (1 to 5 bar) and three different blasting media.

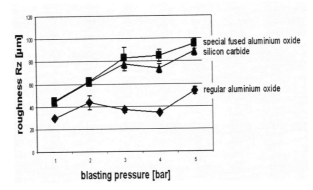

Fig. 7. Development of surface roughness of the aluminium foil with the blasting pressure for three different blasting media

First joining investigations display promising results with a good optimisation capability (Figure 8).

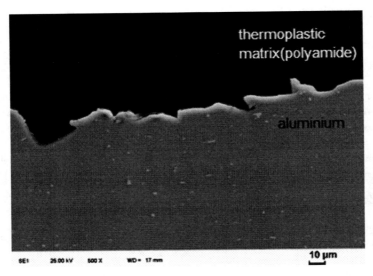

Fig. 8. Joining zone of aluminium and polyamide after mechanical treatment, special fused Al_2O_3, 5 bar (SEM image)

Chemical Treatment
The obtained roughness after the chemical treatment is not so significant. But the scanning electron microscopy (SEM) images display very special surface structures which can be obtained after the chemical treatment. Figure 9 shows two

examples of such different surface structures of the aluminium foil. On the one hand the aluminium was treated with sodium hydroxide and on the other hand with sodium fluoride.

Plasma-electrolytic anodic oxidation
The oxide surfaces which can be obtained by means of the plasma-electrolytic anodic oxidation are characterised by a very homogeneous roughness. Figure 10 displays the SEM image and the 3D-profile of the surface of the aluminium foil after the plasma-electrolytic anodic oxidation.

The objective of this kind of surface treatment is characterised by a very thin coating together with a homogeneous roughness. The current investigations are dealing with the minimisation of thick plasma-anodised layers by means of the combination of appropriate parameters. Figure 11 shows such a thick surface coating which is the initial point of the parameter variation for the current investigations.

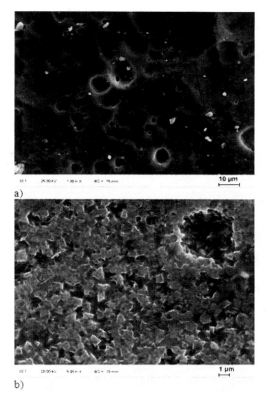

Fig. 9. Aluminium foil after chemical treatment with a) sodium hydroxide and b) sodium fluoride (SEM image)

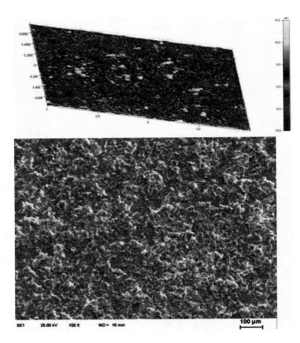

Fig. 10. SEM image and 3D-profile of an aluminium foil after plasma-electrolytic anodic oxidation

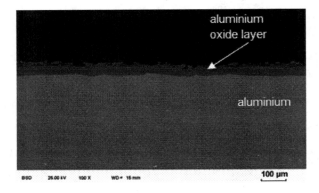

Fig. 11. Thick oxide film after plasma-electrolytic anodic oxidation of an aluminium foil (SEM image)

4 Summary and Conclusions

Already in this early project phase, the results of the single sub-project show that an applicable concept of high-strength and high-stiff hybrid laminates can be achieved by means of an optimally pre-impregnated thermoplastic CFRP together with an optimal interface of the CFRP and the metal foils.

Besides the excellent investigations of each sub-project, an efficient cooperation is necessary.

Acknowledgement

We gratefully acknowledge the cooperation with the project partners and the financial support of the DFG (Deutsche Forschungsgemeinschaft) within the joint research project 3HSL (PAK 413-1).

References

1. N.N.: BONDAL®: das Stahl-Sandwichblech zur effektiven Reduzierung von Körperschall, Produktinformation Thyssen Krupp Steel (2006)
2. N.N.: ALUCOBOND®, Produktinformation ALCAN Singen GmbH (2007)
3. Borchard-Tuch, C.: Chemie Innovativ: Neue Werkstoffe für den Airbus A380. Chemie in unserer Zeit 40, 407–409 (2006)
4. Marissen, R.: ARALL (Aramidfaserverstärkter Aluminiumschicht-Verbundwerkstoff) – Ein neuer Hybrid-Verbundwerkstoff mit besonderen Schwingfestigkeitseigenschaften. Zeitschrift für Materialkunde 17, 278–283 (1983)
5. Westre, W., et al.: Titanium-Polymer Hybrid Laminates, US-Patent, US5866272 (1999)
6. Krishnakumar, S.: Fiber metal laminates – The synthesis of metals and composites. Materials and Manufacturing Processes 9, 295–354 (1994)
7. Lee, J.H.: Herstellung und Charakteristik von Hybrid-Verbundmaterial für die Luftfahrtindustrie. Journal of the Korean Insititute of Metals and Materials 38, 91–97 (2000)
8. Young, K.S., et al.: Spring-back characteristics of fiber metal laminate (GLARE) in brake forming Process. International Journal of Adfanced Manufacturing Technology 32, 445–451 (2007)
9. Sinke, J.: Manufacturing of GLARE Parts and Structures. Applied Composite Materials 10, 293–305 (2003)
10. Jang, K.K., Tong, X.Y.: Forming and failure behaviour of coated, laminated and sandwiched sheet metals: a review. Journal of Materials Processing Technology 63, 33–42 (1997)
11. Wielage, B., Weber, D., Kroll, L., Steger, H., Tröltzsch, J.: Leichte und hochfeste Verbunde aus Kunststoff und Metall, 7. Industriekolloquium des SFB 675, TU Clausthal, 145 f. (2009)
12. Vogel, O., Vogel, H.: Handbuch der Metallbeizerei. Chemie Verlag (1951)
13. Ritupfer, R., von Metallen, B.: Schriftenreihe Galvanotechnik und Oberflächenbehandlung, vol. 24. Leutze Verlag (1993)
14. Wielage, B., et al.: Anodizing – a key for surface treatment of aluminium. Key Engineering Materials 384, 263–281 (2008)
15. http://www.keronite.com 26.03.10

Challenges in the Assembly of Large Aerospace Components

Mozafar Saadat

School of Mechanical Engineering, University of Birmingham, UK
m.saadat@bham.ac.uk

Abstract. Aerospace manufacturing often involves complex large components that are difficult to be assembled. The problems are usually associated with the undesired deformation of large structures when under the influence of jig conditions. Unpredictable pre-assembly manufacturing processes together with the presence of the unaccounted actions of the external forces during assembly often result in the production of geometric errors that exceed the allowable tolerances. A typical example is the assembly of Airbus wing box structure. Often the operative are required to forcefully bring parts together on the jig for joining where there are undesired misalignments. Additionally, the traditional aerospace assembly methods do not lend themselves to the requirements of high throughput, precision assemblies. This paper describes the methodology used to predict the positional variations of wingbox structure during assembly operation. The focus here is on the assembly of ribs under simulated laboratory condition.

Keywords: Assembly, Deformation Analysis, Wingbox Assembly, Variation Prediction, Finite Element Method.

1 Introduction

The civil aerospace manufacturing industry is often faced with major problems in assembly due to their physical size, the flexibility of the components, and the resulting dimensional variation occurring during the assembly process. Defining which variation caused by which assembly processes can be tedious in a complex assembly involving a very large and deformable parts[1]. This is particularly true in the assembly of the wing boxes [2,3,4]. A number of approaches have been presented by researchers to overcome this industrial problem. Transformation vectors were used to represent variation and displacement of feature for modelling assembly of deformable parts. The interaction between parts and tooling are represented by contact chains, which are later used in vector equations [5]. Using a similar cycle for the automobile body assembly, a new simulation procedure for variation analysis by combining the contact and interaction analysis between parts with finite element method was introduced [6]. A method based on structured, hierarchical product description and constraint decomposition for early evaluation of potential flexibility related geometry problems was introduced [7]. A method for simulating fastening assembly taking into consideration of three kinds of variation: positioning, conformity and shape was also developed [8,9]. A study of the variation took place at Airbus UK to measure the actual deformation occurring while assembling the wing panels onto the structure of the wing [10,11]. The study includes identifying the scale and magnitude of these variations, and developing a practical model to implement corrective actions. A model was generated from the results using a complex set of spreadsheets, in predicting the potential variation occurring in assembly depending on various conditions of assembly, especially the profile of the supporting rib prior to the installation of a skin panel.

2 Assembly of Wing Boxes

The principle components of the wing boxes are shown in fig. 1. The stringers are attached to the skin panels. Skin panels are back drilled from inside the wing box, and then removed for deburring. Occasionally, the holes do not match the positions when the panels are re-installed for final bolting. It is also impossible in many locations to make a consistent set of drilling templates. This highlights some dimensional inconsistencies during the assembly process.

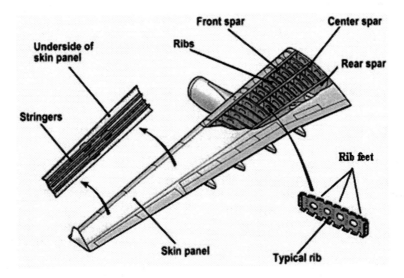

Fig. 1. Wing box assembly detail

The study of the assembly process showed various dimensional variations, which could be inherent to the assembly process itself. Various measuring methods are available to measure over such large distances. Systems such as photogrammetry [12] and Laser Trackers show better solutions. A set of reference points common to the structure and the skin were measured at the three main stages of the final assembly, namely the empty structure, the structure after installation of the top skin, and installation of the bottom skin. Further data were collected on the structure such as the rib profile prior to assembly of the skins.

Despite environmental limitations and industrial restrictions inherent to a real life assembly plant, five sets of aircraft wings were measured [10]. The results showed a certain consistency in the distribution of variation along the wing profile, as well as the amplitude of the variation. However, the rib profiles showed some discrepancies and hinted at a possible cause for the changes in the direction of the variation between aircrafts. Combined with the study of the assembly process, especially the impossibility to manufacture consistent templates, it appears that the differences in the variation between builds can be caused by the initial condition of alignment of the ribs, or more specifically, the spar flanges onto which they are installed.

3 Simulation of Assembly Process

3.1 Experimental Rig

An experimental laboratory rig was constructed, shown in fig. 2, which is capable in supporting the rib such that location, direction and magnitude of applied forces are similar to real life assembly. The forces acting upon the edge of the rib feet are assumed to have resulted mainly from the pressure of the panel onto the rib when it is wrapped over the structure. The rig consists of a steel structure, which holds the rib vertically. It is bolted to two 'spar flanges' attached to the structure. On the side of the rig, twelve flexible load units are mounted to apply the forces on the rib feet, which will generate the deviation recorded during the real life measurements. The force is applied by a turnbuckle and recorded by a force transducer.

The experimental results show the positional variation along each of the orthogonal 3 axes is influenced by the misalignment of the supporting flanges. Fig. 3 shows such variation on the selected rib feet in the Y direction. The X-axis of the graph shows the incremental of 1mm in the Y direction of the rear spar (fig. 2). It becomes clear that the relative position of the two spar flanges has a significant effect on the direction of the variation when the skin panel is installed.

The results achieved from these experiments will provide the basis for the design of a prediction model, which will help in predicting uncertainties in positioning the holes of the skin to match the rib foot. From the data collected, a system can be designed to calculate or interpolate the results for values that were not directly measured from the production jig.

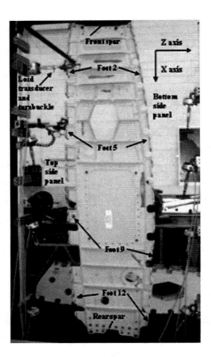

Fig. 2. Experimental rig

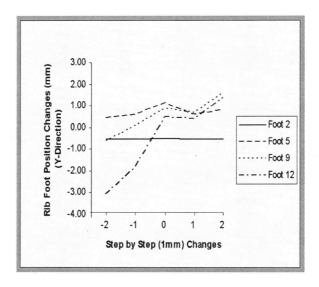

Fig. 3. Displacement of rib feet during bottom skin installation

3.2 Variation Prediction Procedure

A set of database tables were created to show the results according to the initial conditions. Calculation was conducted using proportional interpolation of data for values in-between the steps taken in the measurements. The values for the feet that were not measured were interpolated relative to the position of the neighbouring feet that had been measured.

These data constitute the basis of the system of spreadsheet as a potential format for the model. A resulting set of co-ordinate, measured during the experiments in the laboratory, describe the change of position of the selected rib foot while a certain panel is installed onto the assembly. These results are contained in the set of spreadsheets, which were created to calculate the deformation that can be expected on a particular rib foot depending on which panel is installed and how the two spar flanges are aligned.

This prediction method was verified in order to assess its capabilities. The first estimation of the prediction method's accuracy showed promising results, as given in fig. 4. In either direction of the orthogonal system (X, Y, Z), the variations expected from the prediction method closely match the results of the measurements. The difference between the predicted and measured variation is rarely greater than 0.5mm, which is well within the specification of a maximum 1mm from the measured value. The tests carried out on the simulation rig shows that the prediction method can forecast the variation occurring during the simulation of the assembly, knowing the conditions of alignment of the support flanges. The predictions results match the measurement results within 1mm tolerance, and are acceptable in the assembly environment.

Fig. 5 shows the results of the validation on the real factory jig. Similar to the tests carried out on the test rig, the variation along the X-axis is minimal. The variation along the Y-axis, the axis of larger displacement, is predicted with reasonable accuracy. It is noticeable that the values do not follow a smooth pattern, as the prediction method suggested. This could be attributed to handling errors. The variation occurring along the Z-axis is even greater than in the test rig. As mentioned previously, some local deformation can be attributed to the bolting of the panel onto the rib feet while the overall profile of the rib feet edge does not closely match the profile of the panel. This localised deformation is caused by the bolting of the panel onto the skin, where the amplitude of such forces could not be replicated on the laboratory system.

The tests carried out in the assembly jigs at Airbus UK show that the prediction method can reasonably forecast the variation occurring during the simulation of the assembly, knowing the conditions of alignment of the support flanges.

The resulting predictions match the measurement results within 0.5mm to 1mm, which is acceptable in the assembly environment. Certain discrepancies encountered could be attributed to measurement errors, notably due to some difficulties in using a Laser Tracker in the assembly environment. The Laser Tracker used on the jigs is however much more accurate than the dial gauges used for the tests carried out in the laboratory environment. Despite certain differences in the initial conditions of alignment, the two measurements show similar patterns in deformation, though slightly offset.

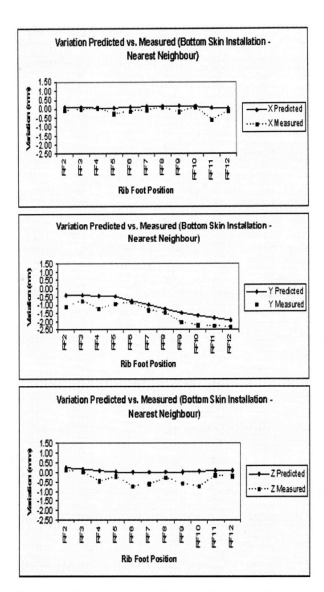

Fig. 4. Verification of the prediction method using the test rig

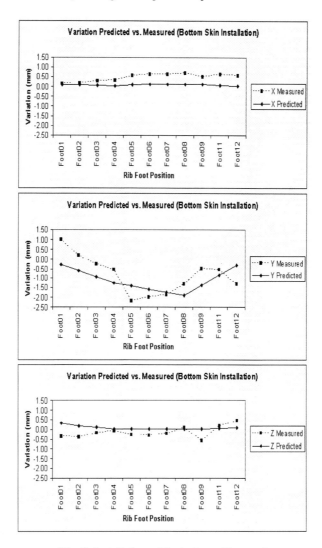

Fig. 5. Installation of the bottom skin panel

4 Single Rib Foot Assembly

The three dimensional geometry for the single rib foot model was developed by ABAQUS CAE finite element software based on the detailed drawing. All the individual components were meshed using hexahedra node and 3D solid elements. This resulted in 600 nodes and 316 elements for the rib foot, and 747 nodes and 408 elements for the skin panel. A small section of the skin panel was modelled from the actual dimension in the assembly operation.

The boundaries for the skin panel are denoted by ABCD as shown in Figure 6. Boundary condition was applied around the line AB in order to be close to the actual assembly condition. No constraint was applied around the line CD as the segment of the skin panel yet to be assembled onto the adjacent rib. However, line AC and BD have the same constraints as AB, so as to show that the skin panel had been coupled to next rib foot. The same boundary condition for the rib foot was also imposed around the line EF to imply a section of the rib structure. A bolt load of 500N is applied on the respective bolt to represent the fastening of the bolt on the skin panel and rib foot.

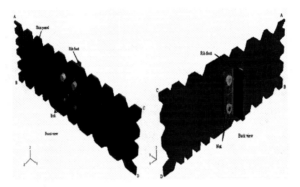

Fig. 6. FE model of single rib foot assembly

Figure 7 illustrates the overall displacement of the RF and a segment of the SP. This simulation focused on the position of the hole on both the RF and SP when the tightening of the first bolt commences. With each tightening of the bolt, the positional variation of the subsequence assembly joint was examined. The effect of displacement on the next assembly joint, where the current joint was fastened, is shown in Figure 8. The abscissa on these graphs represents the bolt tightening sequence between the RF and the SP, with initial condition signifying that both components were not bolted to each other. The next division on the abscissa denotes the fastening of the first bolt between both components. The subsequent divisions represent the remaining bolts to be fastened. The positional displacements of the remaining three holes on the rib foot and skin panel were shown on the ordinate (in millimetre) of the respective graphs for all three axes (xyz).

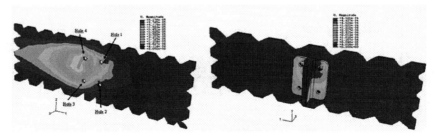

Fig. 7. Displacement distribution of single rib foot and skin panel assembly

Minor displacement was observed on the three orthogonal axes when each bolt and nut is fastened one at a time. Due to the insignificant displacement of the influence of bolt tightening on the single rib foot assembly, the next simulation will concentrate of the displacement effect of the next rib foot when the current rib foot is fastened with four bolts simultaneously.

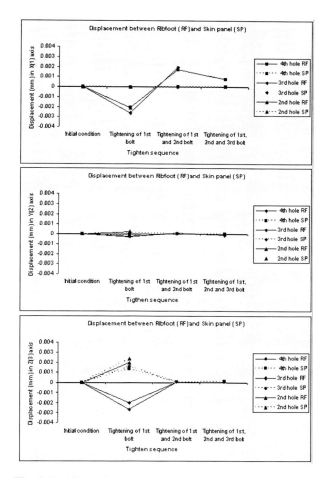

Fig. 8. Positional displacement between rib foot and skin panel

5 Conclusion

This paper presents an outline of a comprehensive study of the actual deformation occurring during part-to-part assembly of the aerospace wing box assembly in Airbus UK. An experimental rig with the capability of representing the condition in the assembly of wing box at the assembly plant was built. A prediction data system was generated with the capability of tackling the positional errors encountered in the assembly of large components, and predicting potential variation occurring in assembly depending on various initial conditions. Results obtained from the prediction model could assist and reduce the assembly time of wing box production processes.

References

1. Ligget, J.V.: Dimensional variation management handbook, A guide for quality, design and manufacturing engineers. Prentice-Hall, Englewood Cliffs (1993)
2. Cretin, L., Saadat, M.: On variation minimisation during aerospace assembly. In: Third World Manufacturing Congress, International Symposium on Manufacturing Technology, Rochester, USA, pp. 24–27 (2001)
3. Jeffreys, D.J., Leaney, P.G.: A dimensional control methodology for the variation analysis of aircraft structure assembly. In: Advanced Manufacturing Technology XII, Proceedings of the Fourteenth National Conference on Manufacturing Research, Derby, United Kingdom, pp. 777–783 (1998)
4. Jeffreys, D.J., Leaney, P.G.: Dimensional control as an integral part of next-generation aircraft development. Proceedings of the Institution of Mechanical Engineers, Part B: Journal of Engineering Manufacture 214(9), 831–835 (2000)
5. Chang, M., Gossard, D.C.: Modelling the assembly of compliant, non-ideal parts. Computer Aided Design 29(10), 701–770 (1997)
6. Hu, M., Lin, Z., Lai, X., Ni, J.: Simulation and analysis of assembly processes considering compliant, non-ideal parts and tooling variations. International Journal of Machine tools and Manufacture 41, 2233–2243 (2001)
7. Dahlstrom, S., Hu, S.J., Soderberg, R.: Identifying variable effects on the dimensional quality of compliant assembly, using computer experiments. In: Proceedings of DETC 2002: 2002 ASME Engineering Technical Conference, Montreal, Canada (2002)
8. Sellem, E., Riviere, A.: Tolerance analysis of deformable assemblies. In: Proceedings of DETC 1998: 1998 ASME Engineering Technical Conference, Atlanta, GA (1998)
9. Sellem, E., Riviere, A., Charles, A.D., Clement, A.: Validation of the tolerance analysis of compliant assemblies. In: Proceedings of DETC 1999: 1999 ASME Engineering Technical Conference, Las Vegas, Nevada (1999)
10. Saadat, M., Cretin, L.: Measurement systems for large aerospace components. Sensor Review: An International Journal 22(3), 199–206 (2002a)
11. Saadat, M., Cretin, L.: Dimensional variation during Airbus wing assembly. Assembly Automation 22(3), 270–276 (2002b)
12. Morrison, R.: Weird and Wonderful Applications of Close-Range Digital Photogrammetry. In: CMSC 1999, Seattle, July 26-30 (1999)

Improvement of Thermal Stability and Fire Behaviour of pmma by a (Metal Oxide Nanoparticles/Ammonium Polyphosphate/Melamine Polyphosphate) Ternary System

B. Friederich, A. Laachachi, M. Ferriol, D. Ruch, M. Cochez, and V. Toniazzo

Centre de Recherche Public Henri Tudor, AMS, 66 rue de Luxembourg,
BP 144, L-4002 Esch-sur-Alzette, Luxembourg
Université Paul Verlaine Metz, LMOPS, 12 rue Victor Demange,
BP 80105, F-57503, Saint-Avold Cedex, France
abdelghani.laachachi@tudor.lu

Abstract. A design of experiment (DoE) was performed on PMMA-(ammonium polyphosphate/melamine polyphosphate/titanium dioxide) ternary system for optimizing both fire-retardancy properties and thermal stability. The software JMP® was used for this purpose. In poly (methyl methacrylate) (PMMA), progressive substitution of titanium dioxide (TiO_2) nanoparticles by melamine polyphosphate (MPP) led to the reduction of the peak of Heat Release Rate (pHRR), whereas the substitution with ammonium polyphosphate (APP) first led to the reduction of pHRR followed by an increase from 9wt% APP onwards. The presence of titanium dioxide led to the increase of the time to ignition (TTI) and thermal stability. Laser Flash Analysis (LFA) measurement showed the heat insulation effect (i.e. low thermal diffusivity) of the residues that have developed during combustion. PMMA-7.5%APP/7.5%TiO_2 showed the major increase of the time of combustion alongside with the decrease of pHRR due to the barrier effect. The high carbon monoxide amount released during the cone calorimeter experiment confirmed an incomplete combustion. That sample presented a ceramized structure and the formation of titanium pyrophosphate (TiP_2O_7) was detected by XRD measurements. TiP_2O_7 resulted from the reaction between APP and TiO_2 upon combustion.

Keywords: Flame retardant, nanocomposites, PMMA, Thermal stability.

1 Introduction

In intumescent systems, the quantity of additives necessary to reach interesting fire-retardancy properties being quite high, it generally leads to a worsening of mechanical properties of the polymer. In order to decrease the level of fire-retardant additives while maintaining mechanical performances, synergistic effects in intumescent systems have been developed [1]. Synergism is observed when two or more additives combined together produce a greater effect than the sum of their individual effects. Flame-retardancy of poly(methyl methacrylate) (PMMA) is significantly improved by the incorporation of additives based on ammonium polyphosphate (APP). Laachachi et al. [2] compared two products marketed by Clariant: AP422 and AP752 mixed with PMMA. They have observed a decrease of the time to ignition (TTI) on samples containing 15% AP422 on the one hand and 15% AP752 on the other hand. But the most noteworthy effect was the decrease of the Total Heat Release (THR) and of the peak of Heat Release Rate (pHRR), especially in the case of AP752 (pHRR = 624 $kW.m^{-2}$ for PMMA, pHRR = 419 $kW.m^{-2}$ for PMMA-15%AP422, pHRR = 300 $kW.m^{-2}$ for PMMA-15%AP752). An intumescent structure was visible after cone calorimeter tests on the sample containing AP752. A comparative X-ray diffraction (XRD) study on both flame-retardants disclosed that AP422 was composed in majority of pure ammonium polyphosphate and that AP752 was composed of APP and of melamine polyphosphate (MPP). Melamine polyphosphate is less stable thermally than APP, but it forms intumescence through the release of ammonia. Schartel et al. [3] also explored the effect of the combination of melamine polyphosphate with phosphorus compounds (aluminium phosphinate) and metallic compounds (zinc borate) in glass-fibre reinforced polyamide 6,6 on flame retardancy. Metal oxides nanoparticles are known for enhancing the thermal stability of polymer. They also help to improve fire-retardant properties. Laachachi et al. [2] therefore combined AP752 and metal oxides nanoparticles (titanium dioxide and alumina) in PMMA so that to benefit from the thermal stability provided by the oxide nanoparticles and from the intumescent behaviour of melamine polyphosphate and of ammonium polyphosphate. The combination of these three additives led to a synergism on flame retardancy. In the present paper, the loading ratio of these fillers (APP/MPP/metal oxide) has been optimized by means of statistical design of experiment, for reaching the best thermal and fire-retardant properties. The statistical software JMP® was used for carrying out that work. Then we have studied the impact of that ternary system of fire-retardants on the heat transfer and on the thermal stability in PMMA. Fire-retardant mechanisms have been investigated by the analysis of cone calorimeter residues.

2 Experimental

2.1 Materials

Poly(methyl methacrylate) (PMMA) (Acrigel® DH LE, Unigel Plàsticos - M_w = 78,000 $g.mol^{-1}$ determined by means of GPC analysis) was used as the matrix. Nanometric titanium dioxide (Aeroxide® TiO_2 P25) with median particles size equal to 21 nm and specific surface area equal to 50 $m^2.g^{-1}$, was provided by Evonik Degussa GmbH. Ammonium polyphosphate (AP 422) was given by Clariant (D_{50} = 0.35 μm) and has a phosphor content of 31-32 wt%. Melamine polyphosphate (Melapur 200, D_{98} = 25μm) was donated by Ciba. It contains 42-44 wt% nitrogen and 12-14 wt% phosphor.

2.2 Nanocomposites Preparation

PMMA pellets were blended with ammonium polyphosphate (APP), melamine polyphosphate (MPP) and metal oxides nanoparticles in a Haake PolyLab 300cm^3 internal mixer at 225°C and 50 rpm. The mixing time was around 7 minutes. The total loading of the additives was 15wt%. Compositions of PMMA-APP/MPP/ metal oxide ternary systems are presented in Table 1. Prior to extrusion, all the materials were dried in an oven at 80°C during at least 4 hours.

Table 1. Compositions of PMMA-based formulations (total loading: 15wt%)

Nom de la formulation	APP	MPP	TiO$_2$
PMMA-15%APP	1	0	0
PMMA-15%MPP	0	1	0
PMMA-15%TiO$_2$	0	0	1
PMMA-7.5%APP/7.5%MPP	1/2	1/2	0
PMMA-7.5%APP/7.5%TiO$_2$	1/2	0	1/2
PMMA-7.5%MPP/7.5%TiO$_2$	0	1/2	1/2
PMMA-5%MPP/5%APP/5%TiO$_2$	1/3	1/3	1/3

2.3 Characterization

Flammability properties of PMMA-based nanocomposites were studied with a cone calorimeter device (Fire Testing Technology). 100×100×4 mm^3 sheets were exposed to a radiant heat flux (35 kW.m^{-2}). Thermogravimetric analysis (TGA) were performed with a STA 409 PC thermobalance from Netzsch operating under gas flow (air) of 100 cm^3.min^{-1} in alumina crucibles (150 µL) containing around 15 mg of sample. The runs were carried out in dynamic conditions at the constant heating rate of 10°C.min^{-1}. The thermal diffusivity was measured using a laser flash technique (Netzsch LFA 457 Microflash™) under inert atmosphere (argon flow: 100 cm^3.min^{-1}). Residues morphology: Residues morphology and chemical composition were ascertained using FEI QUANTA FEG 200 environmental scanning electron microscope (ESEM) coupled with EDAX GENESIS XM 4i energy dispersive X-ray spectrometer (EDS). X-ray diffraction (XRD) measurements were performed using a PANalytical X'Pert PRO Materials Research Diffractometer.

3 Results and Discussion

3.1 Flammability

The cone calorimeter is one of the most effective bench scale methods for investigating combustion properties of polymer materials. Fire-retardancy properties of the PMMA-APP/MPP/TiO$_2$ system have been studied by cone calorimetry under a radiant flux of 35 kW.m^{-2}. Table 2 presents the major parameters measured for all blends of that system (5% standard deviation).

Table 2. Cone calorimeter data of PMMA-APP/MPP/TiO$_2$ ternary system

	PMMA	PMMA-15% TiO$_2$	PMMA-15% APP	PMMA-15% MPP	PMMA-7.5% APP/7.5% MPP	PMMA-7.5% APP/7.5% TiO$_2$	PMMA-7.5% MPP/7.5% TiO$_2$	PMMA-5% APP/5% MPP/5% TiO$_2$
TTI(s)	62	88	63	67	58	75	59	65
TOF(s)	339	608	560	675	710	906	775	935
Combustion time (s)	337	520	497	608	652	831	716	770
pHRR (kW.m^{-2})	533	347	345	260	255	257	278	271
pHRR decrease (%)	/	35	35	51	52	52	48	49
THR (MJ.m^{-2})	117	100	100	99	103	93	99	99
Final residues (%)	0	15	12	6	10	16	11	11
TCOR (g.kg^{-1})	7	10	10	21	7	18	13	10

Figure 1 displays the Heat Release Rate (HRR) and the mass loss as a function of time for PMMA-APP/MPP/TiO$_2$ ternary system under 35 kW.m^{-2}.

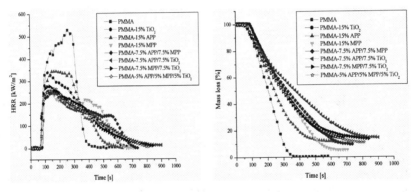

Fig. 1. (a) pHRR and (b) mass loss of PMMA-APP/MPP/TiO$_2$ ternary system measured by cone calorimeter (heat flux: 35 kW.m^{-2})

All formulations based on titanium dioxide showed improved fire-retardancy properties in general compared to PMMA. But the most significant ones were obtained for PMMA-15%MPP, PMMA-7.5%APP/7.5%MPP, PMMA-7.5%APP/7.5%TiO$_2$, PMMA-7.5%MPP/7.5%TiO$_2$ and PMMA-5%APP/5%MPP/5%TiO$_2$ (synergy effects for pHRR). The substitution of a part of ammonium polyphosphate by titanium dioxide led to a significant decrease of pHRR whereas the substitution of a part of melamine polyphosphate by TiO$_2$ resulted in no major changes for that parameter. A reduction of the pHRR means that the fire spread is slowed down. Total Heat Release (THR) decreased of the same factor (about 15%) for all blends. Time of combustion (TOF-TTI) increased for all loaded mixtures, especially for those containing TiO$_2$ and APP and/or MPP. When replacing polyphosphate additives with TiO$_2$, a major increase of the time of combustion was observed alongside with the pHRR reduction. This was the result of the barrier effect and PMMA-7.5%APP/7.5%TiO$_2$ was the most representative sample showing these features.

The increase of the time of combustion can be associated to the diminution of the mass loss rate observable in Figure 1(b), meaning there is a modification of the kinetics of polymer combustion. Moreover PMMA-7.5%APP/7.5%TiO$_2$ presented 16% residues after combustion, which suggested that products stemming from the degradation of PMMA were protected from the flames by the barrier effect. Besides the high amount of carbon monoxide released (18 g.kg^{-1}) compared with PMMA-15%TiO$_2$ and PMMA-15%APP, it also ascertained the incomplete combustion stemming from that barrier. Time to Ignition (TTI) remained constant when taking into account the standard deviation, except for PMMA-15%TiO$_2$ and PMMA-7.5%APP/7.5%TiO$_2$. Two factors can be invoked for explaining that rise: (i) increase of the thermal stability of the polymer in the presence of oxides, confirmed by Thermogravimetric Analysis, (ii) increase of the thermal diffusivity of samples when adding metal oxides [2]. Melamine polyphosphate had a prevailing effect on the reduction of the time to ignition, because it systematically resulted in the fall of TTI even for formulations containing titanium dioxide which used to lead to the highest TTI value (88s). A decrease of TTI is often pointed out for intumescent systems; it comes from the upsurge of the polymer degradation by the additives [4]. pHRR and TTI data have been gathered in the design of experiment presented in Figure 2 for comparing the variations within the ternary system. DoE approach allows minimization of the number of experiments, investigation of the influence of each additive and the mutual interactions between additives.

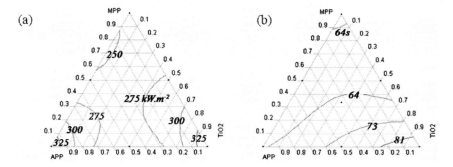

Fig. 2. Variation of pHRR (a) and TTI (b) within PMMA-APP/MPP/TiO$_2$

- PMMA-15%TiO$_2$ showed the highest pHRR value from the system and the progressive substitution of the metal oxide nanoparticles with melamine polyphosphate caused a pHRR decrease. This statement is valid for PMMA-15%APP too. From Figure 2(a), one can therefore say that similar pHRR-performances were reached for a purely intumescent sample (PMMA-15%MPP) and for a ceramized one (e.g. PMMA-5%APP/5%MPP/5%TiO$_2$).
- TTI was at its highest for PMMA-15%TiO$_2$. Replacing metal oxide with APP and/or MPP resulted in its decrease. Formulations containing melamine polyphosphate led to a lowering of TTI, because that compound promoted the development of intumescence through the release of

ammonia (i.e speed up of the polymer degradation) [5] while ammonium polyphosphate had a catalytic effect through polyphosphoric acid [6].

Figure 3 displays photographs of residues of PMMA-APP/MPP/TiO$_2$ after cone calorimeter tests.

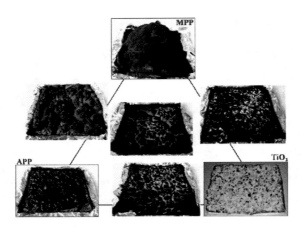

Fig. 3. Cone calorimeter residues of PMMA-APP/MPP/TiO$_2$

One can say, from Figure 3, that the sample showing the more important intumescent effect was that containing 15%MPP which sheet expanded until reaching 4 cm thickness. An intumescent material begins to swell and then to expand when heated beyond a critical temperature. The result is a charred and expanded layer on the surface, which protects the underlying material from the action of heat and of flames [7]. Char formation generally reduces the smoke release [8].

PMMA-15%APP and PMMA-15%TiO$_2$ have similar fire-retardant properties (pHRR, time of combustion) and they did not show an intumescent behaviour (no swelling). Their differences lied in the residues aspect: residues of PMMA-15%TiO$_2$ formed a white crust on the surface of the sample, whereas residues of PMMA-15%APP consisted of a carbonaceous dough containing very likely liquid phosphoric acid. On the contrary, PMMA-15%MPP showed better fire-retardancy properties, because of the water and ammonia released during the decomposition of the sample. Formulations containing TiO$_2$ and APP and/or MPP exhibit a ceramized structure, like PMMA-15%TiO$_2$, combined with a slight swelling observed for PMMA-15%MPP.

Melamine-based fire-retardants normally employ more than one mechanism to flame retard a polymer (condensed phase reactions, intumescence) [9]. Melamine polyphosphate is a salt of melamine and polyphosphoric acid. It decomposes endothermically above 350°C, acting as a heat sink to cool the polymer. The released phosphoric acid forms a coating which shields the polymer substrate. The phosphoric acid further reacts with the polymer to form a char and inhibits the release of free radical gases into the oxygen phase. Simultaneously, nitrogen species (ammonia) released from the degradation of melamine intumesces the char to further protect the polymer from the heat [10]. Phosphor can act in the vapor phase through radical mechanism but also in the condensed phase promoting char formation on the surface which insulates the substrate from heat and air and also interferes with the loss of decomposition products to the flame zone [11].

3.2 Thermal Properties

The investigation of thermal properties aims to enhance the knowledge about the impact of fire-retardants on the thermal properties and on fire-retardant properties in composites. Thermal degradation of all studied materials was performed using Thermogravimetric Analysis (TGA) upon heating under air from ambient temperature to 900°C. Figure 4 presents TG curves of the PMMA-APP/MPP/TiO$_2$ system.

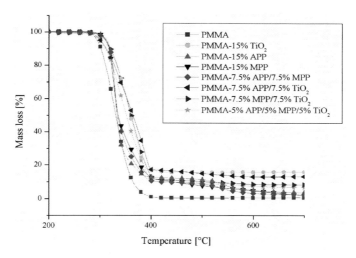

Fig. 4. TG curves of PMMA-APP/MPP/TiO$_2$ under air (10°C.min^{-1})

Mixtures containing APP and/or MPP presented no great thermal stability enhancement compared with PMMA at half degradation temperature (T$_{50\%}$). The replacement of APP and MPP by titanium dioxide resulted in an increase of T$_{50\%}$ by 33°C with a leveling from 7.5wt% TiO$_2$ onwards, under air. The enhancement of thermal stability of PMMA in the presence of titanium dioxide nanoparticles can be explained by: the restriction of mobility of polymer chains, the trapping of radicals by that filler and by the reaction of PMMA with the filler surface via the methoxycarbonyl group [2, 12].

➢ Thermal diffusivity (Laser Flash Analysis)
Thermal diffusivity (α) is the ability of a material to transmit heat rather than to absorb it. Measurements of thermal diffusivity performed by Laser Flash Analysis (LFA) of PMMA-APP/MPP/TiO$_2$ versus the temperature are presented in Figure 5.

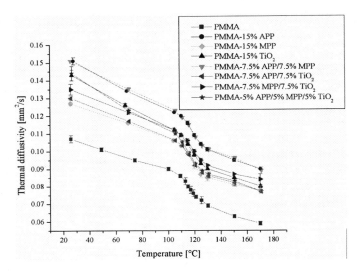

Fig. 5. Thermal diffusivity in function of temperature by Laser Flash Analysis (LFA)

In the temperature range studied (25°C-170°C), all formulations have thermal diffusivity values higher than neat PMMA. PMMA-15%TiO$_2$ exhibited the lowest increase of α after PMMA (26%-increase compared to PMMA). Ammonium polyphosphate-based formulations showed higher values. Thermal diffusivities measured at 25°C have been gathered in the design of experiment presented in Figure 6 for comparing the different mixtures.

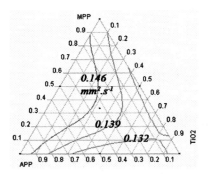

Fig. 6. Design of experiment of thermal diffusivity data for PMMA-APP/MPP/TiO$_2$

The lowest values have been observed for PMMA containing at least 7.5wt% TiO$_2$ and mixed with APP and/or MPP. Substitution of metal oxide nanoparticles with APP and/or MPP resulted in an increase of α with a maximum of 41% for PMMA-7.5%APP/7.5%MPP compared with neat PMMA. According to the design of experiment (Figure 6), similar thermal diffusivity (> 0.146 mm^2.s^{-1}) was obtained for formulations containing few amount metal oxide (< 5wt%) combined with APP and MPP.

3.3 Analysis of Cone Calorimeter Residues

➢ Thermal diffusivity (Laser Flash Analysis)

Thermal diffusivity of polymeric samples evolves upon heating but also during their combustion. It is, at the present time, technically speaking not possible to measure the thermal diffusivity during the samples' degradation, because their thickness changes during that process. Therefore, thermal diffusivity was measured after combustion on cone calorimeter residues using Laser Flash Analysis. Thermal diffusivity of PMMA-7.5%MPP/7.5%TiO_2 residues is $\alpha = 0.089 \pm 0.001$ $mm^2.s^{-1}$ at 25°C. That value was lower than all formulations of the ternary system before combustion and even lower than neat PMMA (0.107 ± 0.002 $mm^2.s^{-1}$). Therefore, the carbonaceous layer which progressively appeared during degradation exhibited an increased heat insulation effect. That layer protected the material from the heat, slowing down the rate of degradation.

➢ Scanning Electron Microscopy

The analysis of the top side of the residues crust was performed by Scanning Electron Microscopy (SEM). SEM images highlighted the presence of holes on the crust surface stemming from the gases released during the combustion. Moreover a flower-like structure and a more square-like one appeared regularly throughout the foamed structure. They are displayed on Figure 7 . The flower-like structure contained Ti and P elements along with C and O, which were homogeneously dispersed into the residues. The other structure (Figure 7) showed an enrichment of P, C and O, but also a shortage of Ti. No other chemical element (especially nitrogen) was detected by SEM-EDS since it was released during combustion (ammonia).

Fig. 7. Mapping of C (a), O (b), P (c) and Ti (d) on the residues crust of PMMA-5%APP/5%MPP/5%TiO_2

A structure chemically comparable to Figure 7(c, d) was also detected on PMMA-7.5%APP/7.5%TiO_2.

➢ X-ray diffraction (XRD) analyses

X-ray diffraction (XRD) analyses were performed on residues crust obtained after cone calorimeter experiments in order to detect the formation of crystalline phase. The combination of APP and TiO$_2$ led to the formation of titanium pyrophosphate (TiP$_2$O$_7$) upon combustion following $2H_3PO_4 + TiO_2 \rightarrow TiP_2O_7 + 3H_2O$ [13]. Therefore, titanium pyrophosphate (TiP$_2$O$_7$) was detected in both PMMA-5%APP/5%MPP/5%TiO$_2$ and PMMA-7.5%APP/7.5%TiO$_2$. Some part of titanium dioxide remained after combustion. XRD diffractogram of PMMA-7.5%APP/7.5%TiO$_2$ is presented in Figure 8.

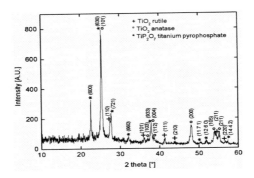

Fig. 8. XRD pattern of PMMA-7.5%APP/7.5%TiO$_2$ residues showing TiO$_2$ (rutile and anatase phases) and TiP$_2$O$_7$ peaks

TiP$_2$O$_7$ formation was also reported by Laachachi et al. [2] in a similar mixture and these authors showed the formation of TiP$_2$O$_7$ from 400°C onwards. The interaction between APP and TiO$_2$ led to a heterogeneous distribution in PMMA even before combustion as shown by Bourson et al. [14] by using Raman spectroscopy: TiO$_2$ presented a heterogeneous distribution (concentration on the surface of the rod), whereas APP was well dispersed.

4 Conclusion

In PMMA-APP/MPP/TiO$_2$ ternary system, progressive substitution of TiO$_2$ nanoparticles by MPP led to the reduction of pHRR. The same trend was obtained when replacing TiO$_2$ with APP until 9wt% APP according to the design of experiment performed with the software JMP®. Moreover the presence of APP, MPP and TiO$_2$ enhanced the heat transfer through the sample. The highest thermal diffusivity was measured for formulations loaded with APP, MPP and with maximum 5 wt% TiO$_2$ nanoparticles. In these mixtures, titanium dioxide was also helpful for increasing the thermal stability of PMMA. Nevertheless, no direct relationships could be highlighted for all mixtures showing enhanced fire-retardancy properties and high thermal diffusivity, as this was shown for PMMA-metal oxide nanocomposites in a previous paper [15]. This could come from the fact that the system presented in the present paper encompassed more compounds, thus involving more fire-retardant mechanisms. Fire-retardants improved the heat transfer in neat PMMA, but it was noted that during combustion, the material got more insulating because the thermal diffusivity of residues was much lower than for mixtures before combustion. Therefore not only the structure of the char plays a significant role in fire resistance, but also its thermal properties. That aspect is under progress.

PMMA-7.5%APP/7.5%TiO_2 showed the major increase of the time of combustion alongside with the decrease of pHRR (a synergy effect was also noted), which came from the barrier effect. The high carbon monoxide amount released during the cone calorimeter experiment confirmed an incomplete combustion. That sample presented a ceramized structure and the formation of titanium pyrophosphate (TiP_2O_7) was detected by XRD measurements. TiP_2O_7 resulted from the reaction between APP and TiO_2 upon combustion. Moreover in the ternary system, the presence of titanium dioxide led to the increase of TTI and of the thermal stability.

Acknowledgements

The authors gratefully acknowledge the financial support of the Fonds National de la Recherche (FNR) from Luxembourg. Evonik-Degussa, Clariant and Ciba are acknowledged for giving respectively TiO_2, ammonium polyphosphate and melamine polyphosphate.

References

1. Duquesne, S., et al.: Synergism in intumescent systems - a review. In: Advances in the flame retardancy of polymeric materials, Current perspectives presented at FRPM 2005, Berlin, Germany (2007)
2. Laachachi, A., et al.: Effect of Al2O3 and TiO2 nanoparticles and APP on thermal stability and flame retardance of PMMA. Polymers for Advanced Technologies 17, 327–334 (2006)
3. Braun, U., et al.: Flame retardancy mechanisms of aluminium phosphinate in combination with melamine polyphosphate and zinc borate in glass-fibre reinforced polyamide 6,6. Polymer Degradation and Stability 92, 1528–1545 (2007)
4. Bourbigot, S.: Intumescence-based fire-retardants. In: Fire retardancy of polymeric materials, pp. 136–149. CRC press, Boca Raton (2000)
5. Bourbigot, S., et al.: Recents developments in intumescence-application to polyolefins. In: 9th European Meeting on Fire Retardancy and Protection of materials, Lille, France (2003)
6. Bourbigot, S., et al.: Recent advances for intumescent polymers. Macromolecular Materials and Engineering 289, 499–511 (2004)
7. Bourbigot, S., et al.: Intumescence and nanocomposites: a novel route for flame-retarding polymeric materials. In: Morgan, A.B., Wilkie, C. (eds.) Flame retardant polymer nanocomposites, Wiley-Interscience, Hoboken, ch.6 (2007)
8. University of Central Lancashire. 2008. How do fire retardants work? (updated September 15, 2008),
   `http://www.uclan.ac.uk/scitech/fire_hazards_science/`
   `sci_fz_firerets.php` (accessed April 21, 2010)

9. Lomakin, S.M.: Modern Polymer Flame Retardancy. VSP, Boston (2003)
10. Ciba: Ciba Melapur (2006) (updated 20 April 2010), http://www.ciba.com/melapur (accessed April 21, 2010)
11. The Fire Retardant Chemicals Association: Flame Retardants - 101: Basic Dynamics Past Efforts Create Future Opportunities in 'Spring Conference. Baltimore, USA' (1996)
12. Jahromi, S.A.J., et al.: Failure analysis of 101-C ammonia plant heat exchanger. Engineering Failure Analysis 10, 405–421 (2003)
13. Bamberger, C.E., et al.: Synthesis and characterization of titanium phosphates, TiP_2O_7 and $(TiO)_2P_2O_7$. Journal of the Less-Common Metals 134, 201–206 (1987)
14. Bourson, P., et al.: L'application de la spectrométrie Raman portable à l'identification des polymères in 'SCF (Société Chimique de France) groupe "Dégradation et comportement au feu des matériaux organiques. Alès, France' (2003)
15. Friederich, B., et al.: Tentative links between thermal diffusivity and fire-retardant properties in PMMA-metal oxides nanocomposites. Polymer Degradation and Stability (accepted for publication, April 2010)

Investigation of Mechanical Properties and Failure Behaviour of CFRP, C/C and C/C-SiC Materials Fabricated by the Liquid-Silicon Infiltration Process in Dependence on the Matrix Chemistry

Bernhard Wielage, Daisy Nestler, and Kristina Roder

Chemnitz University of Technology, Institute of Materials Science and Engineering,
Erfenschlager Str. 73, 09125, Chemnitz, Germany
bernhard.wielage@mb.tu-chemnitz.de,
daisy.nestler@mb.tu-chemnitz.de,
kristina.roder@mb.tu-chemnitz.de

Abstract. Fibre-reinforced ceramics have a higher fracture toughness compared to monolithic ceramics. Therefore, they are an attractive material for lightweight structural components in high-temperature applications. The liquid-silicon infiltration (LSI) process is a cost-efficient manufacturing route for fibre-reinforced ceramics consisting of three processing steps. A carbon-fibre-reinforced plastic (CFRP) composite is fabricated, which is converted in a porous C/C composite by pyrolysis. Liquid silicon is infiltrated to form a dense C/C-SiC composite. The performance of the composites strongly depends on the raw materials. The matrix chemistry in particular plays a key role in developing composites with tailored functional properties. The aim of this work is to investigate the mechanical properties and the failure behaviour of CFRP, C/C and C/C-SiC composites in dependence on the matrix chemistry. The composites are fabricated by the liquid-silicon infiltration process. Different phenolic resins as matrix polymers are used. The mechanical properties are characterized by bending tests and the failure behaviour is observed in-situ. Additionally, microstructural and fractrographic analyses are done. It can be shown that the formation of the microstructure and therefore the mechanical properties and the failure behaviour are strongly influenced by the matrix chemistry over all processing steps.

Keywords: ceramic matrix composites, liquid-silicon infiltration process (LSI), phenolic resin.

1 Introduction

The application of metallic materials in high-temperature environments is limited. Ceramic materials offer a higher service temperature. They possess a low density, a high elastic modulus, a high hardness and a good corrosion and wear resistance. The disadvantages of monolithic ceramics are their brittle failure behaviour and their low damage tolerance, therefore their usage in structural components is restricted [1], [2]. The fracture toughness of ceramics can be improved by embedding fibre reinforcement in the ceramic material. So additional energy-dissipating mechanisms such as crack deflection, crack branching, crack bridging and fibre pull-out can be activated during crack propagation in the ceramic matrix [3]. Due to the tailored properties, fibre-reinforced ceramics are an interesting material for lightweight structural components in high-temperature applications, where conventional materials have an insufficient performance. Typical applications are aircraft engine parts, thermal protection systems for spacecraft and friction systems like brake disks [4].

To produce non-oxide fibre-reinforced ceramics, a lot of different processing techniques can be used, but long processing times, multiple reinfiltration steps and expansive raw materials lead to high manufacturing costs. To overcome these drawbacks, the liquid-silicon infiltration (LSI) process was developed in the German Aerospace Centre (DLR). The processing route can be split into three steps. In the first step, a carbon-fibre-reinforced plastic (CFRP) composite is fabricated. A polymer matrix with high carbon content is used. The CFRP composite is pyrolysed in the second step, whereas the polymer matrix is thermally converted into carbon. The shrinkage of the matrix during this step is hindered by the stiff fibres leading to a defined crack pattern in the composite. A porous carbon-fibre-reinforced carbon (C/C) composite is the result. The third step represents the one-shot infiltration of the porous C/C composite with silicon melt. The silicon reacts with the carbon to silicon carbide in the area of the cracks. A dense heterogeneous C/C-SiC composite with intact C/C segments surrounded by silicon carbide and residual silicon is formed [5], [6], [7], [8].

The performance of the composites strongly depends on the raw materials. The matrix chemistry in particular plays a key role in developing composites with tailored functional properties. Usually phenolic resins are used due to their good processability during CFRP manufacturing, their high carbon yield during pyrolysis and their low costs. Phenolic resins are produced by a condensation between phenol and aldehyde, in many cases formaldehyde, and can be generally divided into resoles and novolacs. Resoles are base-catalysed with a molar ratio of formaldehyde to phenol greater than one. Novolacs are obtained by using an acid catalyst with a molar ratio of formaldehyde to phenol less than one [9]. The microstructure formation and thus the mechanical properties of the composites are affected by the resin chemistry. The pore formation during the manufacturing process of the composite can be particularly influenced by the chemical properties [10]. The aim of this paper is to characterise the mechanical properties and the failure behaviour of CFRP, C/C and C/C-SiC composites in dependence on the used phenolic resins.

2 Materials and Methods

A two-dimensional plain weave fabric made of PAN-based carbon fibres (HTA, manufactured by Toho Tenax Europe GmbH) was used as reinforcement. The textile fabric was made of a 3K filament yarn. The fabric weight was 160 g/m^2 and the yarn count in warp and weft direction was 4 yarns/cm. The fabric was thermally desized by a temperature of 600 °C. Three different types of resoles were used as matrix polymers. Resin 1 is the commercial resol EXP 5E2904U supplied by Dynea Erkner GmbH. Resin 2 and 3 were prepared in our laboratory with a molar mixing ratio between formaldehyde and phenol of 2 : 1. The basic catalysts were NaOH and KOH. Only in resin 3, the additional agent (EtO)$_4$Ti was added to improve the cross linking in the structure during curing.

The processing of the investigated composites consisted of three processing steps. The CFRP composites were fabricated by hand lay-up. 12 plies were used. The curing was carried out in an autoclave by a controlled time-temperature profile with an end temperature of 160 °C and a pressure of 45 bar. Cross linking reactions during curing produce volatile substances resulting in a porous matrix. The high pressure prevents the high amount of closed porosity and leads to a dense matrix. A fibre volume content of circa 60 % could be achieved. In the second step, the CFRP composite was pyrolysed under argon atmosphere at a temperature of 960 °C. The time-temperature profile was adjusted to the decomposition of the polymer matrix to reduce the damage of the composite caused by diffusing volatile decomposition products. In the third step, liquid silicon was infiltrated using a vacuum atmosphere and a temperature of 1575 °C. The silicon dosage was 45 % related to the weight of the C/C sample. In the sample with resin 1, a relative silicon uptake of circa 15 %, and in the samples with resin 2 and 3, a relative silicon uptake of circa 40 % could be achieved.

The mechanical properties were investigated using a three-point bending test. The sample dimensions were 1.8 mm x 8 mm x 45 mm. The support span was 40 mm. The speed of loading was 10 µm/s. The failure behaviour was observed in-situ by a stereomicroscope and was documented by a digital image acquisition system. Four samples were investigated for each resin and material state. The microstructure of the specimens was investigated by optical microscopy. The fracture surfaces of the samples were characterised by scanning electron microscopy.

3 Results and Discussions

The microstructure formation over all processing steps is shown in Figure 1. The composites made from resin 1 and 2 have been selected because the microstructures of the composites manufactured from resin 2 and 3 are very similar. The CFRP composite made from resin 1 shows a homogeneous structure, whereas cracks are found in the CFRP composite fabricated from resin 2. Transversal cracks divide the fibre bundles in several segments. Furthermore, there are partial delaminations in the interfaces between the warp and weft fibre bundles. The crack development can be explained by a higher solvent content in the self-made resins. During pyrolysis, the polymer matrix is converted into carbon. A defined crack pattern is formed in all C/C composites. The composite manufactured from resin 2 shows the lowest crack density. The crack formation during pyrolysis is influenced by the amount of volume shrinkage of the polymer matrix and by the fibre/matrix bonding (FMB). A weak FMB leads to debondings at the interface of fibre and matrix. A strong FMB results in more segmentation cracks and partial delaminations [11]. During the silicon infiltration step, the melt penetrates the C/C composites only through the segmentation cracks and the partial delaminations. The C/C-SiC composite made from resin 1 shows the desired microstructure. Intact C/C segments are surrounded by silicon carbide. The microstructure formation of the C/C-SiC composite fabricated from resin 2 is different. The silicon reacts completely with the whole matrix carbon. In addition, fibre degradation is discernible. The carbon fibre partly reacts with the silicon. The higher conversion rate can be explained by a higher porosity of the carbon matrix after pyrolysis, leading to improved infiltration conditions with silicon. After the silicon infiltration, most of the pyrolysis cracks are closed, but in the composite made from resin 2, additional cracks can be determined. They emerge during cooling down from the infiltration temperature. The higher silicon carbide content leads to a higher coefficient of expansion.

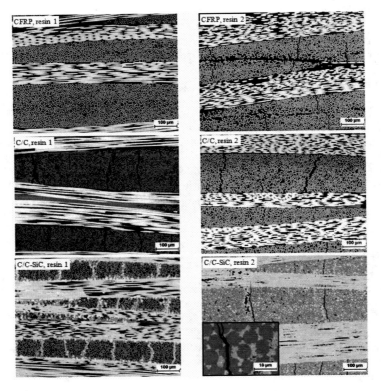

Fig. 1. Structure of the CFRP, C/C and C/C-SiC composites

In Figure 2, the flexural stress-strain response of different composites is demonstrated. The flexural strength and modulus are analyzed statistically. The mean values and the 95 % confidence intervals are plotted in Figure 3. Figure 4 illustrates the failure behaviour during the bending tests for the composites manufactured from resin 1 and 2. All investigated samples show linear elastic behaviour up to the point of failure. The fibres have a higher stiffness and strength compared to the matrix. The main load is carried by the fibres which are aligned in load direction. The macroscopic behaviour of the composites is fibre-dominated.

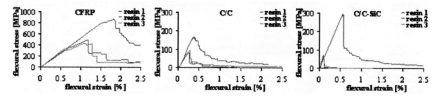

Fig. 2. Flexural stress-strain response of the CFRP, C/C and C/C-SiC composites

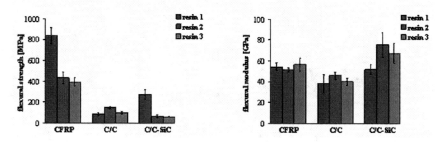

Fig. 3. Flexural strength and flexural modulus of the CFRP, C/C and C/C-SiC composites

The resin 1 composite in the CFRP state has the highest flexural strength. At first, the failure occurs due to compressive stress. With higher loading, the composite fails at tensile stress and shear stress. The composites made from resin 2 and 3 have similar behaviour. The first failure is also caused by compressive stress followed by failure due to shear stress. Delaminations between the textile plies are visible in the fracture picture. The tensile strength is considerably lower because the cracks in the microstructure have already damaged the composite. All composites in the CFRP state show a residual stress after fracture and the failure is successive. There is no difference between the flexural modulus because this value is dominated by the fibre properties.

In the C/C state, the composites made from resin 1 and 3 have similar mechanical properties. The first fracture originates from tensile stress. The additional delaminations between the plies indicate a fracture based on shear stress. The C/C composites made from resin 2 possess the highest flexural strength in this state. The reason for this is the lower crack content in the microstructure compared to the other composites. Shear stress causes the first failure. Only few plies on the sample surface fail due to tensile stress.

In the C/C-SiC state, the composite made from resin 1 has the highest flexural strength. The first failure occurs due to tensile stress. Partial delaminations between the textile plies indicate an additional failure due to shear stress. The behaviour of the composites made from resin 2 and 3 is similar. Their failure which is initiated by tensile stress is brittle. No residual stress can be found because there is no load redistribution between the different textile plies. The brittle failure is caused by the degraded fibres. The interfacial reaction with the silicon melt leads to a high FMB and a hindered crack deflection on the interface. Figure 5 illustrates the fracture surfaces of the composites. The fibre fracture of the resin 1 composite is uncorrelated. Fibre pull-out is visible. The resin 2 composite shows correlated fibre fracture leading to a planar fracture surface with no crack deflection and fibre pull-out. In this case, structurally weak points (pores or cracks) lead to a premature failure. The flexural strength is reduced. The composite has a higher silicon carbide content, which results in a higher flexural modulus compared to the composite fabricated from resin 1.

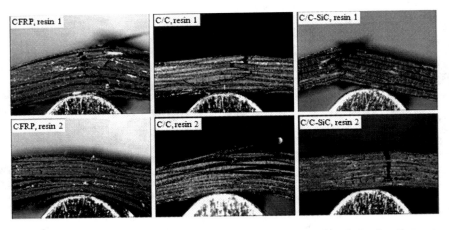

Fig. 4. Failure behaviour of the CFRP, C/C and C/C-SiC composites during bending tests

Fig. 5. Fracture surfaces of the C/C-SiC composites

4 Conclusions

CFRP, C/C and C/C-SiC composites were prepared by using different phenolic resins. The microstructure formation and the properties of the composites were investigated.

The different resins influence the structure formation in the composites in the following way:

- different crack formation in the CFRP state during curing and in the C/C state during pyrolysis,
- different carbon-matrix conversion to silicon carbide during silicon infiltration of the C/C composites.

Different mechanical properties can be achieved in the three processing steps. In the CFRP state, the resin 1 composite, in C/C state, the resin 2 composite and in C/C-SiC state, the resin 1 composite shows the best mechanical properties. In the C/C-SiC state, the resin 2 and resin 3 composites fail in a brittle manner and the achieved flexural strength is low. Therefore, an application in structural components is inappropriate. Resin 2 and 3 are interesting raw materials if composites with a complete silicon carbide matrix are to be fabricated. Furthermore, an additional fibre coating can be used to protect the fibres and achieve higher fracture toughness.

The manufactured composites have a heterogeneous structure resulting in anisotropic properties. During flexural load, a complex stress condition arises. Depending on the used resins, different failure modes result.

Acknowledgements

The authors express their gratitude to the Deutsche Forschungsgemeinschaft (DFG) for supporting these investigations.

References

1. Bansal, N.P.: Handbook of Ceramic Composites. Kluwer, Boston (2005)
2. Warren, R.: Ceramic-Matrix Composites. Springer, Glasgow (1992)
3. Xia, K., Langdon, T.G.: Review - The toughening and strengthening of ceramic materials through discontinuous reinforcement. Journal of Materials Science 29, 5219–5231 (1994)
4. Krenkel, W.: Carbon Fiber Reinforced CMC for High-Performance Structures. International Journal of Applied Ceramic Technology 1, 188–200 (2004)
5. Krenkel, W., Schanz, P.: Fiber Ceramic Structures based on Liquid Silicon Impregnation Technique. Acta Astronautica 28, 159–169 (1992)
6. Krenkel, W.: From Polymer to Ceramics: Low Cost Manufacturing of Ceramic Matrix Composite Materials. Molecular Crystals and Liquid Crystals 354, 353–364 (2000)
7. Krenkel, W.: Cost effective processing of CMC composites by melt infiltration (LSI-Process). Ceramic Engineering and Science Proceedings 22, 443–454 (2001)

8. Heidenreich, B.: Manufacturing methods and typical applications of C/C-SiC materials based on the liquid silicon infiltration process (LSI). VDI-Berichte 2028, 321–336 (2008)
9. Pilato, L.: Phenolic resins: A century of progress. Springer, New York (2010)
10. Mucha, H., Kim, Y.-E., Kuroda, K., Krenkel, W., Wielage, B.: Untersuchungen zur Entstehung von Porosität in Phenolharzmatrices bei der Härtung. In: Krenkel, W. (ed.) Symposium Verbundwerkstoffe und Werkstoffverbunde. Verbundwerkstoffe, vol. 17, pp. 226–232. Wiley-VCH, Weinheim (2009)
11. Krenkel, W.: Entwicklung eines kostengünstigen Verfahrens zur Herstellung von Bauteilen aus keramischen Verbundwerkstoffen. Stuttgart: Deutsches Zentrum für Luft- und Raumfahrt e.V., Institut für Bauweisen- und Konstruktionsforschung (2000)

Investigation of Polymer Melt Impregnated Fibre Tapes in Injection Moulding Process

Jürgen Tröltzsch and Lothar Kroll

Chemnitz University of Technology, D-09107 Chemnitz, Germany
`juergen.troeltzsch@mb.tu-chemnitz.de,`
`lothar.kroll@mb.tu-chemnitz.de`

Abstract. The polymer melt impregnation of glass fibre tapes or rovings is of vital importance for manufacturing textile reinforced thermoplastic composites. In the present paper, the polymer melt impregnation of fibre bundles in injection moulding process is investigated by a specific pull-out test which allows the preparation of the investigated specimen near to process. Several glass fibre roving types with varying appearance and yarn count were placed as a tape insert in a mould for thermoplastic injection process and were impregnated with the injected polypropylene melt. The fibre bundle pull-out test then was used to determine the quality of melt impregnation which could be verified by magnified photomicrograph. A clear relationship between maximal load in pull-out test and impregnated fibre fraction has been observed. Also a strong influence of the fibre bundle geometry and yarn count on impregnation quality was detected.

Keywords: injection moulding, melt impregnation, fibre bundle pull-out test.

1 Introduction

Textile reinforced polymer composites feature outstanding mechanical properties through the possibility of adjusting the textile structure on existing stress. However, missing technologies for fast and fully automated production of thermoset resin based textile composites still limit the applications on fields with low volume production like aerospace or sports engineering. High efficient production technologies are available for fibre reinforced composites with thermoplastic matrix systems like injection moulding or press technology. Traditionally, textile fibre reinforced thermo-plastic composites are melt processed by compression moulding. While press technology is suitable for large and plain shaped structures, injection moulding technology offers complex formed and highly function-integrated parts in combination with short cycle times, high repeat accuracy and relatively low material costs.

Due to the plastification process, fibre reinforced injection moulded parts are characterized through irregular orientated fibres and short fibre length, which is a restriction to mechanical properties. Especially the impact toughness and damage tolerance can be improved by using longer fibres in thermoplastic composite materials [1-3]. This is a crucial fact of interest for crash-sensitive structures, which are primarily needed in high volume automobile production [4]. To combine good mechanical properties and in particular high impact strength with effective injection moulding technology, long fibres or a textile reinforcement has to be placed in the mould cavity. Therefore the melt impregnation, which means the encapsulation of every single fiber by the molten polymer, takes place within the injection process, which is a critical fact due to the high viscosity of thermoplastic melts.

The polymer melt impregnation of textile reinforcements can be classified in a macro and micro impregnation process irrespective of the kind of resin. The macro impregnation describes the melt flow between the fibre bundles of the textile macrostructure. Because of the existing gaps and expanded flow channels it runs with lower pressure gradient and is no limitation for the impregnation process. Much more complicated is the micro impregnation of each roving bundle. Inside the flow gaps are only of a few micrometers in diameter. To keep the melt flow running, high pressure gradients are necessary due to the high viscosity of thermoplastic polymer melts. Caused by the heat transfer from the polymer melt in the cold cavity walls and fibres, the early freezing of the flow front is also a problematic fact.

The impregnation process of textile fibre bundles with a thermoplastic melt was first investigated for the production of long fibre reinforced thermoplastic pellets or tapes. Several authors determined thermoplastic impregnation by using the common method of pulling continuous unidirectional glass fibre bundles over several parallel pins within a polypropylene or polyamide melt pool [5-7]. The pins act as deflexion rollers, whereby the pressure drop in the melt film between roving and roller impregnates the glass fibres. The optimization of process and material parameters like fibre tension, pin diameter, contact angle or melt viscosity resulted in high degrees of impregnation [8, 9].

Melt impregnation of roving is commonly verified evaluating the ratio of number of melt encapsulated filaments to the total number of filaments in the roving. This so called degree of impregnation [5] can be determined in a direct way by preparing photomicrographs of the cross section of the roving and counting the monofilaments. For sufficient confidence level a major number of cross section must be taken into account which can be time consuming. Therefore several authors developed indirect methods to determine voids in the cross section of the roving, which marks not impregnated areas. The degree of impregnation can be determined through measuring opacity of the cross section in relation to the reflectance with a high reflective and completely dark background. The opacity is than significantly higher in areas with voids [9]. In related manner the grounded cross section was treated at one end with colouring agent, which penetrates between the fibres if unimpregnated regions exist, and appeared as dark areas in microscopy [5]. An advantageous melt impregnation assessment technique is the determination of mechanical properties of the composite, which are related to the degree of melt impregnation and avoid time-consuming preparing of polished photomicrographs. The flexural test was used to measure the resistance of the impregnated roving strand to interlaminar shear stress, which is related to the magnitude of unimpregnated regions in the roving. An impregnation assessment number was defined which represents the ratio between the measured force at strand rupture and the maximum measured rupture force, which was assumed to have maximum impregnation [7].

In the present paper a new test method for indirect mechanical measurement of fibre bundle melt impregnation in injection moulding process is proposed, which is currently developed at Chemnitz University of Technology. For the melt impregnation assessment technique the fibre bundle pull-out test was used and adopted on process related test specimen. In a first step the mould had to be developed, which especially should allow the integration of glass fibre rovings in the cavity for thermoplastic melt impregnation. For the determination of the melt impregnation through the mechanical fibre bundle pull-out test the geometry of the moulded part had to be suited for pull-out test set up. In a second step than the injection moulded parts were tested and the correlation of mechanical properties with the grade of impregnation was verified.

2 Fibre Bundle Pull-Out Test for Injection Moulded Test Specimen

The pull-out test of single fibres or fibre bundles is a common test method for investigating fibre matrix adhesion by interfacial shear strength. While the single fibre pull-out test allows the measurement of failure load which can be directly converted in an interfacial shear strength, the fibre bundle pull-out test only indirectly reflects the shear strength of the hole embedded roving [10]. However, the fibre bundle technique is easier to handle and practicable to model the real shear behavior of textile reinforcements. By use in injection molding process the impregnation behavior can be taken into account, which is necessary in this study.

The common single fibre or fibre bundle technique allows the pullout of pure fibres out of a block of resin. Thereby the embedded fibre length must be short enough, to preclude fibre fracture before the fibre debonding of the matrix. The curve of measured load over displacement (Figure 1) has a typical characteristic for both single fibre and fibre bundle pull-out tests. With load build-up the displacement starts linear in the phase of elastic material behavior. With crack initiation and transition in ductile material behavior, displacement increases nonlinear to maximal load, which leads to sudden debonding of the fibre or fibre bundle.

To maintain the typical injection moulding process for impregnation of the fibre bundles the original test set-up of pull-out test and specimen needs to be adapted to the process. For partial insert moulding of the roving, which is necessary to have a free end for pull out, the roving has to lead through the closed and sealed mould, which is critical due to the brittleness of glass fibres. To avoid fibre fractures, the test set-up was modified in such a way, that the whole test specimen was integrated in the cavity. A geometrical form of a T-beam was chosen, whereas the transition of the cross beam in the lengthwise beam was notched deliberately (Figure 2a). The thereby caused crack during load increase led to separation into a pulled-out roving and pulled-out resin block (Figure 2b). The width of the cross beam, which corresponds with the pull-out length was adjusted to 10 and 6 mm. The pull-out length of 10 mm was optimized for rovings with 2400 tex and 1200 tex whereas the 6 mm cross beam especially was developed for lower yarn counts. This reduction avoided fibre fraction before pull-out. The injection mould contains a cold runner system with a runner gate at each cross beam of the part. Sealing rings fasten and seals the roving bundles in the mould.

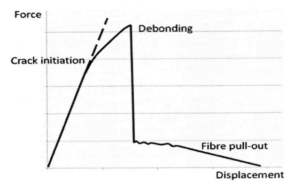

Fig. 1. Load-displacement curve for single fibre pull-out test [11]

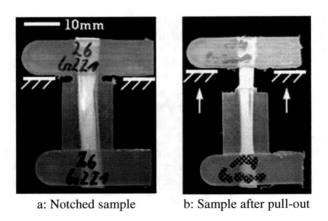

a: Notched sample b: Sample after pull-out

Fig. 2. Injection moulded samples for fibre bundle pull-out test

3 Experimental

3.1 Materials

Polypropylene homopolymer HP500V, supplied by LyondellBasell, was used for melt impregnation of the rovings. The high fluidity polymer was measured with a

Investigation of Polymer Melt Impregnated Fibre Tapes 71

Melt Flow Rate (MFR) of 120 g/10 min (ISO 1133; 230 °C/2.16 kg). The zero shear rate viscosity was measured at 190 °C with 220 Pa*s and at 230 °C with 85 Pa*s. A maleic anhydride modifier (PP-g-MAH), supplied by Kometra (Schkopau, Germany), was added to the polymer with 2 %w/w to improve the fibre-matrix adhesion.

Several commercially available E-glass fibre rovings with 2400 tex, 1200 tex, 600 tex and 300 tex were used for impregnation tests. Besides the variation of yarn count the differentiation in spreading width was considered. The fibre diameter of all types was 16-17 µm. The sizing type for all rovings was an silane/silicone sizing with concentration between 0,4 and 0,64 %w/w which was determined by IR-Spectroscopy and Thermogravimetric Analysis. For further identification of the tested roving types the numeration in Table 1 is introduced.

3.2 Processing Methods

For the impregnation tests a standard injection molding machine with 800 kN clamp force was used. The glass rovings were fit in the ejection side of the open mould. Through closing of the mould the sealing rings fastened and fixed the integrated rovings. For all test series the mould was tempered with 25 °C. The adjusted injection process parameters are summarized in Table 2. At least 15-20 samples of each roving type were produced.

Table 1. Numeration of the tested roving types

Roving No.	Yarn count [tex]	Spreading width [mm]	Sizing concen-tration [%w/w]
Ro1	2400	6,4	0,63
Ro2	2400	5,6	0,41
Ro3	2400	5,5	0,64
Ro4	2400	4,7	0,52
Ro5	1200	4,1	0,60
Ro6	1200	2,9	0,54
Ro7	600	1,8	0,53
Ro8	300	1,1	0,52

After demoulding the symmetrical samples were cut in two halves, whereof one side was used for pullout-test and the other side for preparing the photomicrographs. For the fibre bundle pull-out test a tensile testing machine with a 10 kN load cell was used. Before the pull-out test the spreading width of the roving in each sample was measured in the area of the pulled-out resin block. The resin block than was fixed with form closure clamps to avoid clamping stress which would adulterate the pull-out load. The crosshead moved with a velocity of 2 mm/min, while the load-displacement curve was monitored. The polished specimen for the photomicrographs were taken from the cross section of the roving in the cross beam and lengthwise beam of the injection moulded probes. The photomicrographs were then analyzed in image processing software to detect fraction of impregnated fibers.

Table 2. Process parameters of injection moulding test set-up

Process parameter	Value	Unit
Melt temperature	240	°C
Injection time	1,8	s
Injection Pressure	500	bar
Packing Pressure	600	bar
Part weight to be filled	7,2	g

4 Results and Discussion

The injection moulded samples were processed successfully with all roving types in the above described method. The shifting of the rovings inside the cavity as a result of the high injection pressure could be avoid through the sealing rings, which fastened the fiber bundles. The rovings were in a stretched position in the injection moulded samples which was important for the fibre bundle pull-out-test. A floating of the rovings on the part surface appeared occasionally, especially for the 2400tex rovings. These defective samples were not taken for the pull-out tests.

The failure behavior of the tested fibre bundles in the pullout test agreed very well with the observed failure modes, reported in literature. For ductile matrix systems the typical non-linear slop to the maximum load peak in the load-displacement curve could be observed. While in single fibre bundle test a distinctive kink occurs in the slope which marks the beginning of debonding and interfacial crack initiation, we monitored a continuous slope due to the gradual and superposed fracture and debonding of many single fibres in the bundle. At maximal load the remaining embedded fibre bundle debonded suddenly and the measured force dropped to a lower value. The remaining force level resulted from the friction between fibre bundle and matrix and decreased nearly linear with the pulled out length. By trend samples with high debond load showed lower pull-out loads and vice versa. So the drop in force was higher and with a sharp kink in transition to pull-out force for strong bonds but lower for weak bonds. Typical experimental load-displacement curves are shown in Figure 3.

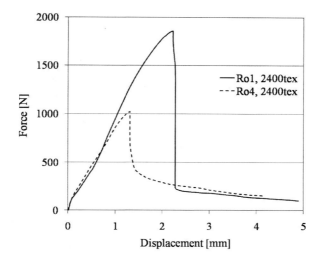

Fig. 3. Typical load-displacement curve for tested pull-out samples

Examination of the tested specimen showed, that the outer fibres of the roving were broken and remained in the cross beam of the probe. A correlation between debond load and quantity of remained fibres was detected, which in detail means that higher debond loads results in more broken and therefore remaining fibres. This is also the reason for the higher drop in force, since the broken fibres did not cause friction force through pull-out any more. Impregnated fibres with good matrix adhesion in the outer area of the roving and a decrease of impregnation in the roving could be observed in the magnified photomicrographs. Therefore the quantity of broken fibres is an indicator for the level of impregnation and fibre-matrix adhesion.

Among all roving types distinctive differences in maximum load, which corresponds to bundle debond load could be observed. Also between rovings with same yarn count a wide range of debond loads was detected (Figure 4). Besides the sizing, which differs between the rovings with regard to sizing type and concentration, the differences in maximum load can be attributed to the spreading width of the rovings.

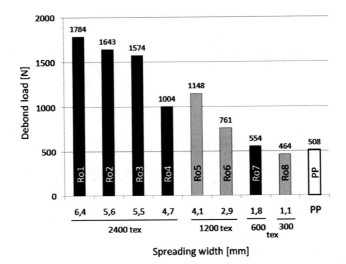

Fig. 4. Debond loads of the tested rovings, sorted to average spreading width in compare to PP-samples without roving

The impregnation of the fibre rovings starts with the melt overflow and is perpendicularly adjusted to overflow direction (Figure 5). Due to the limited flowability of the high-viscosity polymer melt, a better spreading of the roving leads to shorter flow channels and therefore better impregnation with higher debond load values. This therefore allows within same yarn count the defining of a critical spreading width which is necessary for complete impregnation of the rovings.

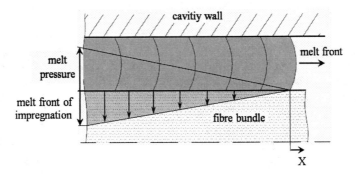

Fig. 5. Model of impregnated fibre bundles during melt overflow

Assuming the fibre bundle geometry with a rectangular cross section, the critical spreading width also leads to a maximum of fibre layers, lying upon another, which the polymer melt can impregnate. This can be clarified through the differences in debond loads between same yarn counts and also between different yarn counts but nearly same spreading width. Within the 2400tex and 1200tex rovings the increase of debond load with higher spreading width was clearly detected. But instead observing a halving of the debond load, when halved the yarn count from 2400 tex to 1200 tex, a smooth transition depending on spreading width was detected. This is because full impregnated 1200 tex rovings with high spreading have the same number of impregnated fibres as fractional impregnated 2400 rovings with poor spreading. This could be verified by the polished photomicrographs, which were taken from the cross section of the embedded rovings (Figure 6). The poor impregnated rovings shows a boundary area with impregnated fibres while the inner fibres remain unimpregnated and therefore without reinforcement effect. With better spreading the number of impregnated fibre layers stays constantly while the inner unimpregnated area becomes smaller. The critical number of fibre layers for melt impregnation then could be estimated which an average value of 10. For rovings with yarn count of 1200 tex a minimum spreading width of 3,5 mm and for 2400 tex a spreading width of 7 mm is necessary.

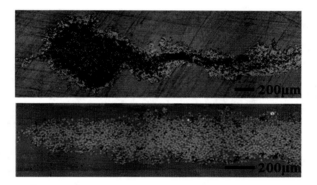

Fig. 6. Photomicrographs of melt impregnated 2400tex glass rovings; above: poor impregnation through low spreading width; below: well impreg-nated roving with high spreading width

To distinguish differences in debond load between the rovings, attributed to sizing and impregnation quality, the spreading width was compared to bundle debond load of each tested sample of several roving types (Figure 7). This clearly shows the increase of debond load with higher spreading width over all roving types. For the 600tex roving (Ro7) a complete impregnation can be assumed through nearly constant debond loads despite higher spreading. Linear interpolation of all points except roving Ro4 and Ro7 leads to a function with a slope of 330 N/mm. Roving Ro4 shows a distinctive deviation from this trend over all tested samples. An increase of debond load with higher spreading width can be observed but on a lower level as the other 2400tex rovings. This can be attributed to poor fibre-matrix adhesion due to not definable sizing properties in the IR-Spectroscopy and Thermogravimetric Analysis.

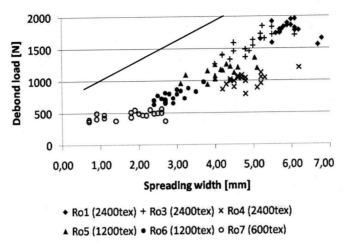

Fig. 7. Debond loads of the different roving types as a function of spreading width

5 Conclusion

The thermoplastic melt impregnation of textile reinforcements within the injection moulding process is an essential requirement to obtain good mechanical properties. Therefore the fibre bundle pull-out test is a suitable near-to-process test method to determine the impregnation quality of the used fibre and polymer materials. In the performed pull-out tests the debond loads correlate well with the resulting fracture of impregnated fibres, which was verified through magnified photomicrographs. Subject to the existing material and process parameters a critical number of fibre layers can be determined, which can be impregnated through the polymer melt. This leads to a necessary minimum spreading width of the roving in order to achieve full impregnation.

Acknowledgements

This work was performed in the project PAFATHERM (Partial Textile Reinforcement of Thermoplastic Injection Moulded Parts). Financial support by the BMBF within the initiative INNOPROFILE is gratefully acknowledged.

References

1. Thomason, J.L.: The influence of fibre length, diameter and concentration on the impact performance of long glass-fibre reinforced polyamide 6,6. Composites, Part A: Applied Science and Manufacturing 40(2), 114–124 (2009)

Investigation of Polymer Melt Impregnated Fibre Tapes

2. Senthil-Kumar, K., Bhatnagar-Naresh; Ghosh-Anup, K.: Mechanical properties of injection molded long fiber polypropylene composites, Part 2: Impact and fracture toughness. Polymer Composites 29(5), 525–533 (2008)
3. Thomason, J.L., Vlug, M.A.: Influence of fibre length and concentration on the properties of glass fibre-reinforced polypropylene: 4. Impact properties. Composites, Part A: Applied Science and Manufacturing 28(3), 277–288 (1997)
4. Schemme, M.: Long-fibre reinforced thermo-plastics (LFT)-development status and pers-pectives. In: 10th International AVK Conference for reinforced Plastics and Thermosets Conference Proceedings, Stuttgart, Germany (2007)
5. Peltonen, P., Lähteenkorva, K., Pääkkönen, E.J., Järvelä, P.K., Törmälä, P.: The influence of melt impregnation parameters on the degree of impregnation of a Polypropylene/ Glass Fibre Prepreg. Journal of Thermoplastic Composite Materials 5(4), 318–343 (1992)
6. Gaymans, R.J., Wevers, E.: Impregnation of a glass fibre roving with a polypropylene melt in a pin assisted process. Composites Part A: Applied Science and Manufacturing 29(5-6), 633–670 (1998)
7. Bates, P.J.M., Charrier, J.M.: Effect of process parameters on melt impregnation of glass roving. Journal of Thermoplastic Composite Materials 12(4), 276–296 (1999)
8. Xian, G., Pu, H.T., Yi, X.S., Pan, Y.: Parametric optimisation of pin-assisted-melt impregnation of Glass Fiber/Polypropylene by Taguchi Method. Journal of Composite Materials 40(23), 2087–2097 (2006)
9. Nygard, P., Gustafson, C.: Continuous Glass Fiber–Polypropylene Composites Made by Melt Impregnation: Influence of Processing Method. Journal of Thermoplastic Composite Materials 17(2), 167–184 (2004)
10. Herrera-Franco, P.J., Drzal, L.T.: Compa-rison of methods for the measurement of fibre/matrix adhesion in composites. Composites (23), 2–27 (1992)
11. Zhandarov, S., Mäder, E.: Determination of interfacial parameters in fiber-polymer systems from pull-out test data using a bilinear bond law. Composite Interfaces 11, 361–391 (2004)

Numerical Analysis of Relaxation Test Based on Prony Series Material Model

Mahdi Mottahedi, Alexandru Dadalau, Alexander Hafla, and Alexander Verl

ISW institute, Stuttgart University, Germany
com73499@stud.uni-stuttgart.de,
alexandru.dadalau@isw.uni-stuttgart.de,
Alexander.Hafla@isw.uni-stuttgart.de,
Alexander.Verl@isw.uni-stuttgart.de

Abstract. In performing an experimental analysis it is always important to take into account different parameters influencing the results. Boundary conditions in addition to the specimen size in case of nonlinear materials could be even more effective. Hence, a full analysis of these parameters before doing any test on the real material seems necessary. In this paper, these effects are numerically analyzed, using ANSYS APDL coding. Samples are cylindrical elastomers becoming ready for a relaxation test in order to get Prony series of the material. The influences of friction and applied displacement in addition to the area-length ratio have been investigated and the optimal values in order to do an appropriate relaxation test are proposed.

Keywords: Prony series, Relaxation test, Boundary condition effect, Hypoviscoelastic material, FEM test modeling.

1 Introduction

The history of natural rubber use comes from 1600 BC when ancient Mayans made a kind of latex material from a specific plant in order to make playing balls. Since then different application of this material from household till industrial and military equipments was extended [1, 2, 3]. The scientific use of synthetic rubber originates after the Second World War when England imposed an embargo on natural rubber against Germans. It then led to first production of synthetic rubber and categorizing them by their physical characteristics. The classical way of categorizing elastomers is differentiating them by their shore hardness and tensile stress. However, this kind of categorizing would not give all necessary information for modeling or investigating the behavior of these kinds of materials [4, 5]. After improving FEM and extending several phenomenological material models for modeling the real responses, it was found that the materials like elastomers could be classified as viscous like materials which react elastically as well. In other words, their stiffness depends on how fast the load is applied to them. They have actually time dependent material properties. It was then realized that such viscoelastic materials could justify the creep and relaxation behaviors. Some more complicated potential energy functions like Ogden and Mooney-Rivlins then helped engineers to model also large strain effect of these materials, namely hyperviscoelastic, where the elastic deformation is large. Moreover, when the whole continuum body has small strain deformation these elastomers would be named hypoviscoelastic and could be modeled by a series of MAXWELL branches were arranged parallel to each other [6, 7].

2 Maxwell Model

Viscoelastic materials could be modeled by different rheological models. One of these basic rheological models for modeling viscoelasticity was proposed by James Clerk Maxwell in 1867. This model includes both elastic and viscous property of the material and consists of a linear ideally viscous Newtonian damper and linear ideally elastic Hookian spring in series. If we put the elements in parallel then the model would be named as Kelvin-Voigt. Figure 1 shows the Maxwell model. The model can anticipate relaxation behavior but may not model the creep [8]. As the elements are in series, the force is equal in each of the parts. However, the total strain would be equal to the summation of strain in elastic and viscous elements. In applying instant displacement, namely relaxation test, the viscous part needs some time to move, however, the spring could move instantaneously. Hence, the whole displacement will be compensated by the spring at time zero. As the time goes on, the dashpot will have also enough time to move, and then the displacement would decrease from the spring and increase in the damper. As the time goes infinity, the strain in elastic part would be zero and therefore its stress will be zero as well. It means the stress in the whole Maxwell branch is then zero. It is but the problem of simple Maxwell model in modeling complicated elastomers. Nevertheless, stress in these materials even in the long time would not lead to zero. Thus, a general Maxwell model for modeling complex behavior of materials in relaxation is needed [9].

By applying instantaneous force onto this model and beginning of dashpot movement, we would see that the dashpot would be elongated till ever. Hence, the Maxwell model has a basic problem in modeling of creep. Instead, we should use Kelvin-Voigt model in creep analyzing [10].

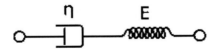

Fig. 1. Maxwell rheological model

3 Generalized Maxwell Model (Prony Series)

As Maxwell model has the problem of giving zero stress at time infinity in relaxation test, the general Maxwell model was proposed to solve the problem and to be used in appropriate modeling of complicated viscoelastic materials. It takes into account that the relaxation does not occur at a single time, but at a distribution of times. Figure 2 shows the rheological model of "General Maxwell".

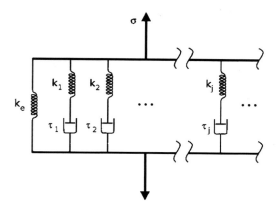

Fig. 2. Rheological model of General Maxwell

In this part Prony series as one of the best functions for modeling the behavior of the elastomers as viscoelastic materials is explained.

As mentioned before the viscoelastic behavior can be divided into large and small deformation. In large deformation modeling one of the hyperelastic models should also be added in order to get a hyperelastic material at the end. However, in case where the strain is small just initial values of shear and Bulk modulus would be enough as the starting values of the material properties over the time. These shear and bulk modules are representative of deviatoric and volumetric parts of the stress respectively -equations 1 and 2.

$$\sigma = \sigma_{deviatoric} + \sigma_{volumetric} \tag{1}$$

$$\sigma = \int_0^t 2G(t-\tau)\frac{de}{d\tau}d\tau + I\int_0^t K(t-\tau)\frac{d\Delta}{d\tau}d\tau \tag{2}$$

Where σ is Cauchy stress, e and Δ are deviatoric and volumetric part of the strains. And G (t) and K (t) are shear and Bulk modulus functions, t and τ are current and past time and I is identity matrix.

Prony series was then proposed by the following formulas 3 and 4 relating shear and bulk modulus over the time.

$$G = G_0\left[\alpha_\infty^G + \sum_{i=1}^{n_G} \alpha_i^G \exp\left(-\frac{t}{\tau_i^G}\right)\right] \tag{3}$$

$$K = K_0 \left[\alpha_\infty^K + \sum_{i=1}^{n_K} \alpha_i^K \exp\left(-\frac{t}{\tau_i^K}\right) \right] \tag{4}$$

Where superscript shows belonging to shear or bulk modulus, and subscript indices the number of series component.

$\alpha_i = \dfrac{G_i}{G_0}$ and τ_i is relaxation time constant for each Prony series component.

α_∞ could be simply calculated by t equal to zero. Then equation 3 changes to equation 5;

$$G_0 = G_0 \left[\alpha_\infty^G + \sum_{i=1}^{n_G} \alpha_i^G \right] \tag{5}$$

where the term $\alpha_\infty + \sum_{i=1}^{n} \alpha_i$ should then be equal to 1. It means that $\alpha_\infty = 1 - \sum_{i=1}^{n} \alpha_i$.

Hence, the only constant of the formula are α_i and τ_i which should be determined by a relaxation test. By choosing the number of Prony components and performing a curve fitting coming from the experimental results of relaxation test, we can get the unknown coefficients and model our viscoelastic material based on Prony series [11].

The initial values of G and K would be taken into account in the series at time equal to zero. In the case where the material is incompressible like elastomers the Poisson ratio could be assumed constant equal to 0.5, on the other words infinite bulk modulus. However, in the case interested, the procedure for finding the coefficients of the bulk function would be the same as shear function. Although there are also some other additional functions which could be implemented for modeling the material behavior in different temperatures, like William-Landel-Ferry shift function, in most cases isothermal procedure assumption in room temperature could be held. Depending on how exact the material parameters are asked, different number of components of the function could be used, but nevertheless most of the time 4 or 5 would be enough [12].

4 Relaxation Test

Relaxation test is becoming more and more important as material models and their application get advanced. A couple of years ago only scientist in universities were trying to do such a test for their purposes. However, nowadays different applications of rubbers and elastomers in industry have increased the importance of categorizing the characteristics of these materials, and relaxation test as a general standard method of determining viscoelastic characteristics has been dominated in industry as well. Using different sealings in industry and tires in automobile companies led to application increase of the test. In the following some points regarding to standard set up of the test will be mentioned.

When a constant strain is applied to a rubber sample, the stress necessary to keep the strain would be decreased over the time. It means that the material resistant against the strain is decreasing gradually. This behaviour is named "stress relaxation". On the other hand, when the elastomeric specimen is subjected to constant stress, the strain is gradually increasing leading to a phenomenon named as "creep" [13, 14, 15].

Chemical processes within the material could lead to relaxation as well. However, in low temperature this effect could be neglected and the most dominant sort of relaxation comes from physical processes.

As mentioned above the appropriate experiments in order to get Prony series of a viscoelastic material are creep and relaxation tests. However, as the Prony series are based on the relaxation, this kind of test would lead to better results and most of the proposed experiments mentioned in standards are therefore based on the relaxation test. The relaxation test is, in simple words, analysis of the material properties when a constant strain is applied. It would then lead to a stress decrease over the time. Some standard methods starting from ISO TC 45, ISO 3384 measuring relaxation test have been also published. Some others like ISO 6056 was related to relaxation test of a ring specimen in liquids. The standard relaxation test in tension was also explained by ISO 6914. The new ISO 3384 is now a mixture of classical ISO 3384 and ISO 6056 for testing both in open air or liquid space. It is still under supervision to improve the exactness of the test. Standard ISO 3384-1999 could be performed in two versions A and B depending on the specimen shape. Both of the tests could be done in either air or liquid. In version A a cylindrical disc of height 6.3 mm, diameter 13 mm is used. In version B a ring of square cross section 2x2 mm and internal diameter of 15 mm is chosen. ISO 6914 which predict the relaxation behaviour by a tension test could be done in two versions. It could be done either by applying a continuous or cyclic strain. The test specimen is 1 mm wide strip elongated 50% [16, 17].

In order to perform this test in the proposed work a cylindrical specimen with length 100 and diameter of 50 mm is considered (figure 3). It would be constrained in one side, although it could be deformed in its radial direction. On the other side, a constant displacement should be applied to the specimen, and then the effect of two different material constants over the time should be considered to have a general idea about the material parameters. Prony series is based on the shear and bulk modulus over the time, as it consists of the two stress terms, namely deviatoric related to the shear and volumetric related to the bulk modulus; as elastomers are mostly behaving incompressible it would be possible to take into account a constant Poisson ratio about 0.5 for the analysis. Nevertheless from the practical point of view in order to prevent unconverged FEM because of volumetric locking, it is needed to take into account the Poisson ratio very close to 0.5, likely 0.4999. The effect of shear modulus could also be considered using modulus of elasticity and using the formula $G=E/(2(1+v))$. Calculating E is simply coming from dividing the stress by strains. However, as in the case of friction we have a constant stress neither in length nor in radial direction of the specimen, a kind of mean stress should be considered in order to give mean stress in the model. Getting the stresses is based on dividing the contact forces over the contact area. By using this method it is also then simple to get the stress when the testing apparatus is working. For calculating the area, it is beneficial to neglect the effect of friction and use the incompressibility characteristics of elastomers as Area= (pi x Radius x Radius)x Length/ (Length-displacement). On the other hand, we are doing a one dimensional analysis to acquire the material parameters in the specific direction here in length of the cylinder, which could simply lead to the strain in that direction as well. As the applied displacement on the material is constant, strain is then calculated as displacement over the initial length of the specimen. At the end, the modulus of elasticity could be calculated by dividing the gained mean stress in length direction dividing by the strain in the same direction, and by using the mentioned formula it could be afterwards transferred into the shear modulus.

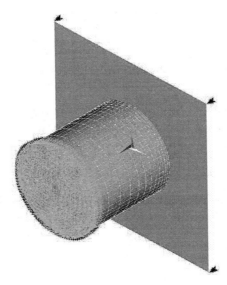

Fig. 3. Topology of the considered cylinder and the boundary conditions applied to it

By having the shear modulus over time and using one of the global optimization algorithm e.g. genetic algorithm, optimal parameters fitting the Prony series for shear could be found. It should be still considered that the ideal conditions where no external effect could influence the experiment would hardly happen. Thus these external parameters should be first of all be detected and then their effects in order to minimize them should be analyzed. In this work these different conditions were detected and analyzed, and as a first step in performing a practical relaxation test, needed parameters for the specimen and external effects in order to perform the best test and consequently getting the material parameters which realistically anticipate the analyzed material were proposed.

These parameters are coming from contacts in constraint and the condition of applying load. Effect of coefficient of friction as an boundary condition, plus the ratio of cross section area over length as the specimen effect in addition to the amount of applied displacement as the experiment characteristic were considered and would be explained in the following sections.

Convergence problem in performing all numerical analysis when nonlinear materials in addition to nonlinear boundary conditions namely contact exist is dominant. The problem frequently comes from volumetric locking effect exists in the case of incompressible behavior and hourglassing problem coming from using reduced integration formulation for the elements. The problem were coming from having both of these troubles simultaneously because of using the full integration method in order to solve hourglassing problem could lead to locking effect on the other hand. Increasing the degree of freedom of the system by refining the mesh still in some cases could not solve the problem. The toughest situation was in the case using no friction and small deformation in order to get convergence. To solve the problem, a small change was done in the FEM model. The contact and target were excluded and the displacement was simply applied to the surface. The result was not only getting faster convergence but also very exact results in fitting the Prony series curve. Compared with using mixed FEM formulation for elements, this method has also benefit from the computational time point of view.

The used elements were 3D solid elements with mid side nodes and the same with quadratic ansatz function for the contact elements. Contact stiffness update was based on each solution iteration according to current mean stress of the underlying individual elements. The contact algorithm were mostly augmented, however, in the case no convergence could be gained Lagrange multipliers in contact and penalty method in target elements were used. It should be considered that the specimen and target element could not be separated however are allowed to be slipped over each other. This effect in addition to giving a logical tolerance value, here 0.01, for the penetration between target and contact elements was also included in the FEM model.

The implemented material was 2 components of the Prony series for modeling hypoviscoelastic materials $(\alpha_1^G = 0.5, \alpha_2^G = 0.25, \tau_1^G = 2.0, \tau_2^G = 4.0)$, in addition to the initial Young's modulus 20e5 Pa and constant Poisson ratio 0.4999.

5 Results

5.1 Effect of Friction Coefficient

In this part five different coefficients of friction were considered. These coefficients were 0.001, 0.005, 0.025, 0.625, and 3.125. The experiment was performed for different displacement as well. 0.001, 0.002, 0.004, 0.008, 0.016 mm were the displacement used for doing the experiment. Getting convergence for the whole 25 experiments because of the locking and hourglassing problem was a tough task. Using an explicit solver could probably simplify the simulation; however, the in hand implicit solver of ANSYS had lots of convergence problems where both friction coefficient and displacement were high or low. Hence, by increasing the displacement the number of DOF were also increased enormously. In addition, full integration method and quadratic shape function elements were used in order to run the project. Using reduced integration method parallel to mixed formulation could also lead to more accurate results. Using tetrahedron element solid 187 which was designed in ANSYS in order to model the elastomers behaviour could also be helpful. However, hexahedral solid 186 was used in the model in order to have well formed mesh. The results also showed that applying displacement on area of the cylinder in the case no friction exists compared to the case where pilot node for target element is used could lead to a better result.

Figures 4 till 7 show the effect of different friction coefficients when the displacement is constant. As it is expected from a relaxation test, as the time is going on the shear modulus is decreasing. It means that the material resistance against the displacement is decreasing. This effect is coming from viscous effect existing in the dashpot of the rheological Prony model. However, this decrease is not continuing for ever, but became closer and closer to its asymptote. As it is expected in the case no friction exists, less G is gained. It means that the material is behaving softer. And as the friction coefficient is increasing it leads to less accurate results because then the material behaves stiffer. On the other hand, this value for high friction is not also going to infinity but even for high values of friction coefficient leads to a specific value. By taking a look at the asymptote where friction is zero, we can conclude that when friction is less than 0.005 we could still get good accurate results as the curve more or less fits the frictionless one. It is also noticeable that all the curves starting from the more or less the same values, because not only the dashpots need time to react against applied displacement but friction could also not affect at time zero. The only parameters here are the initial modulus of elasticity and Bulk which were given to the material at the beginning, and could be gained by the real relaxation test as initial properties of the material.

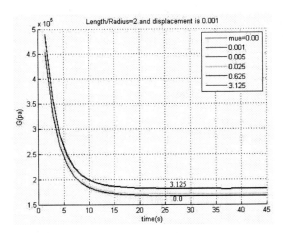

Fig. 4. Relaxation behaviour for different friction coefficients when displacement is 0.001

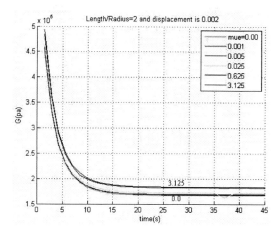

Fig. 5. Relaxation behaviour for different friction coefficients when displacement is 0.002

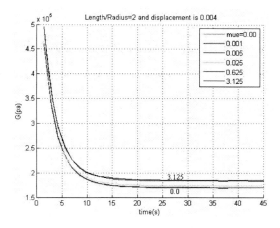

Fig. 6. Relaxation behaviour for different friction coefficients when displacement is 0.004

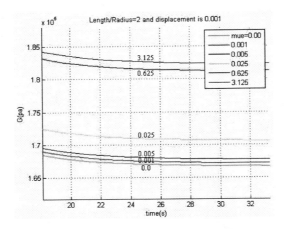

Fig. 7. Relaxation behaviour for different friction coefficients when displacement is 0.001 in a zoomed period of time

5.2 Effect of Applied Strain

It can be deduced from figures 8 and 9 that the applied displacement has some effect on the shear modulus. It is seen that for all cases the curve is decreasing over the time and get close to its asymptote. Comparing different displacements shows that as the applied displacement is increased, the shear modulus is gained from relaxation test is also increasing. Hence, in a specific case we should know how much strain is going to be applied on the system in order to determine the curve and hence the material parameters in that range.

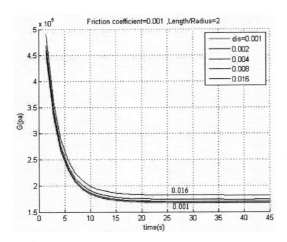

Fig. 8. Relaxation behaviour for different displacements

Numerical Analysis of Relaxation Test Based on Prony Series Material Model

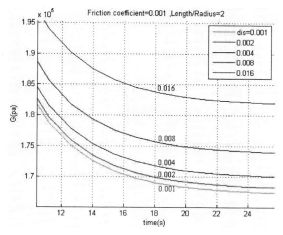

Fig. 9. Relaxation behaviour for different displacements in a zoomed period of time

5.3 Area Effect

As it was said in previous sections by using friction coefficient of 0.005 or less it could be guaranteed that the external effect of friction could be negligible. Here it is shown that by using this friction coefficient, effect of area could be also omitted. By taking a look at figure 10 we see that when applied displacement is 0.005 and friction coefficient is also low and 0.005 there is almost no difference between the curves using different cross section area. As it is obvious, as the cross section area with respect to length is increasing the effect of friction becomes dominant. However, in small amount of the friction this effect has no main influence. By cutting a zoomed area of figure 10 and putting it in figure 11 it is found that by increasing the area, measured stiffness is also increasing, but nevertheless the intensity of its effect depends on the coefficient of friction.

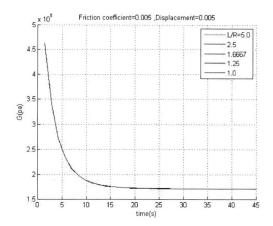

Fig. 10. Relaxation behaviour for different L/R

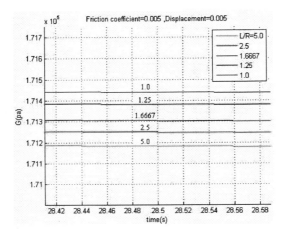

Fig. 11. Relaxation behaviour for different L/R in a zoomed period of time

6 Conclusion

Prony series is one of the appropriate models in order to model the viscoelastic behaviours of elastomers. By using relaxation test the unknown Prony coefficients can be determined. In performing a relaxation test boundary conditions are affecting the coefficients which were noticed and analyzed here. This article showed that as the coefficient of friction between the setup and material is increasing, the material depicts also stiffer characteristics. However, when the coefficient of friction is less than 0.005 no specific change in the stiffness is taking place. Moreover, the measured material behavior is dependent on applied strain. More applied displacement leads to stiffer behavior of the elastomer. So the range of applied strain on the material should be determined in advance in order to get more realistic material parameters. Furthermore, the stiffness increase is directly dependent on L/R factor. Nevertheless when friction coefficient is less than 0.005, this effect could also be neglected.

Acknowledgment

Helpful discussion with scientific staff of Institute for applied mechanics – Stuttgart University, specifically Dr. J. Mendez and providing the project fund by ISW institute is highly acknowledged.

References

1. Martinez, J.M.M.: Natural rubber by a rubber man. Materials Today 9(3), 55 (2006)
2. Mendez, J.: Finite Elastic and Inelastic Behavior of Rubbery Polymers: Experiment and Modeling, Diez, M.Sc. thesis, Institute for applied mechanics, chair 1, Stuttgart University (August 2004)

3. Okwu, U.N., Okieimen, F.E.: Preparation and properties of thioglycollic acid modified epoxidised natural rubber and its blends with natural rubber. European Polymer Journal 37(11), 2253–2258 (2001)
4. Siddiqui, A., Braden, M., Patel, M.P., Parker, S.: An experimental and theoretical study of the effect of sample thickness on the Shore hardness of elastomers. Dental Materials (in Press, Corrected Proof, available online March 4, 2010)
5. Meththananda, I.M., Parker, S., Patel, M.P., Braden, M.: The relationship between Shore hardness of elastomeric dental materials and Young's modulus. Dental Materials 25(8), 956–959 (2009)
6. Park, S.W.: Analytical modeling of viscoelastic dampers for structural and vibration control. International Journal of Solids and Structures 38(44-45), 8065–8092 (2001)
7. Park, S.W., Schapery, R.A.: Methods of interconversion between linear viscoelastic material functions. Part I. a numerical method based on Prony series. International Journal of Solids and Structures 36(11), 1653–1675 (1999)
8. Miehe, C., Linder, C.: Computational mechanics of material lecture note, Institute for applied mechanics, chair 1, Stuttgart University
9. Lewandowski, R., Chorążyczewski, B.: Identification of the parameters of the Kelvin–Voigt and the Maxwell fractional models, used to modeling of viscoelastic dampers. Computers & Structures 88(1-2), 1–17 (2010)
10. Rajagopal, K.R.: A note on a reappraisal and generalization of the Kelvin–Voigt model. Mechanics Research Communications 36(2), 232–235 (2009)
11. ANSYS manual
12. Del Nobile, M.A., Chillo, S., Mentana, A., Baiano, A.: Use of the generalized Maxwell model for describing the stress relaxation behavior of solid-like foods. Journal of Food Engineering 78(3), 978–983 (2007)
13. Zhou, Z.: Creep and stress relaxation of an incompressible viscoelastic material of the rate type. International Journal of Solids and Structures 28(5) (suppl. 1), 617–630 (1991)
14. Cortés, F., Elejabarrieta, M.J.: Modelling viscoelastic materials whose storage modulus is constant with frequency. International Journal of Solids and Structures 43(25-26), 7721–7726 (2006)
15. Haneczok, G., Weller, M.: A fractional model of viscoelastic relaxation. Materials Science and Engineering A 370(1-2), 209–212 (2004)
16. Brown, R.P., Bennett, F.N.B.: Compression stress relaxation. Polymer Testing 2(2), 125–133 (1981)
17. Bassi, A.C., Zerbini, V.: Pressure relaxation. Polymer Testing 6(2), 121–123 (1986)

Powder Metallurgy of Particle-Reinforced Aluminium Matrix Composites (AMC) by Means of High-Energy Ball Milling

Daisy Nestler, Steve Siebeck, Harry Podlesak, Swetlana Wagner, Matthias Hockauf, and Bernhard Wielage

Chemnitz University of Technology, Faculty of Mechanical Engineering,
Institute of Materials Science and Engineering, D-09107 Chemnitz, Germany
`steve.siebeck@mb.tu-chemnitz.de`

Abstract. This paper deals with the production of aluminium matrix composites through high-energy milling, hot isostatic pressing and extrusion. Spherical powder of the aluminium alloy AA2017 (grain fraction > 100 µm) was used as matrix material. SiC and Al_2O_3 powders of submicron and micron grain size (< 2 µm) where chosen as reinforcement particles with contents between 5 and 15 vol.% respectively. The high-energy milling process was realised in a Simoloyer mill (Zoz). The milling time was about 4 hours. Hot isostatic pressing (HIP) was used to convert the compound powder into compact material. The extrusion process realises semi-finished products with different geometrical shapes.

The stages of the composite powder formation during high-energy ball milling will be discussed by means of metallographic studies. The SEM and TEM results show the microstructure of the compact composites obtained by HIP and extrusion. This study focuses on the dispersion and embedding of reinforcement particles and matrix/reinforcement interfaces. Typical appearances like ferrous contaminations through abrasion of the steel balls and the formation of the spinel phase $MgAl_2O_4$ during the subsequent powder-metallurgical processing will be discussed. The semi-finished product exhibits a good particle dispersion and low microporosity. There are no accumulations of the brittle phases nor any microcracks at the interface between reinforcement particles and matrix.

Keywords: Aluminium, silicon carbide, alumina, high-energy ball milling.

1 Introduction

The particulate reinforcement of aluminium alloys is carried out with the main aim of improving mechanical properties such as modulus, strength, and creep resistance (even at elevated temperatures). In addition, the wear resistance of the material increases, too [1]. Type, size, quantity and distribution of the introduced reinforcements have great influence on the intensity of property changes. A high degree of dispersion, the full integration of the hard material in the metal matrix and the formation of a suitable interface condition are the key to achieving the desired range of properties [2, 3]. The particle size exerts significant influence on the distribution and dispersion of the reinforcements. Coarser particles tend to have a worse embedding quality. Moreover, they tend to break during the powder-metallurgical processing or post-deformation [4–6]. Aluminium matrix composites (AMC) can be produced by the powder route or by the melting route. The powder route has some advantages. Because of the lower temperatures, there are no chemical reactions between matrix and reinforcements as they occur in the molten state [2, 7, 8]. With decreasing particle sizes of the reinforcement components, this aspect gains importance. The composite powders with a high degree of dispersion are produced in the first process step and then compacted in the second step. High-energy milling (HEM) is a suitable method for the production of composite powders [9–11].

This paper deals with the production and characterisation of aluminium composite powders with a fine-scaled hard-material component as the starting material for the powder-metallurgical production of massive AMC materials. First results were presented in [12, 13]. A number of further investigations on the manufactured materials will be carried out. Besides the determination of mechanical properties and equal-channel angular pressing (ECAP) for further structural optimisation [14], among others the machinability of the material for further processing is of great interest [15].

2 Experiments

The aluminium alloy of the 2000 series, which was used as matrix material, is characterised by good toughness and high resistance to crack propagation under cyclic loading in the cold-hardened state (T3, T4) [16, 17]. Furthermore, this alloy has excellent high-temperature strength [16]. Its composition is similar to that of the alloy AA2017, only the silicon content is slightly lower (Table 1). Gas-atomised spherical powder with a fraction size < 100 μm (Fig. 1) was used as feedstock. The cellular cast structure contains the primary Al solid-solution phase and intermetallic phases.

The experiments presented in this paper were carried out with small-grained alumina powders with a fraction of $0.2 - 2$ μm as reinforcing components (Fig. 2). The reinforcement amount was varied between 5 vol.% and 15 vol.%. The high-energy milling was carried out with a ball mill Simoloyer® CM08 (Zoz) with steel equipment in air atmosphere.

Table 1. Composition of matrix powder as-received

Al	Cu	Mn	Mg	Fe	Si	Zn
> 94.10	4.10	0.84	0.58	0.22	0.08	< 0.01

Because of the high energy density during milling at high rotational speeds, in the case of metals, the effect of welding predominates. In contrast, the shear forces that occur at low speeds break the material. However, for ductile materials such as aluminium, the comminution is very limited. It occurs only in already highly flattened particles [9]. To limit the welding of the particles and any undesirable build-up on the rotor, the balls and the chamber wall, small amounts of stearic acid are added as process control agent (PCA).

Metallographic sections of the powder and the bulk material were prepared and characterised in the unetched condition by light microscopy and scanning electron microscopy (LM, SEM). In order to optimise the process parameters, cross-sections of the composite powder after different milling times are necessary. The increasing integration and distribution of particles in the matrix is well represented by means of light microscopy. For SEM, a secondary-electron detector (SE) and a backscattered-electron detector (BSE) are used in parallel. The SE detector predominantly records the topographical contrast. With this method, a way to map the SiC particles is given. In contrast, the BSE detection is used in the material contrast mode. This way is particularly well-suited for the imaging of intermetallic phases and small pores.

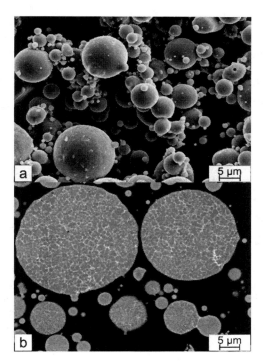

Fig. 1. AA2017 powder in as-received state, SEM micrographs of: a) topography; b) cross-section

Fig. 2. Alumina in as-received state, SEM micrograph, topography

Transmission electron microscopy (TEM) is additionally used to investigate the compact material. The interpretation of the TEM images is relatively complicated because of the dominant diffraction contrast. For the visualisation of the particle phases and the internal interfaces, the observation of several tilting angles is often required. However, the situation is difficult for the present material. Due to the irregular shape of the SiC particles, sample sites with a vertically truncated interface are rarely found. In addition, they show similarities with other phases. In this respect, it is necessary to use the energy-dispersive x-ray spectroscopy (EDXS) for phase identification of the individual particles.

3 Results

3.1 High-Energy Ball Milling

The investigation of cross sections allows much profounder statements compared to the method of powder characterisation using powder surface imaging [11, 18, 19]. Apart from the shape of the powder particles, changes in the structure and the input and distribution of the reinforcements are detectable. Although it is impossible to depict fine-scaled SiC particles individually in light-microscopic investigations, a rough characterisation of the distribution of reinforcements by means of bright-field light microscopy is successful. While the areas without reinforcements appear bright, the areas with reinforcements are dark. SEM images were used for the evaluation of the agglomeration of the reinforcements and the microporosity.

The formation of the composite powder can be divided into typical stages (Fig. 3). Initially, the spherical aluminium particles are deformed into flat particles. Simultaneously, the reinforcements attach on the surface of these flakes. In the next stage, the effect of cold welding starts and leads to the formation of larger composite particles with lamellar structure. The resulting structure is a mixture of alternating reinforced and unreinforced lamellas. Free particles are no longer existent at that stage. A steady deformation of composite particles causes an increase in mixing and thus an improvement of the dispersion degree. The chronological procedure of the composite powder formation depends on the milling parameters. Strong welding effects due to high rotational speeds lead to premature formation of large, poorly mixed particles. Additionally, a material adhesion on the surfaces of the milling tools leads to losses of grinding stock. Therefore, a cyclical change of speed seems to make sense. Good results will be achieved by varying between short intervals at high speed (700 rpm) and long intervals at low speed (400 rpm).

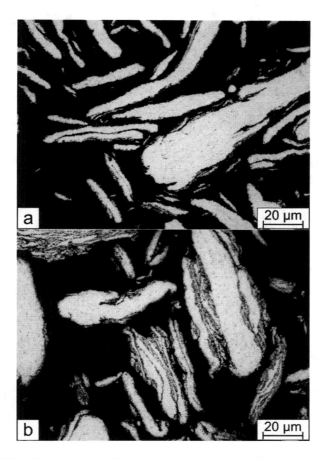

Fig. 3. AA2017 with 10 vol.% Al_2O_3; optical micrographs of polished sections of composite powder after different milling times: a) 1 h, b) 2 h, c) 2 h, d) 4 h

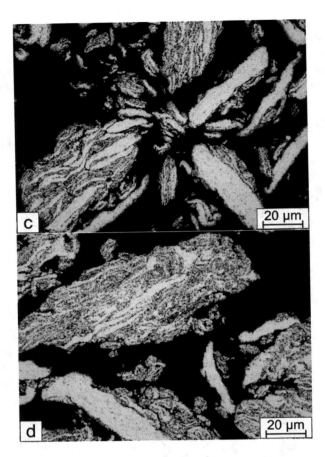

Fig. 3. (*continued*)

Not all powder particles pass through the stages simultaneously. Consequently, there is no sharp separation between the individual stages, but rather they go into each other and overlap. A small but not insignificant number of particles contain little or no hard material even in the final stage of the powder. The particle size distribution is rather broad.

The extension of the grinding time in favour of the "homogenisation" of the composite powder is limited. With increasing grinding times, the cold welding causes a progressive coarsening of the composite powder and increasing deposits on the grinding tools. This loss of grinding stock can be reduced by the use of larger amounts of PCA. Longer milling times are therefore possible in principle. However, it has been shown that this measure increases the degree of contamination of the powder by means of steel particles. Such particles were identified using

EDXS measurements as residues from grinding balls (detection of iron and chromium) and rotor wings (detection of iron, vanadium and chromium).

The entry of steel particles with dimensions up to a few 10 micron sizes is due to the erosion at the surface of the grinding tools (Fig. 4). The increased PCA amount causes a decrease in the adhesion of the aluminium on the metal surfaces and thus reduces the protection effect of the build-up against erosion.

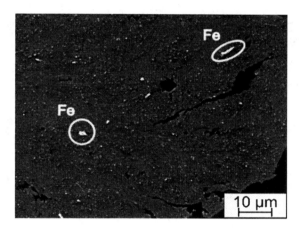

Fig. 4. AA2017 composite powder particle with iron contaminations, SEM micrograph in material contrast mode

The contamination of the grinding stock by milling tools is known in principle but has not yet been much discussed in specialist publications about AMC production [18, 20]. Our own studies have shown that the steel particles are transformed during the powder-metallurgical processing of the composite powder. Because of the high temperatures during hot degassing (450 °C), hot isostatic pressing (450 °C) and extrusion (355 °C), aluminium and copper atoms diffuse into the steel particles, which causes the creation of intermetallic phases of the type $Al_3(Fe, Cu)$. With particle sizes of 1-50 microns, the uniformity of the AMC structure is strongly affected [12, 13]. Similarly, the incorporated stearic acid must be regarded as a contaminant. The stearic acid is liquid till about 69 °C. Personal thermogravimetric studies show that stearic acid evaporates in an argon atmosphere at about 300 °C and burns in air at about 186 °C. Although these temperatures cannot be achieved in the milling container, a decomposition of the PCA is possible due to the local but very severe thermo-mechanical conditions at the moment of ball collision [21]. Because a residual content of stearic acid in the composite powder is likely to be present, a hot degassing of the powder must be carried out before the powder-metallurgical processing. Otherwise, an undesirable pore formation during extrusion or during a subsequent thermal post-treatment of the AMCs cannot be excluded. Thus, the present results can be seen as a compromise between the degree of dispersion and the percentage of impurities from steel particles and stearic acid. The amount of well-dispersed areas is roughly 80 %. The remaining unmixed areas occur in the form of individual metal particles or lamellae of the composite powder particles. Insofar, the above parameters have led to satisfactory results. Scanning-electron-microscopic studies show an adequate embedding of the hard material in the metal matrix without significant agglomeration. In contrast to experiments with coarse reinforcements [4–6], there are no indications for the destruction of the reinforcements during the milling. The size of the reinforcements does not differ significantly from the original size fraction. It can be concluded that the SiC powder is pressed rapidly into the aluminium particles and thus protected against breakage.

3.2 Microstructure in the Extruded State and after T4 Treatment

After the hot isostatic pressing (450 °C) of the composite powder and the subsequent extrusion (ca. 350 °C) to rods, a compact material is obtained. The cross-section shows a nearly homogeneous distribution of the particles (Fig. 5). Only a few areas (light) contain no reinforcements. However, the longitudinal section clearly reveals certain lamellarity (Fig. 6) which is typical for extruded materials and can be attributed to the anisotropic arrangement of unmixed, rod-shaped areas in longitudinal direction. The shaping of the lamellarity is characterised by the inhomogeneity of the composite powder. So, a better mixing of the powder leads to fewer lines.

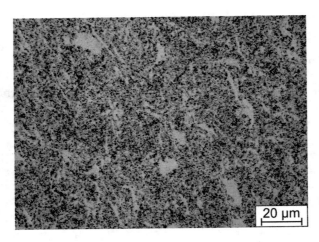

Fig. 5. LM microsection, Al + 15 vol.% Al$_2$O$_3$ in extruded state, lateral to extrusion direction

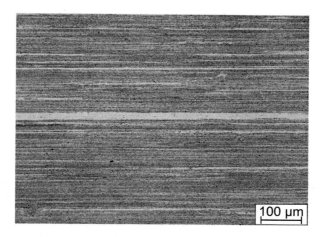

Fig. 6. LM microsection, Al + 15 vol.% Al$_2$O$_3$ in extruded state, longitudinal to extrusion direction

Scanning-electron-microscopic investigations allow the high-resolution imaging of the reinforcements. The images of the AMC with Al$_2$O$_3$ in Fig. 7 are representative for the structure in mixed areas. In addition to the image with SE electrons, the image with backscattered electrons is shown. The latter is suitable for the detection of micropores (black) and intermetallic phase particles (white) in the metal matrix. The extensive distribution of the reinforcements without marked agglomeration is clearly determined by the images. The extruded material shows low microporosity in the form of individual pores of about 1 µm in the maximum.

The number of micropores is lower for a hard-particle content of 5 vol.% than for 15 vol.%. The necessary embedding of the reinforcements into the matrix takes place. Neither microgaps nor differently formed cavities occur at the reinforcement/matrix interface. The mentioned structural characteristics apply for the AMCs with Al$_2$O$_3$ as well as for the AMCs with SiC particles. These statements are confirmed by the TEM investigations in the T4-heat-treated state (505 °C). In TEM overview images (Fig. 8), the image interpretation is much more difficult because the mapping of grey values to phases in the microstructure is not easily feasible, and some of the particles are superimposed on the image plane in the projection of the irradiated volume with a final thickness of about 200 nm.

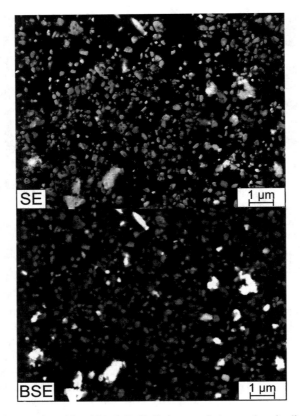

Fig. 7. SEM microsection, Al + 15 vol.% Al$_2$O$_3$ in extruded state, longitudinal to extrusion direction

Fig. 8. TEM micrograph, AA2017 with 15 vol.% Al$_2$O$_3$/T4-aged

The aluminium solid solution/matrix phase is visible as a bright phase. The thin black lines represent the grain boundaries. The particle size varies between 0.5 and several microns. The embedded particles appear mainly medium to dark grey. These include both Al_2O_3 and SiC particles as well as intermetallic phases in the metal alloy.

The identification of individual particles is successful on the basis of the particle form and the additional use of X-ray microanalysis (EDS). The intermetallic particles can be assigned to the Al_2Cu phase. The particle size reaches about 1 micron in maximum. Al_2O_3 and SiC particles have an irregular shape and a locally varying stripe contrast (Fig. 9, Fig. 10). In addition, there are oxide particles of some 10 nm in the metal matrix which are not visible in the SEM investigations of the microsection. Frequently, they are present as thin polyhedral tiles (Fig. 9, Fig. 10). This morphology, but in particular the contrast for different sample angles as well as in the dark-field mode, confirms the presence of a crystalline state. Phase identification by means of fine-area diffraction was impossible because of the small particle size and the mixing with other phases. However, the EDXS measurements revealed an oxide with Al and Mg contents (Fig. 11). The determination of the chemical composition is difficult because the analytical results may be distorted by the surrounding Al matrix. For this reason, the determined element concentrations are given in the following form: >50 O; >10 Mg; <40 Al in at.%, so the composition is close to spinel $MgAl_2O_4$ (57 O; 14 Mg; 28.5 Al in at.%). Because of the crystallinity, it is assumed that it is spinel. The oxide particles are distributed at the reinforcement/matrix interface as well as in the Al matrix.

In the case of SiC as a reinforcing component, there are single rod-shaped crystallites at the interface in addition (Fig. 12). Because of the shape and the results of the EDS analysis, it is assumed that it is aluminium carbide. The frequency of these particles is very low.

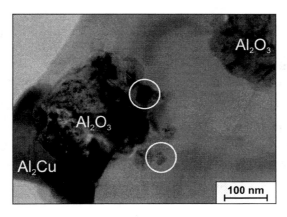

Fig. 9. TEM micrograph, AA2017 with 15 vol.% Al2O3/T4-aged, circles: crystalline Mg-Al oxides

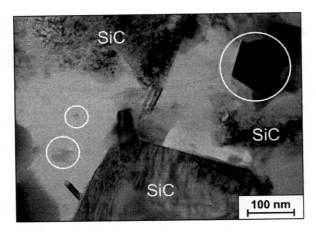

Fig. 10. TEM micrograph, AA2017 with 15 vol.% SiC/T4-aged, circles: crystalline Mg-Al oxide particles

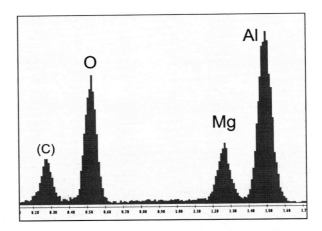

Fig. 11. EDS point analysis of an oxide particle (spinel)

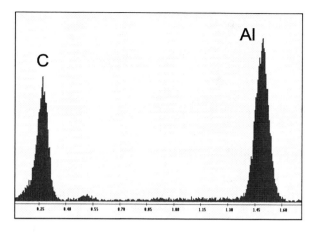

Fig. 12. EDS point analysis of an aluminium carbide particle

4 Conclusions

The high-energy ball milling has been tested and optimised successfully for the production of aluminium composite powders with up to 15 vol.% alumina with reinforcement sizes between 0.2 and < 2 µm. Therefore, the formation of the composite powder was characterised in detail by optical micrographs. Scanning-electron-microscopic examinations have confirmed the statements and contributed to the assessment regarding agglomeration and embedding in the metal matrix. The process parameters rotor speed, grinding time and the proportion of the added stearic acid partly influence the formation of the composite powders, the contamination and the loss of grinding stock in the opposite direction so that the obtained adjustment constitutes a compromise. From this perspective, the achieved degree of dispersion is acceptable.

After the consolidation of the compound powder, the composite material shows the necessary embedding of the reinforcements. The interface reinforcement/Al is designed without voids and without significant Al carbide phases in the case of SiC. Because of little deficiencies in the reinforcement distribution, there is a relatively small band-type formation in the direction of extrusion. In this respect, a possibility for improving the HEM process can be seen. By detailed TEM studies, the formation of an oxide phase due to the process conditions is shown. Because of its small particle size and wide distribution in the metal matrix, no negative but rather a positive effect on the mechanical properties is expected.

Acknowledgements

The authors would like to thank the "Deutsche Forschungsgemeinschaft" (DFG) for supporting the research project SFB 692 A2.

References

1. Drossel, G.: Umformen von Aluminium-Werkstoffen, Gießen von Aluminium-Teilen, Oberflächenbehandlung von Aluminium, Recycling und Ökologie. Aluminium-Taschenbuch, vol. 2, 15th edn. Aluminium-Verlag, Düsseldorf (1999)
2. Beffort, O., Long, S., Cayron, C., Kuebler, J., Buffat, P.A.: Alloying effects on microstructure and mechanical properties of high volume fraction SiC-particle reinforced Al-MMCs made by squeeze casting infiltration. Compos. Sci. Technol. 67(3-4), 737–745 (2007)
3. Cheng, N.P., Zeng, S.M., Liu, Z.Y.: Preparation, microstructures and deformation behavior of SiCP/6066Al composites produced by PM route. J. Mater. Process. Technol. 202(1-3), 27–40 (2008)
4. Ozdemir, I., Ahrens, S., Mücklich, S., Wielage, B.: Microstructure Characterization of Al-Al2O3p Composites Produced by High Energy Ball Milling. Prakt. Met. 44(3), 103–112 (2007)
5. Ozdemir, I., Ahrens, S., Mücklich, S., Wielage, B.: Nanocrystalline Al-Al2O3p and SiCp composites produced by high-energy ball milling. J. Mater. Process. Technol. 205(1-3), 111–118 (2008)
6. Ozdemir, I., Ahrens, S., Mücklich, S., Wielage, B.: The Production of Ultra Fine Grained Al-SiCp Composites Produced via High Energy Ball Milling. Prakt. Met. 45(3), 136 (2008)
7. Torralba, J.M., Da Costa, C.E., Velasco, F.: P/M aluminum matrix composites: an overview. J. Mater. Process. Technol. 133(1-2), 203–206 (2003)
8. Shorowordi, K.M., Laoui, T., Haseeb, A., Celis, J.P., Froyen, L.: Microstructure and interface characteristics of B 4 C, SiC and Al 2 O 3 reinforced Al matrix composites: a comparative study. J. Mater. Process. Technol. 142(3), 738–743 (2003)
9. Suryanarayana, C.: Mechanical alloying and milling. Prog. Mater. Sci. 46(1-2), 1–184 (2001)
10. Maneshian, M.H., Simchi, A., Hesabi, Z.R.: Structural changes during synthesizing of nanostructured W-20 wt% Cu composite powder by mechanical alloying. Mat. Sci. Eng. A 445-446, 86–93 (2007)
11. Tjong, S.C.: Novel Nanoparticle-Reinforced Metal Matrix Composites with Enhanced Mechanical Properties. Adv. Eng. Mater. 9(8), 639–652 (2007)
12. Wielage, B., Mücklich, S., Podlesak, H.: Gefügecharakterisierung von hochenergiegemahlenen Verbundpulvern und Verbundwerkstoffen mit EN AW 2017-Matrix. In: Krenkel, W. (ed.) Symposium Verbundwerkstoffe und Werkstoffverbunde. Verbundwerkstoffe, vol. 17, pp. 52–58. Wiley-VCH, Weinheim (2009)
13. Podlesak, H., Siebeck, S., Mücklich, S., Hockauf, M., Meyer, L., Wielage, B., Weber, D.: Pulvermetallurgische Erzeugung von SiC- und Al2O3-verstärkten Al-Cu-Legierungen. Materialwiss. Werkstofftech. 40(7), 500–505 (2009)
14. Hockauf, M., Schönherr, R., Wagner, S., Meyer, L., Podlesak, H., Mücklich, S., Wielage, B., Krüger, L., Hahn, F., Weber, D.: ECAP-Umformung mittel- und hochfester ausscheidungshärtbarer Aluminiumknetlegierungen. Materialwiss. Werkstofftech 40(7), 540–550 (2009)
15. Schubert, A., Funke, R., Nestler, A.: Oberflächenausbildung beim Drehfräsen von Aluminiummatrix-Verbundwerkstoffen mit Diamant-Werkzeugen. wt Werkstattstechnik online 99(11/12), 858–864 (2009)

16. Totten, G.E.: Physical metallurgy and processes. Handbook of aluminum, vol. 1. Dekker, New York (2003)
17. Brandes, E.A., Brook, G.B.: Smithells light metals handbook. Butterworth-Heinemann, Oxford (1998)
18. Prabhu, B., Suryanarayana, C., An, L., Vaidyanathan, R.: Synthesis and characterization of high volume fraction Al-Al2O3 nanocomposite powders by high-energy milling. Mat. Sci. Eng. A 425(1-2), 192–200 (2006)
19. Hesabi, Z.R., Simchi, A., Reihani, S.M.: Structural evolution during mechanical milling of nanometric and micrometric Al2O3 reinforced Al matrix composites. Mat. Sci. Eng. A 428(1-2), 159–168 (2006)
20. Cleary, P.W.: Predicting charge motion, power draw, segregation and wear in ball mills using discrete element methods. Min. Eng. 11(11), 1061–1080 (1998)
21. Kleiner, S., Bertocco, F., Khalid, F.A., Beffort, O.: Decomposition of process control agent during mechanical milling and its influence on displacement reactions in the Al-TiO2 system. Mater. Chem. Phys. 89(2-3), 362–366 (2005)

Initial Stress Behaviour of Micro Injection-Moulded Devices with Integrated Piezo-Fibre Composites

Lothar Kroll, Marco Walther, Wolfgang Nendel, Michael Heinrich, Matthias Klärner, and Jürgen Tröltzsch

Chemnitz University of Technology, Kompetenzzentrum Strukturleichtbau e.V., Chemnitz, Germany
lothar.kroll@mb.tu-chemnitz.de,
marco.walther@slb.tu-chemnitz.de,
wolfgang.nendel@hrz.tu-chemnitz.de,
michael.heinrich@mb.tu-chemnitz.de,
matthias.klaerner@mb.tu-chemnitz.de,
juergen.troeltzsch@mb.tu-chemnitz.de

Abstract. Based on an increasing demand for function integration in small components, micro injection moulding offers a highly productive solution to combine plastic structures with additional electronic and mechatronic features. Particularly two-component micro injection moulding allows embedding of active elements as well as the application of electric contacts in one process step by using insolating and conductive compounds, respectively. The study outlines major effects on the thermo-mechanical compatibility of active modules comprising several stacked piezo fibre composite beams. With regard to material properties, geometry of the module, machine set up and processing parameters a structural and strength analysis including process induced residual stress is used to predict favourable material combination and composite design.

Keywords: two-component micro injection moulding, piezo fibre composite beams, electrical conductive thermoplastics.

1 Introduction

The increasing miniaturisation of products including electronic and mechatronic assemblies leads to a maximal integration of functions and developments of manufacturing technologies for very small multi-component devices. For this purpose the micro injection moulding technology for plastic based micro parts is outstandingly appropriated. This technology is able to produce plastic parts with a high degree of shape-freedom in a high quantity without post processing. The flexibility of this process allows the combination of different materials and the integration of metallic and ceramic inserts as well as electronic or magnetic functional parts [1, 2].

Particularly the in the electrical industry required increase of the functional density of injection moulding parts with small installation space is a specific challenge, wherein the internal stress effect and the shrinkage effect have a high weight [3]. For this purpose are basic numerical and verifying experimental analysis of the mould-filling characteristics, of the warpage as a result of thermal stress to the anisotropic mechanical structural behaviour and the dependence between thermal mechanical material stress and process related parameters essential [4-6]. The usage of plastic based piezo ceramic modules as actuator, sensor and as energy supplier have already been discussed in theoretical and experimental studies with different objective targets [7-10].

The ideal method for the economical bonding and integration of piezo elements in plastic based parts are compounded polymers with conductive additives, because of their cheapness in large quantities. By means of the two-component micro injection moulding process it is possible to injection-mould the technical embedding and bonding of a piezo ceramic component in use of an electrical conductive thermoplastic (E-plastic), as well as to enable defined mould cavities with an electrical insulated plastic (I-plastic), so that the electrodes and piezo elements lie electrical separated. The considered piezo fibre composites are made of uniaxial piezo fibres (lead zirconate titanate (LZT), Sonox® P53), which are embedded in epoxy resin (circa 70% vol. fibre content), wherein the piezo fibres are arranged perpendicular to the rod axis. One possible example of a micro injection moulding produced µIM-Piezomodule (micro injection moulding piezo module) is shown in figure 1.

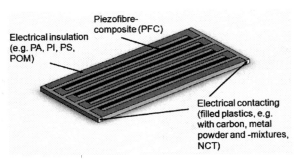

Fig. 1. Design example of the µIM-Piezomodul

2 Mechanics of Materials

For the numerical calculation of the internal stress the material parameters of the single components are needed. Therefore the E-plastic (TinCo50) and the I-plastics (PA6, PA6-GF20, PA6-GF50) can be assumed as isotropic material, whereas the piezo-fibre-stick (PFS) component shows, because of their unidirectional alignment of the ceramic fibres, a distinctive orthotropy (transversal isotropy). The analysed material properties of the regarded plastics are subsumed as averages in tab. 1. Basically high volume contents of fibres result in increasing of the stiffness and strength and to drop of the tensile stretch and thermal expansion. Without regarding the strength this is similar for TinCo compounds. The tensile strength of TinCo50 compound reaches less than a half of the tensile strength of the matrix of PA6. Furthermore, the hybrid plastics show a clearly lower ductility. The tensile stretch drops to less than tenth of the initial polymer.

The strength of the TinCo50 produced μ-tension members is lower than the strength of standard tension members. From this it follows that the inhomogeneous micro structure of the filled polyamide 6 (distribution of the particle size, effects of alignment and side effects) causes a decidedly debilitation of the material at small profiles (μ tension member: 2 x 1,5 mm).

The unidirectional configuration of the piezo ceramic fibre in epoxy resin implicates intensely different mechanical and thermal attributes in direction of the rotational axis of the material. Whereas the piezo fibres are vertical to the rod axis. Thereby the rotational axis of the material coincide with the axis of the squarish profile of the PFS. The bonding takes place along the oppositely stick's longitudinal surface, which complies with the effective direction across the stick's longitudinal axis.

Table 1. Averaged material properties

Material	Young's Modulus [MPa]	Tensile Strength [MPa]	Thermal Expansion $[10^{-6}$ K$^{-1}]$
TinCo50 (standardtest member)	3,950	35.4	20
TinCo50 (μ tension member)	3,700	34.3	-
PA6	3,000	85.0	80
PA6-GF20	6,600	140	50
PA6-GF50	12,200	180	30

For a fail-safe actuatory reliable operational performance of such piezo converters, a preload pressure of about 30 MPa is aspired.

The engineer constants and basic strength of the piezo fibre composites (composite material PF-EP01) are given in table 2.

Table 2. Engineer constants and basic strength of the transversal isotropic PFS *(PF-EP01)*; legend: t:tension, p: pressure, no superior indices at R: shear strength; ||: parallel, ⊥: normal, #: parallel/normal

Young's Modulus [MPa]	$E_{\|}$	40,000
	$E_{\perp}$	8,000
Shear Modulus [MPa]	$G_{\#}$	3,500
	$G_{\perp\perp}$	3,600
Transversal contraction [-]	$\nu_{\|\perp}$	0.3
	$\nu_{\perp\perp}$	0.11
Thermal expansion $[10^{-6}\,K^{-1}]$	$\alpha_{\|}$	6.9
	$\alpha_{\perp}$	38
Strengths [MPa]	$R_{\|}^{t}$	35
	$R_{\|}^{p}$	100
	$R_{\perp}^{t}$	20
	$R_{\perp}^{p}$	80
	$R_{\perp\|}$	25
	$R_{\perp\perp}$	27

3 Process Conditioned Internal Stress

The analysed material characteristic values of TinCo50 and the basic properties of the orthotropic PFS build together with the characteristic values of the I-plastics the necessary material pool (material cards) for FE calculation of the internal stress and the involved warpage conditions. To estimate the behaviour of the whole module in a raise of stiffness and strength of the I-plastic component PA6 with the short-fibres filling degrees: 0-, 20-, 50-vol. % were prospected. The modelling and simulation were accomplished with ANSYS for three building variations (see also figure 1):

    I.   active layer (without top layer),
    II.   asymmetric structure (one top layer),
    III. symmetric structure (two top layers),

whereas the top layer dimensions amount 20 mm x 10 mm x 1 mm and the PFS dimensions amount 20 mm x 1 mm x 1 mm.

In the first FE analyses initially a stationary cooling process was assumed (ΔT= -200 K) and the flow-induced local anisotropy of the PA20 or PA50 component was neglected. To calculate failure critical stresses of the PF compound the fibre reinforced UD compounds proven failure criterion according to CUNTZE will be used. The single failure criterion are given by courtesy of a special invariant equation (1) (for more information see [11, 12]):

$$\text{FF1:} \quad F_\parallel^\sigma = \frac{I_1}{R_\parallel^t} \leq 1$$

$$\text{FF2:} \quad F_\parallel^\tau = \frac{-I_1}{R_\parallel^p} \leq 1$$

$$\text{IFF1:} \quad F_\perp^\sigma = \frac{I_2 + \sqrt{I_4}}{2R_\perp^t} \leq 1 \tag{1}$$

$$\text{IFF2:} \quad F_{\perp\parallel} = \frac{I_3^{3/2}}{R_{\perp\parallel}^3} + b_{\perp\parallel}\frac{I_2 I_3 - I_5}{R_{\perp\parallel}^3} \leq 1$$

$$\text{IFF3:} \quad F_\perp^\tau = (b_\perp^\tau - 1)\frac{I_2}{R_\perp^p} + \frac{b_\perp^\tau I_4 + b_{\perp\parallel}^\tau I_3}{(R_\perp^p)^2} \leq 1$$

with F_i as failure index of the respective failure mode (FF1: fibre fracture pull, FF2: fibre fracture pressure, IFF1: inter fibre fracture pull, IFF2: inter fibre fracture thrust vertical/parallel, IFF3: inter fibre fracture thrust vertical/vertical) and

$$\begin{aligned}
I_1 &= \sigma_1 \\
I_2 &= \sigma_2 + \sigma_3 \\
I_3 &= \tau_{31}^2 + \tau_{21}^2 \\
I_4 &= (\sigma_2 - \sigma_3)^2 + 4\tau_{23}^2 \\
I_5 &= (\sigma_2 - \sigma_3)(\tau_{31}^2 - \tau_{21}^2) - 4\tau_{23}\tau_{31}\tau_{21}
\end{aligned} \tag{2}$$

as well as the material dependent free curve parameters

$$\begin{aligned}
b_{\perp\parallel} &= 0,1 \\
b_\perp^\tau &= 1,3 \\
b_{\perp\parallel}^\tau &= 0,2
\end{aligned} \tag{3}$$

The constitution of the piezo module out of three different material components requires optimal connection properties for a good force transmission and functional efficiency of the piezo sticks. Especially the connection between PFS and the E-plastic for bonding and transmission as well for the I-plastic for electrical insulation is significant for a reliable function of the hybrid compound. Also the adhesive strength of the surrounding E- and I-plastics among each other essentially determine the total structure behaviour of the piezo converter module. Basically the interlaminar tensile strength will be distinguished from shear strength, which were ascertained in tensile or shear tests. The experimental determined strength values are summarized in table 3 and will be used further as basics to interpret the FE results.

Table 3. Averages of the inter laminar strength

Material combination	Tensile test, tensile strength [MPa]	Shear test, shear strength [MPa]
PA6/TinCo50	19	21
PA6-GF20 /TinCo50	26	29
PA6-GF50 /TinCo50	29	31
PFS/I-plastics	18	22
PFS/TinCo50	15	20

In first numerical computations the preselected principal structures of hybridal piezo modules composed of electrical conductive thermoplastic TinCo50, insulated thermoplastic (PA6, PA6-GF20 or PA6-GF50) and the PFS were analysed.

The very different thermal expansion coefficients and the shrinkage behaviour of the single components the cooling causes a heterogeneous and global "anisotropic" warpage of the piezo module, whereby in case of asymmetric structures occur relatively high deformations vertical to the layer plane. This process caused deformation involves additional bending stresses in the thermoplastic and ceramic components. In contrast to this a global flexure of the plate will be avoided because of the symmetric structure, so that no bending stresses will be added to the usual internal stresses.

Exemplarily results of the numerical calculation as Von Mises yield criterion of the E- and I-plastic components are shown in figure 2 for the structure with one top layer and in figure 3 for the structure with two top layers; each for the I-plastic PA6-GF20. Herein are well identifiable the strong strain exaggerations in arrays of gradations of material and contour discontinuities.

Initial Stress Behaviour of Micro Injection-Moulded Devices 115

Fig. 2. Von Mises yield criterion in MPa of the E-plastic components (above, TinCo50) and I-plastic components (below, PA6-GF20) for structures with one top layer

Fig. 3. Von Mises yield criterion in MPa of the E-plastic components (above, TinCo50) and I-plastic components (below, PA6-GF20) for structures with two top layers

To interpret the stresses of the orthotropic PFS, the single stress components in direction of the material rotational axis have to be collected in pursuance of equation (2). Therefore in figure 4 and 5 are given the particular normal and shear stress curves for the build options with one top layer (PA6-GF50), whereas the x-direction conforms to the orientation of the piezo fibres.

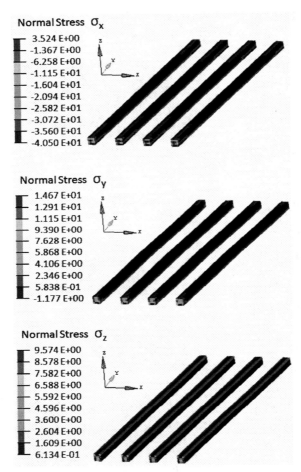

Fig. 4. Normal stress in MPa of the PFS for the structure with one top layer at PA6-GF50

The estimation of the complex spatial stress, in particular in pull in the plastic components and pressure in the PFS, happened by usage of material-specific failure criterion. In this regard, the Von Mises yield criterion was used for the I- and E-plastics and for PFS the CUNTZE-Criterion was consulted. Due to the small tensile and shear strength of the PFS the corresponding normal and shear stress are of special interest.

Initial Stress Behaviour of Micro Injection-Moulded Devices 117

Furthermore the analysis of stress at gradations of material (interface) was realized with help of accordingly interlaminar tensile or shear strengths in pursuance of table 3.

At the strength test of the piezo modules the uniform assessment measure for failure of the so called failure index F_i will be determined as function value of the particular failure criterion. The F_i value complies at homogeneous failure equations of a known material strain A_R or the reciprocal value of the standby factor as the rate of breaking stress and load induced stress vector ($f_{res} = |\sigma_B|/|\sigma_L| = 1/A_R$). Except for the phrase IFF3 of the CUNTZE-Criterion (see eq. (1)) this case applies to all of here chosen fracture conditions, also at the Von Mises condition.

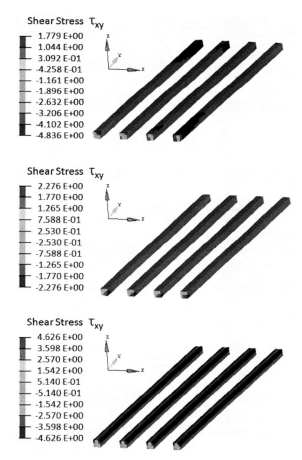

Fig. 5. Shear stress in MPa of the PFS for the structure with one top layer at PA6-GF50

The determination of critical state of stress conforms to the Von Mises criterion for the thermoplastic components (inter digital electrode structure and insulation components) is shown summarized in table 4. For analyses especially the high strain exaggerations were captured, which naturally occur only in local areas. Such peaks at visco-elastic thermoplastics will be reduced by restoring of stress to neighbouring, lower loaded areas (Neubauer hypothesis, cf. e.g. [13]). Therefore it is assumed, that merely at the build options with unreinforced I-plastic the risk of failure is given. (see therefore tab. 4).

Table 4. Failure index (material strain) in pursuance of Von Mises criterion for E- and I-plastic components

Number of layers	GF-content in PA6	Failure-Index	
		E-plastic	I-plastic
0	0	2.55[*)	0.55
	20	1.49[*)	0.25
	50	0.96	0.08
1	0	2.89[*)	0.60
	20	1.76[*)	0.29
	50	0.73	0.13
2	0	2.75[*)	0.64
	20	1.14[*)	0.28
	50	0.41	0.11

[*) failure

The failure index for the single FF and IFF failure mode of the PF composite is given in tab. 5. Due to the lower reduction of stress peaks of the brittle piezo fibre composite, a failure is expected for the build options with one or two top layers as well for the case of PA6-GF20 as for PA6-GF50. By varying the process parameters the internal compressive stress of PFS can be proportioned so that an optimal working area of the piezo elements can be adjusted for the present application.

Table 5. Failure index in pursuance of CUNTZE- Criterion (eq.(1)) fotransversal isotropic PFS

Number of layers	GF-content in PA6	Failure-Index, CUNTZE-Criterion				
		FF1	FF2	IFF1	IFF2	IFF3
0	0	0.28	0.13	0.58	0.80	0.45
	20	0.14	0.04	0.51	0.42	0.46
	50	0.26	0.04	0.45	0.17	0.40
1	0	0.57	0.66	1.28*)	1.02*)	1.23*)
	20	0.22	0.31	0.70	0.55	1.05*)
	50	0.24	0.03	0.57	0.18	0.77
2	0	0.40	0.83	0.46	0.69	1.40*)
	20	0.22	0.40	0.44	0.15	1.29*)
	50	0.14	0.06	0.52	0.03	0.93

[*) failure

The failure analyses of the interlaminar stresses arose that, regarding to the shear and compressive strength in pursuance of tab. 3, no critical stress conditions appear. The maximal material exploitation is here at circa 0,8. From the progressed numerical calculation and followed strength analyses by means of material-specific criteria follows all in all that only the structures with I-plastic made out of PA6-GF50 had enough structural performance. An important reason for this is the small shrinkage in combination to high strength of the highly filled plastic compared with PA6 and PA6-GF20, whereby the tensile internal stress can be reduced and higher forces can be taken.

4 Summary

The executed investigations of the process of electrical conductive plastic compounds in (two-component) micro injection moulding shows a fundamental possibility of bonding of piezo ceramic elements. It was possible to demonstrate that the bonding and the connection of piezo elements and the transmission of the generated charge with electrical conductive thermoplastic allows a profitable structure. In this process different thermoplastics with different electrical conductive fillers for their specific electrical resistance were investigated. TinCo50 – a PA6 thermoplastic filled with tin, zinc and copper – proves itself as a high conductive plastic with a specific resistance of less than $1,0 \cdot 10^{-3}$ Ωcm.

In FE analyses different structures of piezo modules with different process parameters and material combinations were tested for their internal stress behaviour during their production. Due to their different coefficients of expansion of the piezo ceramic and the used plastics deformations of course of their strong varying shrinkages of the materials among themselves in asymmetric structures are listed, which appealed with a tensile stress of 220 MPa of the piezo ceramic sticks. Simulations with symmetric structures show a clearly lower internal stress in the module, which has a better compressive stress of 80 MPa on the piezo sticks. As a consequence a symmetric module structure is favoured in pursuing investigations.

However, TinCo shows certain deficits in micro inject moulding in combination with PFS. Especially in view of the flow rating and the adjusting mechanical boundary layer properties in micro scale particle sizes influence negativ. With further compounding a tweak of the boundary layer as well as a homogenization of the material have to be set to avoid failures in the whole system of piezo modules at overlay of internal stress and operating voltage.

The basic realization of micro injection moulding technical production of complex modules can be integrated into related boards, for example: lightweight structures with integrated Structural Health Monitoring, actor module to reduce noise emission and also piezo ceramic generators for Energy Harvesting.

As part of the SFB/TR 39 "Production Technologies for light metal and fiber reinforced composite based components with integrated Piezoceramic Sensors and Actuators" (PT-PIESA) this scientific work had been supported by "Deutsche Forschungsgemeinschaft (DFG)". The authors thank for financial support.

References

1. Heinle, C., Vetter, M., Brocka-Krzeminska, Z., Ehrenstein, G.W., Drummer, D.: Mediendichte Materialverbunde in mechatronischen Systemen durch Montagespritzguss. Kunststofftechnik / Journal of Plastics Technology (8), 428–450 (2009)
2. Ehrenstein, G.W., Drummer, D.: Hochgefüllte Kunststoffe mit definierten magnetischen, thermischen und elektrischen Eigenschaften. Springer-VDI-Verlag (2009)
3. Giboz, J., Copponnex, T., Mélé, P.: Microinjection molding of thermoplastic polymers: a review. Journal of Micromechanics and Microengineering (17), R96–R109 (2007)
4. Giboz, J., Copponnex, T., Mélé, P.: Microinjection molding of thermoplastic polymers: morphological comparison with conventional injection molding. Journal of Micromechanics and Microengineering (19) (2009)
5. Griffiths, C., Dimov, S., Brousseau, E.: Microinjection moulding: the influence of runner systems on flow behavior and melt fill of multiple microcavities. Proc. IMechE, Part B: Engineering Manufacture 222, 1119–1130 (2008)
6. Dormann, B., Jüttner, G.: Hochpräzise Miniaturen. Kunststoffe 2, 34–37 (2009)
7. Heinrich, M., Kroll, L., Elsner, H., Leibelt, J.: Lightweight structures with autarchic functional piezoceramic modules for energy harvesting. In: 17, International Conference on Composites or Nano engineering, Honolulu (2009)
8. Kroll, L., Gelbrich, S., Ulbricht, J., Elsner, H.: Material integrated textile sensors in lightweight structures. In: 4th International Conference on Structural Health Monitoring on Intelligent Infrastructure (SHMII-4), Zürich (2009)
9. Kroll, L., Heinrich, M., Elsner, H.: Fertigungstechnologien für Piezomodule mit Drahtsensorik. In: Neugebauer, R. (Hrsg.) 2. Wissenschaftliches Symposium des SFB/Transregio 39, Großserienfähige Produktionstechnologien für leichtmetall- und faserverbundbasierte Komponenten mit integrierten Piezosensoren und –aktoren, pp. 48–52 (2009)
10. Inman, D.J.: Trans in Adaptive Structures. Neugebauer, R (Hrsg.): 2. Wissenschaftliches Symposium des SFB/Transregio 39, Großserienfähige Produktionstechnologien für leichtmetall- und faserverbundbasierte Komponenten mit integrierten Piezosensoren und –aktoren (2009)
11. Cuntze, R.G., Freund, A.: The predictive capability of failure mode concept-based strength criteria for multidirectional laminates. Composites Science and Technology 64, 343–377 (2004)
12. Kroll, L., Czech, A., Müller, S.: Identification of Anisotropic Damage on CFRP Tubes using Computer Tomography and Automated Analysis Methods. In: 17th International Conference on Composite Materials, ICCM-17, Edinburgh International Convention Centre (EICC), Edinburgh, UK, July 27- 31 (2009)
13. Kroll, L., Kostka, P., Lepper, M., Hufenbach, W.: Extended proof of fibre-reinforced laminates with holes. Z. Journal of Achievements in Materials and Manufacturing Engineering 33/1(3), 41–46 (2009)

Development of Particle-Reinforced Silver-Based Contact Materials by Means of Mechanical Alloying

Bernhard Wielage, Thomas Lampke, Daisy Nestler, and Heike Steger

Chemnitz University of Technology, Institute of Materials Science and Engineering,
Erfenschlager Str. 73, 09125 Chemnitz, Germany
bernhard.wielage@mb.tu-chemnitz.de,
thomas.lampke@mb.tu-chemnitz.de,
daisy.nestler@mb.tu-chemnitz.de,
heike.steger@mb.tu-chemnitz.de

Abstract. To produce silver-based contact materials such as silver/tin oxide different ways are possible. All usual technologies have their typical advantages and disadvantages. Expensive techniques are characterised by a very fine reinforcement distribution. Techniques with a high output quantity display a very limited particle distribution within the matrix material.

By means of the economical mechanical alloying process, particle-reinforced metal-matrix composite powders with a fine distribution of the reinforcement within the matrix material can be produced. This contribution shows the morphology and microstructure of mechanically alloyed silver-based powders by means of the high-energy ball-milling process as well as the microstructure and mechanical properties of the consolidated materials. The objective of the present investigation has been the successful and reproducible production of particle-reinforced metal-matrix composites as well as the reduction of the noble-metal part by increasing the content of the reinforcement. Current investigations present a successful increase of the oxide content to 17 wt.-%.

Keywords: contact materials, high-energy ball milling.

1 Contact Materials

Silver shows the highest electrical and thermal conductivity [1-3] and is therefore appropriate for applications in switches. Because of its bad welding and wear resistance as well as its tendency to material migration in direct current applications, however, silver as a contact material satisfies the requirements only in a limited range of electrical applications. So, silver alloys or silver composite materials are used to meet the requirements. AgNi 0.15 (fine-grain silver) and Ag/Cu alloys are typical examples for silver alloys. Contact materials of silver/metal oxide demonstrate very good anti-welding properties. [1-10] In particular, the material system silver tin oxide with an oxide quota between 8 and 12 and up to 14 wt.-% [3, 10 - 13] have to be mentioned.

1.1 Production of Silver-Based Contact Materials

In the following, the advantages and disadvantages of three typical production processes for silver tin oxide contact materials are briefly described.

1.1.1 Internal Oxidation

By means of the internal oxidation of silver/tin alloys, a very fine tin oxide distribution can be obtained. The disadvantage of this technology is that very expensive indium has to be added to the silver/tin alloy to reach high oxide contents by means of internal oxidation. [7]

1.1.2 Wet-Chemical Precipitation

A very fine distribution is also possible with the wet-chemical precipitation and subsequent powder-metallurgical processing. Unfortunately, this technology is expensive [7, 13, 14].

1.1.3 Powder Blending

The powder metallurgical process of powder blending is distinguished by a high output quantity but a very limited particle distribution within the matrix material [7, 10, 13].

2 Mechanical Alloying

Mechanical Alloying is an appropriate and economical procedure to achieve metal-matrix composite powders. The technology was developed in the 1960 to produce oxide-dispersion-strengthened nickel basis super alloys. [15 - 17] The mechanical alloying process can be achieved by means of different types of mills. To mill the powders, which can be systematised in the systems ductile/ductile, brittle/brittle and ductile/brittle, grinding media such as balls with different diameters are necessary. Other shapes, e.g. cylinders, are also possible. By means of a permanent cold-welding, milling and breaking processes, a very homogeneous distribution of the components can be obtained. [15 - 17]

Different investigations tested the high-energy ball-milling process (HEM) for different material systems and applications. E.g. feedstock for thermal spraying processes or composite powders for the production of compact materials were tested [18 - 22].

This contribution illustrates the system ductile/brittle to reach metal-matrix composites. Milling balls were used as grinding media.

3 Experiments

3.1 High-Energy Ball-Milling Process

The applied high-energy ball mill is characterised by a horizontally arranged, fixed milling chamber and a spinning rotor (Figure 1). By means of this laboratory-scaled high-energy ball-milling unit of the type Simoloyer® CM01 (Fa. Zoz GmbH), ductile silver-matrix material was milled together with the brittle tin oxide reinforcement.

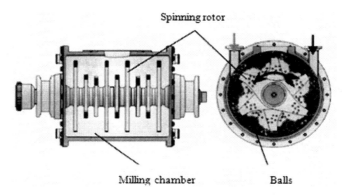

Fig. 1. Scheme of a high-energy ball mill Simoloyer® [Fa. Zoz]

The HEM process was performed in air atmosphere and at various rotation speeds and milling times. A special ceramic chamber and rotor (Si_3N_4) as well as ceramic balls (ZrO_2) were used to avoid erosive impurities which normally result from the steel configuration of the high-energy ball mill. The milling balls are exposed to a very high acceleration by the spinning rotor. Because of this acceleration, two processes can be defined: the milling process on the one hand and the cold-welding process on the other hand.

The matrix material was reinforced with 8, 12, 14 and 17 wt.-% tin oxide. Three different tin oxide (Table 1) fractions were applied to study possible agglomeration effects, in particular for the fine fraction with the highest content.

Table 1. Tin oxide fractions

Fraction	fine	medium	coarse
d_{50} [µm]	0.7	2.6	4.2

The milled powders were studied after different milling times and various rotor speeds using the optical microscopy (OM) and the scanning electron microscopy (SEM). After reaching an appropriate distribution of the reinforcement within the matrix material, the powders were consolidated to semi-finished products (e.g. wires).

The microstructures of the so achieved wires were also characterised by means of OM and SEM in cross-section as well as longitudinal section. Concerning their mechanical properties, tension tests were used.

3.2 Consolidation

The milled powders with the appropriate particle distribution within the matrix material were consolidated to semi-finished products. In a first step, the powders were pressed to a preform by means of cold isostatic pressing (CIP). The extrusion to wires follows after a special heat treatment.

4 Results

4.1 Microstructure

4.1.1 Microstructure Observation of the High-Energy Ball-Milled Powders

The microstructural observations of the obtained composite powders after different milling periods show that the reinforced particles are finely-dispersed after an appropriate milling time. During the milling process, a typical feature can be observed which has already been described in [17]. In a first step, the ductile matrix material is formed to lamellas and plates and the brittle reinforcement particles are ground. After this first step, the small brittle particles are attached to the ductile matrix. Then the plate-forming matrix encloses the reinforcements (step 3) and a cold-welding process follows. Coarse lamella structures develop and another cold-welding and deformation process together with a repeated breaking and milling process takes place.

This refining process of repeated milling leads to the disappearance of the lamellar orientation. So a homogeneous distribution occurs. The characteristic steps for the breaking, milling and cold-welding processes for the combination of a ductile matrix material and a brittle reinforcement are presented in the left column of Table 2. [17]

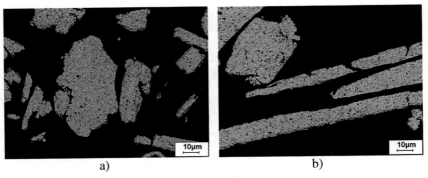

Fig. 2. Ag-SnO$_2$ powder with two different SnO$_2$ fractions: a) fine; b) coarse (cross-section, SEM images)

The development of the distribution of the tin oxide reinforcement in the silver matrix and therefore of the composite powder was observed by means of OM and SEM after different milling times. The process (left column) has been validated and the evolution of the Ag/SnO$_2$ material is presented in the right column of Table 2.

The distribution of step 3 and 4 is reached after a milling time of 10 minutes. After 20 minutes, the state of step 5 and 6 can be identified. The homogeneous distribution of the tin oxide component within the matrix material of step 7 can be reached after a milling time of 60 minutes.

Table 2. Development during the high-energy ball-milling process

Scheme [17]	Observation
Step 1	Starting Mixture
Step 2	
Step 3	Development of the milled powder after 10 miutes
Step 4	

Table 2. (*continued*)

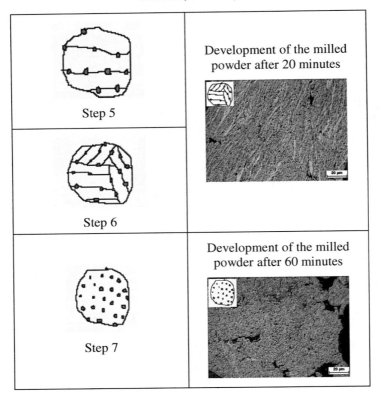

A significant grinding process of the tin oxide which is classified in step 2 (Table 2) cannot be detected in the case of the very ductile silver matrix. The typical distinctive flake form (Figure 3) of the composite powder particles is also caused by the ductile matrix material.

Fig. 3. Morphology of the distinctive flake form of Ag/SnO2 composite particles after the high-energy ball-milling process (SEM image)

The cross-sections (SEM images) in Figure 2 present very dense silver tin oxide powders with a very finely distributed oxide component, both for the very fine fraction (Figure 2a) and the coarse fraction (Figure 2b).

The three different tin oxide fractions (fine, medium, coarse) as well as the four different tin oxide contents (8, 12, 14 and 17 wt.-%) are very homogeneously distributed within the flake-formed composite particles. The distribution of two Ag/SnO$_2$ materials with 12 wt.-% tin oxide content (fine fraction and coarse fraction) is displayed in Figure 2.

4.1.2 Microstructure of the Compact Material

The microstructure of the consolidated composite powders was evaluated. In particular the longitudinal section of the wires can display inhomogeneous reinforcement distributions due to the extrusion process.

Figure 4 displays the polished longitudinal section (SEM images) of three Ag/SnO$_2$ wires with 8 (a), 12 wt.-% (b) and 14 wt.-% (c) tin oxide. The SEM images show a metal matrix composite material with a very fine reinforcement distribution for these typical tin oxide contents. Only a minimum of silver-rich lines is visible.

To reach a maximum of noble-metal substitution, the tin oxide content was increased to 17 wt.-%. The consolidated material with this high oxide content is displayed in Figure 5. The oxide component is well embedded and homogeneously distributed within the matrix material. The silver-rich lines are slightly increased but still insignificant.

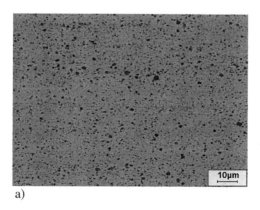

a)

Fig. 4. SEM image of longitudinal sections of Ag/SnO2 HEM material after extrusion with a) 8 wt.-%, b) 12 wt.-% and c) 14 wt.-% tin oxide

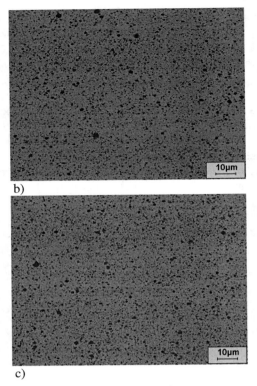

b)

c)

Fig. 4. (*continued*)

Figure 7 shows that the extruded wires are very dense and there are only a few micropores. The oxide component is well embedded in the silver matrix material. The FE-SEM illustrates in particular the integration of very fine oxide particles of a few nanometres.

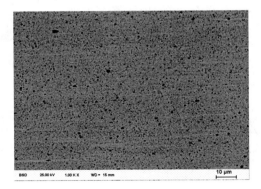

Fig. 5. SEM image of longitudinal sections of Ag/SnO2 HEM material after extrusion with 17 wt.-% tin oxide

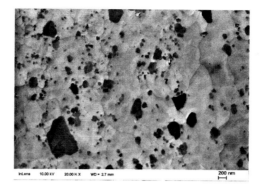

Fig. 6. SEM image of an ion-polished silver/tin oxide wire with 12 wt.-% tin oxide

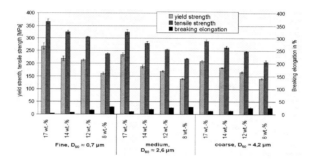

Fig. 7. Tension tests of Ag/SnO$_2$ with 8, 12, 14 and 17 wt.-% tin oxide at room temperature

4.2 Mechanical Properties

4.2.1 Methods

Besides the microstructural observations, the mechanical properties of the composite materials are very important for the interpretation of the results and the applicability of the production process. Different ways are possible to obtain the mechanical properties of the consolidated silver-matrix composite materials. Tension tests at room temperature were used to describe the properties of the achieved materials. These tests were performed with a simple fixation of the untailored specimens.

4.2.2 Results

The results of the tension tests show the expected correlations. In the case of the dispersion strengthening, the tensile strength and the yield strength rise with the increasing quota of the oxide component and the decreasing particle size of the reinforcement. The maximum is reached at 17 wt.-% of the fine tin oxide. The breaking elongation also decreases with the rising oxide component quota. Here the minimum is reached also at 17 wt.-% of the fine tin oxide component.

5 Summary and Conclusions

The investigations have shown that the high-energy ball-milling technology is a suitable production process to produce silver-based composite materials with a wide range of oxide contents (8 – 17 wt.-%) as well as different fractions (fine, medium, coarse). The composite powders and the finished compact materials show a reproducible homogeneous particle distribution. Such fine structures cannot be obtained with other mixing and milling processes. The increased content of the reinforcement can significantly reduce the necessary noble-metal quota.

Acknowledgement

We gratefully acknowledge the cooperation with the project partners and the financial support within the framework of the BMBF project 03X3500C.

References

1. Schröder, K.-H. (hrsg.): Werkstoffe für elektrische Kontakte und ihre Anwendungen: Fertigungsverfahren, Eigenschaften, Verbindungstechniken, Prüfverfahren. Expert-Verlag (1992)
2. Burstyn, W.: Elektrische Kontakte und Schaltvorgänge: Grundlagen für Physiker. Springer, Heidelberg (1956)
3. http://www.amidoduco.com (Stand: 22.03.10)
4. Schatt, W.: Pulvermetallurgie. Springer, Heidelberg (2007)
5. Rieder, W.: Elektrische Kontakte, Eine Einführung in ihre Physik und Technik. VDE-Verlag (2000)
6. Vinaricky, E. (hrsg.): Elektrische Kontakte, Werkstoffe und Anwendungen: Grundlagen, Technologien, Prüfverfahren (2002)
7. Kempf, B., Braumann, P., Böhme, C., Fischer-Bühner, J.: Silber-Zinnoxid-Werkstoffe: Herstellverfahren und Eigenschaften. Metall. 61, 404–408 (2007)
8. Schreiner, H.: Pulvermetallurgie elektrischer Kontakte, Reine und angewandte Metallunde in Einzeldarstellungen. Band 20, Springer Verlag (1964)
9. Blawert, C.: Edel- und Buntmetall-Matrix-Verbundwerkstoffe. In: Kainer, K.U. (ed.) Metallische Verbundwerkstoffe, Wiley-VCH (2003)
10. http://www.technicalmaterials.umicore.com (Stand: 22.03.10)
11. Graff, M.: Einfluss oxidischer Zusätze auf die Phasenbildung und die Schalteigenschaften von Kontaktwerkstoffen auf Silber/Zinnoxid-Basis, Dissertation, TU Darmstadt (2001)
12. Müller, M.: Rühlicke, Dl; Heringhaus, F.; Wolmer, R.; Goia, D.V.: Verfahren zur Herstellung von Verbundpulvern auf Basis Silber-Zinnoxid und deren Verwendung zur Herstellung von Kontaktwerkstoffen, Patentschrift DE 10017282 (2002)
13. Braumann, P.: Silberhaltige Kontaktwerkstoffe für die Energietechnik, Vortag im Rahmen der Veranstaltung Die 7. Todsünden der Kontaktphysik, Windisch, CH (2006)
14. Braumann, P., Koffler, A.: Einfluss von Herstellverfahren, Metalloxidgehalt und Wirkzusätzen auf das Schaltverhalten von Ag/SnO2 Relais (Teil 1 und 2). Albert-Keil-Seminar, Karlsruhe (2003/2005)

15. Lü, L., Lai, M.O.: Mechanical Alloying. Kluwer Academic Publishers, Dordrecht (1998)
16. Surynanarayana, C.: Mechanical Alloying and Milling. Marcel Dekker, New York (2004)
17. Kudashov, D.: Oxiddispersionsgehärtete Kupferlegierungen mit nanoskaligem Gefüge, Dissertation, TU Bergakademie Freiberg (2003)
18. Wielage, B. et al.: Mechanically alloyed SiC composite powders for HVOF applications. In: Proceedings ITSC 2002, DVS, pp. 1047-1051 (2002)
19. Wilden, J., et al.: Performance of HVOF nanostructured diboride composite coatings. In: Proceedings ITSC, DVS, pp. 1033-1037 (2002)
20. Zwick, J., et al.: Hochenergetisches Mahlen von Metall mit oxidischen Dispersionen zum thermischen Spritzen, Proceedings Hagener Symposium für Pulvermetallurgie, Pulvermetallurgie in Wissenschaft und Praxis, Band 18 (2002)
21. Wielage, B., et al.: Hochenergiekugelmahlen und ECAP-Umformung von pulvermetallurgisch hergestelltem EN AW-2017SFB mit Partikelverstärkung – Mikrostruktur und mechanische Eigenschaften. Tagungsband SFB 692 HALS, Chemnitz 25 (2007)
22. Wielage, B., et al.: Kontaktwerkstoffe auf Silberbasis – Gefüge und mechanische Eigenschaften. Materialwissenschaft und Werkstofftechnik 39, 940–943 (2008)

Design of Sports Equipment Made from Anisotropic Multilayer Composites with Stiffness Related Coupling Effect

J. Kaufmann, L. Kroll, E. Paessler, and S. Odenwald

Chemnitz University of Technology, D-09107 Chemnitz, Germany
Joerg.Kaufmann@mb.tu-chemnitz.de,
Lothar.Kroll@mb.tu-chemnitz.de,
Erik.Paessler@mb.tu-chemnitz.de,
Odenwald@hrz.tu-chemnitz.de

Abstract. Sporting goods such as snowboards have to resist enormous loads while using them. Especially freestyle snowboards are extremely stressed during the landing of high jumps or while sliding sideways on a rail. During such a board slide the steel edge of a snowboard tends to cut into the handrail and the motion of the athlete can be interrupted abruptly. This fact leads to crashes and can cause severe injuries. Catching the edge is nearly eliminated by using an all-new anisotropic layer design (ALD)-snowboard which has been developed, simulated and tested by the Professorship of Lightweight Structures and Polymer Technology (SLK) in cooperation with the Professorship of Sports Equipment and Sports Technology (SGT).

Keywords: snowboard, anisotropic layer design, ALD, coupling effects, CLT, fiber-reinforced plastics.

1 Introduction

The market for Fiber-Reinforced Plastics (FRP) shows annual growth rates of 4% to 6% and is strongly influenced by issues for resource-efficient lightweight structures. Currently, FRP are used in almost all economic sectors, whereas the vehicle industry and the aerospace industry are taking the largest market share with about 32%. To compared this, the sports equipment technology has a share of about 7%, which will increase significantly over the coming years in order to application-specific sports equipment design [1]. For many years natural materials e. g. wood and leather have already been substituted successfully, whereby the strong variation of the mechanical properties, such as the component strength and stiffness were noticeably improved. Moreover, the use of composite materials allows a user-specific design in terms of shock- and vibration-damping combined with low weight.

134   J. Kaufmann et al.

By having the objective to improve the handling characteristics, especially for freestyle snowboards, the SLK - in cooperation with the SGT an SLB e. V. - developed an all-new ALD-snowboard [2]. Thereby the development can be divided into three major steps:

1. Determination of the critical load was done by using the physically motivated failure criterion of Puck in combination with analytical and finite element methods [3].
2. Manufacturing and testing of slightly curved test samples to verify the numerical predictions of the ALD-snowboard. These tests demonstrate the usable amount of deformation generated by the coupling effects.
3. Manufacturing of a complete ALD-snowboard for laboratory and field-testing.

This paper presents the results of step 1 to 3 except the field-testing.

2 Structural Behavior of Fiber-Plastic Composites

Thin FRP structures can be treated as a coupled disc-plate problem using the Classical Laminate Theory (CLT). For the calculation of complex geometries a variety of numerical software solutions has been established, which allows the identification of essential stress values. The analysis of the anisotropic coupling effects in floating sports equipment such as skis, snowboards or kiteboards bases on the structure law of the multilayer structure (MLS) [3]. That - unlike isotropic materials - permits a coupling between internal forces and bending moments as well as a coupling between internal moments and distortion [4, 5, 6, 7]. Specifically, the strain and shear coupling from the disc theory and the bend and twist coupling from the plate theory are included.

$$
\begin{Bmatrix} \hat{n}_x \\ \hat{n}_y \\ \hat{n}_{xy} \\ \hat{m}_x \\ \hat{m}_y \\ \hat{m}_{xy} \end{Bmatrix} = \begin{bmatrix} A_{11} & A_{12} & A_{16} & B_{11} & B_{12} & B_{16} \\ A_{12} & A_{22} & A_{26} & B_{12} & B_{22} & B_{26} \\ A_{16} & A_{26} & A_{66} & B_{16} & B_{26} & B_{66} \\ B_{11} & B_{12} & B_{16} & D_{11} & D_{12} & D_{16} \\ B_{12} & B_{22} & B_{26} & D_{12} & D_{22} & D_{26} \\ B_{16} & B_{26} & B_{36} & D_{16} & D_{26} & D_{66} \end{bmatrix} \cdot \begin{Bmatrix} \varepsilon_x \\ \varepsilon_y \\ \gamma_{xy} \\ \kappa_x \\ \kappa_y \\ \kappa_{xy} \end{Bmatrix}_0
\tag{1}
$$

The ABD-Matrix in (1) consists of the disc stiffness matrix [A], the coupling stiffness matrix [B] and the plate stiffness matrix [D], which is crucial in the dimensioning of multilayer components with anisotropic qualified coupling effects.

Under the condition A_{16}, $A_{26} \neq 0$ there is a strain-shear coupling and if condition D_{16}, $D_{26} \neq 0$ is fulfilled, a curvature-torsion coupling exists in the considered sports equipment. The coupling coefficients B_{11}, B_{12} and B_{22} describe the link between strain and curvature of the disc plate problem. In the case of symmetric layer structures independent of the fiber orientation of the individual layers the disc plate problem is decoupled [4, 5, 6, 7].

3 Requirements of Sports Equipment

Floating sports equipment in the field of high performance sport experiences extremely high loads that have high demands on material combination, structure and functionality. Above all, by "sliding" on the edges (e. g. on a handrail, see Figure 1) high attrition and early defects occur in the edge area of floating sports equipment, which often attracts an increased risk of accidents. Due to sliding sideways on handrails consisting of metallic rod an idealized transverse line load is acting on the snowboard. This leads to a high bending moment in the central region of the board. Thereby the snowboard tends to tilt in the peripheral areas and that occasionally leads to laminate failures [3]. With defined lateral inclination of the snowboard the steel edge of the board "cuts" into the material of the handrail and the slide motion is suddenly interrupted, which often causes severe injuries. Due to the combined mechanical and thermal load of the steel edge caused by friction ridges occur, which severely restrict the mode of action of the edge on solidified snow or icy underground. State-of-the-art technology includes various approaches to avoid catching the edge while sliding like the substitution of steel using plastic edges [8], although this leads to inferior handling characteristics of the snowboard on solidified surface.

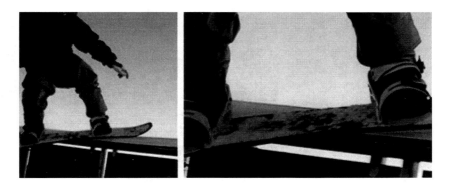

Fig. 1. High structural load during sliding sideways

Using internal coupling effects of FRP is a new and innovative method for anticipating these problems. The coupling effects cause a slight lift of the border areas and thereby the edges of the board if the board is bent perpendicularly to the longitudinal axis.

4 Concept of the All-New ALD-Snowboard

The concept of the ALD-snowboard for specific edge-lift can be performed by an unsymmetrical MLS like an angle balanced composite or a symmetric MLS as an unbalanced composite with specific angles (see Figure 2). Either a bending-curvature coupling in consequence of the coupling coefficient B_{16} or $B_{26} \neq 0$ or a bending-torsion coupling due to the plate stiffness D_{16} or $D_{26} \neq 0$ (see Figure 3) cause unusual deformations in each of the four deformation active zones (DA-zones).

Fig. 2. Individual layers showing the schematically fiber orientation of the four DA-zones

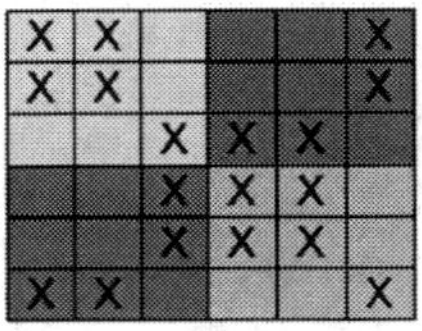

Fig. 3. ABD-matrix of a balanced MLS with unsymmetrical lay-up (DA-zone A: α/core/-α)

For the numerical simulation of the ALD-snowboard an allocation of layers in four DA-zones is useful (see [3] and Figure 2, A and B). These are connected by appropriate conditions of contact with each other. For the optimum transfer of force flow between the DA-zones of the functional samples, five overlapping zones (see Figure 4, C to G) are necessary, which counteract failure of the component in the laminates contact area.

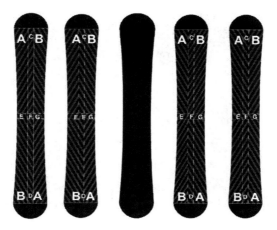

Fig. 4. Individual layers that shown schematically fiber orientation of the four DA-zones and five overlapping zones

5 Numerical Modeling

The boundary conditions for the numerical model are defined to assemble feasibility of that simulation of the real application ("boardslide") and corresponding laboratory experiment. For this purpose, the model is fixed at four points and a displacement-load is applied. Thereby the effective direction of the both displacements faces towards the center of the board (Figure 5).

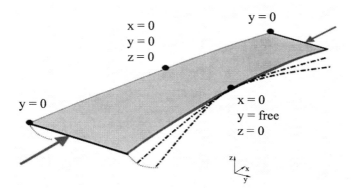

Fig. 5. Boundary conditions and deformation of the FE model

The slight pre-deformation of the snowboard's geometry of approximately 1 mm in z-direction in the central region leads to a realistic deformation constraint. In this way the load case of the transverse line load can be modeled with sufficient accuracy. Commercially available carbon fibers of type T300 and a standardized epoxy resin system as well as wood have been used for building the test-samples. The simulation of the typical sandwich structure of a snowboard has been done by using a popular core with constant thickness and the material data for a typical Carbon FRP (CFRP) with T300 fibers and epoxy resin (see Table 1 and Figure 5).

Table 1. Mechanical properties of the used sandwich components

	E_1	E_2	G_{12}	ν_{12}
CFRP	105	7,5	4,5	0,3
Wood	8,8	0,7	0,6	0,42

For the analysis of the anisotropic coupling-effects the software tool ANSYS Workbench has been used in combination with the ESAComp laminate software application [3]. The comparison of numerical results for the bending-curvature coupling and the bending-torsion coupling confirms large out-of-plane deformations in the peripheral areas, in particular with unsymmetric balanced laminates. The resulting stress peaks are drastically reduced by the steel edges. In this way an interlaminar component failure is not likely to occur.

6 Verification Tests

The experimental setup is similar to the boundary conditions of numerical simulation with the restriction of fixing three respectively two of the displacement degrees of freedom in the middle of the board (Figure 5). Grub screws on an adjustable clamping device perform the displacement along the longitudinal axis. The measurement of the deformation is recorded by a laser measuring system at 14 points along the board transversal axis at a distance of 20 mm each. For a sample with balanced angles and a lay-up of the DA-zone A (α/core/-α) or DA-zone B (-α/core/α) a lifting of the edge area up to 3 mm could be measured at a base deformation in z-direction of 80 mm due to a line load (see Figure 6). Even with relatively small deformations of 40 mm, a lifting of the edge area of approximately 2 mm and thus a resultant angle between the snowboard edge and handrail of 3,3° could be observed. This edge effect is already contributing significantly to the prevention of tilting and reduces the risk of falling.

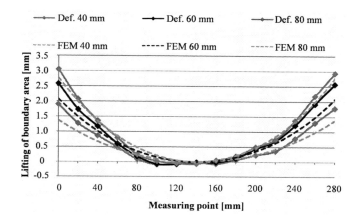

Fig. 6. Deformation of the snowboard cross section due to bending load

The comparison of results from the FE simulations and experiments confirms the high quality of the modeling and simulation. In addition to the mechanical test an analysis of the deformation took also place under thermal stress. For that purpose some of the demonstrators have been stored at -20 °C over a defined period. Due to the selected MLS-Lay-up as balanced respectively unbalanced angle-laminate, the cooling has small effect on the mechanically unloaded state of deformation.

As a result the test-sample verification demonstrates that the ALD-technology leads to a significant edge-lift. Based on this fact a ridable prototype of the all-new ALD-snowboard has been manufactured and tested at the SLK. For this purpose a new test-setup has been designed in order to reproduce the loading condition of a boardslide as realistic as possible. Therefore the board is supported on a metallic square profile in the middle and loaded by a displacement of 15 mm in vertical direction at its binding areas. This load case is comparable to a rider having a mass of 80 kg performing a boardslide. The edge-lift in the center area has also been recorded by laser measuring system and shows maximum values of approximately 2 mm (see Figure 7).

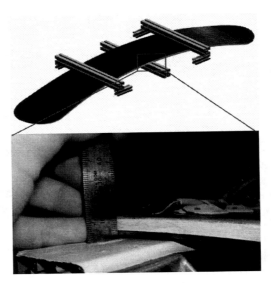

Fig. 7. Test-setup reproducing the load-case "boardslide" and demonstration of the edge-effect

7 Conclusions

The deformation measurements of the test-samples with the selected lay-up show reasonable accurate results for the numerical simulation of the disc-plate problem. Therefore the results could be used for specifying and designing the all-new ALD-snowboard.

The observed angle between edge and handrail obtained from normal bending deformation already contributes to avoid the tilting of ALD-snowboards. By using MLS as anisotropic composite temperature-dependent coupling effects are of secondary importance. Thus corresponding ALD-floating sports equipment can be used without restriction at temperatures down to -20 °C.

Furthermore a prototype of the ALD-snowboard has been laminated and tested with regard to coupling effects using a new test-setup reproducing the loading conditions of a real boardslide. The results show a sufficient edge-lift to avoid cutting into the handrail also under realistic conditions.

As a next step different designs of the snowboard regarding composite structure, core properties etc. should be investigated by using the developed methods for both numerical simulation and mechanical testing of the manufactured specimen. The major aims are maximizing the coupling effects and a further improvement of the resulting edge-lift by performing sensitivity analysis for the different designs. Also a mechanical test device has to be created for spatial measurement of the multiple deformation snowboards. In addition strain gages will be applied on the ALD-prototype in order to investigate the influence of the coupling effects on the snowboards handling and drivability within the scope of field tests.

References

1. Flemming, M., Roth, S.: Faserverbund-bauweisen. Springer, Heidelberg (2003)
2. Kroll L, Heinrich M, Kaufmann J.: DE 102007055532 A1, Sport- und Spielgerät - Anisotrope Koppeleffekte im Snowboard. (November 21, 2007).
3. Kroll, L., Heinrich, M., Müller, S., Kaufmann, J.: FE-Analyse anisotroper Koppeleffekte mit der ESAComp-Schnittstelle zur belastungsgerechten Auslegung von Snowboards. ANSYS Conference and 25. CADFEM Users' Meeting, Dresden (2007)
4. Wiedemann, J.: Beitrag zum Problem orthotroper Platten ohne allgemeine Neutralebene. Z. Luftfahrttechnik 8 (1962)
5. Wiedemann, J.: Leichtbau Bank 1: Elementen. Springer, Berlin (1986)
6. Kroll, L.: Zur Auslegung mehrschichtiger anisotroper Faserverbundstrukturen. Dissertation, TU Clausthal (1992)
7. Kroll, L.: Berechnung und technische Nutzung von anisotropiebedingten Werkstoff - und Struk-tureffekten für multifunktionale Leichtbauan-wendungen. Habilitationsschrift, TU Dresden (2005)
8. Quattro, J. S., Eberhardt, J.: WO 03/086554 A1, Non-metallic edge gliding board (March 20, 2003)

Structural Optimization of Fibre-Reinforced Composites for Ultra-Lightweight Vehicles

Bernhard Wielage, Tobias Müller, Daisy Nestler, and Thomas Mäder

Chemnitz University of Technology, D-09107 Chemnitz, Germany
bernhard.wielage@mb.tu-chemnitz.de,
tobias.mueller@mb.tu-chemnitz.de,
daisy.nestler@mb.tu-chemnitz.de,
thomas.mäder@mb.tu-chemnitz.de

Abstract. The potential of lightweight construction with composite materials is frequently not fully tapped. The reason is that the complex anisotropic properties are not fully reflected in the design. Instead, to be on the safe side, more or thicker material has been used. The numerical structural optimization is a way to calculate the optimal load-dependent material distribution or fibre orientation in complex fibre-reinforced components. This method was used to design the chassis of the ultra-lightweight, extremely energy-saving vehicle "Sax 3" of the student project "fortis saxonia" for the Shell Eco-marathon. Thus it has become possible to keep the weight of the chassis under 10 kg. This shows the high potential of the implementation of this optimization approach for fibre-reinforced composites.

Keywords: Optimization, Composite structures, FEM, energy-saving vehicle, lightweight construction.

1 Introduction

Each year, the Shell companies organize a special car racing: the Shell Eco-marathon. The principle of this contest is simple: to design and build a vehicle that uses the least amount of fuel to travel the farthest distance. At all events, teams can enter futuristic prototypes: streamlined vehicles where the only design consideration is reducing drag and maximizing efficiency. Conventional fuels such as diesel, petrol/gasoline and liquefied petroleum gas, as well as alternatives like GTL, solar energy, ethanol, hydrogen and biofuels can power the vehicles. As long as teams adhere to safety rules, the design of their vehicles is limited only by their imagination.

In 2005, students of the Chemnitz University of Technology set up the research project Fortis Saxonia (Latin for "Powerful Saxony") [1] to participate in the Shell Eco-marathon. To complete the marathon successfully, all components of the vehicle had to be optimized fundamentally. One of the difficult tasks was to engineer an ultra-light chassis that meets all the requirements for stiffness and strength. The use of composite materials alone did not yield the desired weight savings. The material had to be used sparingly and precisely in the right place. In the complex form of chassis and the many different stress situations, an optimal design could be achieved only with the help of numerical optimization methods. The vehicle Sax 3 (Fig. 1) was constructed by using different ways of numerical optimization as described below. The software system for optimization was Hyperworks from Altair.

Fig. 1. Vehicle Sax 3 at the Shell Eco-marathon 2008 (source: www.fortis-saxonia.de)

However, in a lot of cases, particularly in the aerospace industry and the automotive industry, an efficient use of material is essential. In addition to savings in material, the weight of a component can be significantly reduced by optimizing design, shape and size. The mathematical approaches of optimization are not new, but with increasing application of FEA (Finite Element Analysis) methods and the use of powerful computing technology, the possibilities have been expanded considerably. A historical review goes beyond the scope of this article. Detailed information can be found in [2, 3].

2 Numerical Optimization

Numerical optimization is an iterative process. From the results of numerical simulation, conclusions of necessary changes will be generated. In a defined domain, the optimization tool can accomplish changes automatically and restart the numerical simulation for the next optimization step.

For a better understanding, some important terms relating to optimizing are explained in the following. At first, there are different common definitions. In the second part, some types of optimization are described. An FEA model consisting of geometry, boundary conditions and load cases is a prerequisite for successful optimization.

2.1 Definition of Some Terms Related to Numerical Optimization

To optimize a design or a structure, there must be some freedom to change it. The potential for change is typically expressed as a group of parameters called *design variables*. Different ways to express design variables exist in the literature. In this paper, the vector of design variables b_j is called $\mathbf{b} = \{b_j\}$, $j = 1, \ldots, n$ based on [3, 4, 5]. In every iterative step, the optimizer changes the value of the design variables to $\mathbf{b}^{(s+1)} = \mathbf{b}^{(s)} + \delta \mathbf{b}$ and detects in which direction a local destination of optimization is oriented. Design variables such as the thickness of a plate, fibre orientation, or a virtual density are to be decided and whether there is need for material at a specific position or not.

An *objective* is the variable to be optimized. An objective is expressed as the function $\psi_0(\mathbf{b})$ of the vector of design variables that can be either maximized or minimized. In a perfunctory approach in the case of the vehicle Sax 3, the weight is to be minimized. But as described below, in many cases, an indirect optimization shows better results. For structural optimization problems, weight, displacements, stress, vibration frequencies, buckling loads and cost, or any combination of these can generally be used as objective.

In many cases, design variables run to zero or to infinity, so that no meaningful optimization is found. To prevent this, *design constraints* are defined. An example for this problem is a beam in bending. A minimization of mass without constraints leads to a complete disappearance. The definition of the maximum displacement as constraint of the point where the force is connected produce optimal material distribution. Design constraints can also be described as a function of design variables $\psi_i(\mathbf{b})$.

The entire model is not always involved evenly in the optimization. Often there are areas the geometric properties of which will remain unchanged. A *design room* defines the area where the optimization takes place.

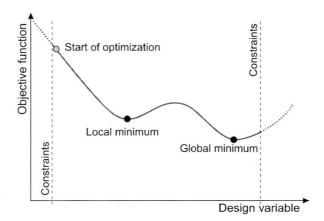

Fig. 2. Correlation between different terms of optimization

Fig. 2 shows an example of an objective function with two minima. The best solution is always the global minimum but in practice, it is difficult to decide if the found minimum is the absolute minimum or if there is a better solution. A

problem is to determine the changes in the design variables $\delta \mathbf{b}$. This search direction can be determined directly from the result of the FEA. Objective and constraints $\psi_i(\mathbf{b})$, $i = 0,...,n$ can be approximated for each design $\mathbf{b}^{(s)}$ using series expansion:

$$\psi_i(\mathbf{b}) = \psi_i(\mathbf{b}^{(s)}) + \sum_{j=1}^{n} \frac{d\psi_i}{db_j} \delta b j \qquad (1)$$

The gradient $d\psi_i / db_j$ can be obtained directly from the results of the numerical analysis. If the gradient is known, the search direction $\delta \mathbf{b}$ can be obtained from the solution of an approximate optimization problem. Details can be found in [3, 4].

To find a global minimum or maximum, the start configuration (comparable to the start of optimization in Fig. 2) of different optimization steps can be chosen with genetic, random or incremental algorithm.

2.2 Types of Optimization

One of the most important types of Optimization is the *topology optimization*. This optimization determines the distribution of the material in the design room, so that a maximal stiffness of the component for given boundary conditions will be achieved. The typical result of this optimization is a truss-like structure.

The design variable for topology optimization is a virtual density ρ^*. The range of the virtual density goes from zero (no material) to one (full material). The local stiffness of the material is a function of ρ^*. A linear correlation of stiffness and virtual density generates primary regions with average density. Therefore, the stiffness of finite elements E^* is calculated as:

$$\frac{E^*}{E_0} = (\rho^*)^n \qquad (2)$$

with the material stiffness E_0 and the polarization exponent n, which is automatically increased from the used solver.

The result of topology optimization is a design proposal for a reconstruction of the component, not a finished construction. The reason is the rugged surface of the calculated structure. Moreover, there are often insular materials that have no connections to other material accumulations.

A fundamental mathematical description of topology optimization can be found in [6] and, also with a lot of applicatory examples, in [7].

The *shape optimization* can be applied in the final stage of construction. Existing forms are not changed fundamentally. The marked FEA nodes are optimized according to certain criteria, for example, to reduce stress peaks and notch effects.

If the shape of a component is defined, a fine adjustment can be done with *size optimization*. This includes among others the thickness of shell elements or the diameter of bar elements, and the configuration of laminated composite materials. The latter is particularly meaningful for the development of the ultra-lightweight vehicle Sax 3.

3 Model Assembling

3.1 Material Composition

The forerunner model of Sax 3 is the initial point for the design process. The subshell is built as a sandwich structure. The sandwich core consists of an aluminum comb (10 mm thickness) capped on both sides with a ($0_f/\pm 45_f/0_f$) laminate of carbon fibre fabric (twill-weave, approx. 0.75 mm per lamina). For aerodynamic efficiency and preservation of shape, one fabric lamina has to be on the outside of the vehicle. All other sandwich components can be optimized for the following loading cases.

3.2 Loads Caused by Mass of the Components

All components of the vehicle and also the driver produce loads to the chassis. Figure 3 shows the position of the driver, with a mass of 50 kg the main load on the chassis. Other components which must be taken into account are the fuel cell (6 kg) and the fire drencher (2 kg). The motor of the vehicle is integrated in the rear wheel. The loads are transferred through the wheel suspension. The modeling of the load with masses has the advantage that the loads do not need to be redefined among the various driving. The loads will be assigned by definition of inertia.

Fig. 3. Wheel and driver position in Sax 3

In most cases, point masses (sometimes connected with damping elements) are sufficient to model the loads. But in the case of the large distributed load caused by driver, the modeling with ballistic gelatin in volume elements was a good alternative. Ballistic gelatin has almost the same density as the human body and a low Young's modulus so that the chassis is not stiffened. The head and extremities were additionally modeled with mass points connected to the volume elements via rigid elements. The position of the volume elements is shown in Figure 4.

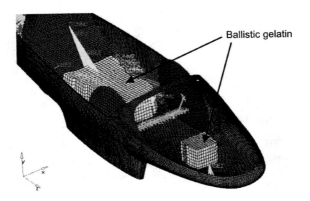

Fig. 4. Position of ballistic gelatin to model the driver

3.3 Loading Cases

A lot of loading cases have to be considered in the optimization process. These cases are described in detail in the following. First are figure some driving related load cases, the second part deals with some accident cases and at last, there are some specific load cases related to this ultra-lightweight vehicle.

The most commonly used driving condition is *straight ahead*. On exactly waveless roadways, only the gravitation takes an effect on the vehicle. No dynamic loads are possible in the current model. To include small potholes, threefold gravity is assumed.

By *cornering ability* additionally the radial acceleration takes an effect. It is different between the right and left curve. The radial acceleration can be specified directly in the simulation software. With a maximal deflection of 10° of the rear wheel, a minimum curve radius of 8.6 m is achieved. The maximum velocity of this deflection is 30 km/h. For the simulation of the forces during cornering, the mass of the motor (approx. 10 kg) is included in the model.

During the racing any *braking* is undesirable. In dangerous situations, the energy-saving vehicle has to be stopped very quickly. The rules of the Shell Eco-marathon dictate two independent brake systems. One of the brake systems is attached to the rear wheel; the other is attached to both front wheels. Two different loading cases model the brake application. Therefore the wheels were stopped. A maximum deceleration of 8.8 m/s^2 will be obtained with a maximum friction coefficient $\mu_0 = 0.9$.

For protection of the driver in case of an accident there are three loading cases included. The small design of the vehicles in the class of prototypes is susceptible to roll-over. Therefore, the *roll-over bar* is very important. The marathon rules stipulate that the roll-over bar can bear a minimum weight of 70 kg without bending. In the model, a vertical force of 1 kN is induced at the top of the roll-over bar in the direction to the bottom.

A second safety rule is the *seatbelt*. This five-point belt has to bear 1.5 times the weight of the driver. This force is induced in the five anchors (1 kN per anchor) in the direction of the expected tension.

Structural Optimization of Fibre-Reinforced Composites 149

The last loading case for the driver protection is the *frontal crash*. To stop the vehicle at 30 km/h at the distance between the front of the vehicle and the driver's feet, a deceleration of 17 g_e would be necessary. This loading case would dominate all others. However, a frontal crash is an absolute case of misusage. In the model, a deceleration of 3 g_e is assumed.

Two loading cases have to be considered for the safe handling of the vehicle. During getting into the vehicle, the lateral boarders are especially loaded and the seat of the vehicle has to bear a kick with the foot. Because of such a kick, delamination took place at the forerunner model Sax 2.

4 Optimization Cycle

The optimization process consists of different parts to be executed step by step:

1. Topology optimization
2. Redesign based on the topology optimization
3. Shape optimization
4. Size optimization

4.1 Topology Optimization

The first step of the optimization cycle is to find out regions where the aluminum comb and the inside laminate have to be placed. Topology optimization is the right tool to do that. As recommended by Altair, the optimization is executed with isotropic material properties. For some anisotropic materials like the comb and the laminate, comparable isotropic approximations must be found. Table 1 outlines the material parameter for all used Materials.

Table 1. Isotropic elastic material properties for all model components

Material	Young's Modulus in GPa	Lateral Contraction	Density in kg/m^3
Aluminum Comb	0.91	0.400	8200
Laminate	50.62	0.307	1630
Ballistic Gelatin	$2.3 \cdot 10^{-05}$	0.499	1100
Makrolon	2.4	0.400	1200
PETG	1.7	0.400	1270
Steel	210	0.300	7900

Two objectives for optimization are possible, the minimal mass and maximal stiffness. But only one objective can be set, the other one has to be a constraint. Whereas very precise ideas existed for the mass to be reached caused by the competitive situation, no evidence was available for the necessary stiffness of the structure. Therefore, the mass was chosen as a constraint and the stiffness was maximized in the form of the minimization of the objective strain energy. The mass of the whole chassis including the cap should be less than 10 kg. That is why the subshell had to be lighter than 8 kg. The optimization was started with different mass constraint values. From a mass of 6.5 kg, the optimization runs stable. At the end, a mass of 7 kg shows the best results.

Figure 5 shows the result of the topology optimization with a mass constraint of 7 kg. All elements with a virtual density less than 0.45 are hidden. In the red-marked areas, reinforcement with the aluminum comb is required. A large connected area was formed in the region of the side and the partition wall. At the sides, the typical truss-like structure is shown.

Fig. 5. Result of topology optimization

4.2 Redesign

The results of the topology optimization are the basis for the redesign of the chassis. With the help of the CAD software CATIA, the structure shown in figure 6 was designed. The optimization output was directly used in this software to create this structure.

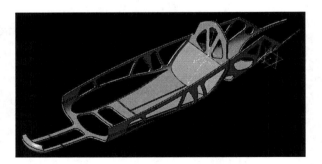

Fig. 6. Redesigned structure without the outer aerodynamic casing

4.3 Shape Optimization

Now, the new design had to be checked with a new simulation. Potentially occurring stress peaks can be diminished by the shape optimization. This simulation was also run with isotropic material properties.

Some changes of the comb material were made during the time of redesign. The reason for that was the bad processibility of the aluminum comb. It was not possible to create such fine framework structures. Furthermore, the adhesion between aluminum and the composite casing was not good enough in Sax 2. For the new vehicle Sax 3, an aramid comb was used. The thickness of this comb is 12.7 mm.

With the aluminum comb and also with the new aramid comb, no critical stress peaks were found. A shape optimization did not have to be accomplished.

4.4 Size Optimization

For the construction of the vehicle, two semi-fabricated materials were available: the low-temperature prepreg KGBX 2508 with carbon-fibre-woven fabric (250 g/m^2) and unidirectional carbon fibre material KGBX 2707. The latter is used for targeted introduction of preferential directions. The elastic properties of these materials are presented in table 2.

Table 2. Anisotropic elastic material properties of lamina

Material	KGBX 2508	KGBX 2707
E_1 in GPa	64.5	131
E_2 in GPa	64.5	10.3
v_{12}	0.032	0.213
G_{12} in GPa	3.8	4.4
ρ in kg/m^3	1630	1630

To insert anisotropic materials in the simulation model, a defined orientation of the elements is a basic requirement. As figure 7a shows, the default element orientation is heavily dependent on the mesh and more or less accidentally. For the correct alignment of the element coordinate systems, the draping analysis of CATIA was used. The resulting grid of this analysis describing local material orientations is assigned to the Hyperworks model by a specially developed macro (Figure 7b).

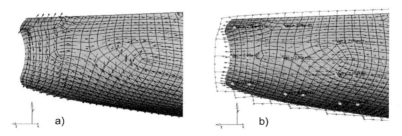

Fig. 7. Alignment of the material orientation: a) default alignment by Hyperworks, b) the oriented direction with the help of draping analysis by CATIA

The size optimization for laminated composites has two different modes. Either a thickness is predefined and the orientation angles are depending on the load, or the orientation angles are defined and the thickness of each layer is calculated by the optimization tool. For the described optimization process, the last modus was used. The load-bearing areas consist of a (0°/±45°/90°) laminate with KGBX 2707. The woven fabric material was only used for the outside shell in 0° and 45° direction.

The size optimization itself optimizes the laminae for all elements with the same value. To get different lamina thicknesses for every element, a method called free-size optimization is included in Hyperworks. This method is used for the following results. The order of the laminae has no effect on the result of the simulation. During the simulation, all laminae are considered as projected to the center of the shell.

5 Results

There is not enough space available in this article to show all results in detail, but the two examples demonstrate the effectiveness of optimization. All scales of the shown pictures are divided in the same way. The minimum of lamina thickness is 0 mm and the maximum is 2 mm. Therefore, the results can be compared easily.

Figure 8 shows the first example of size optimization. The wall behind the driver with the roll-over bar at the top is illustrated. The position of this component is viewed in the Figure 3. Figure 8 refers to the front side of the sandwich structure. The back side is separated by topology optimization.

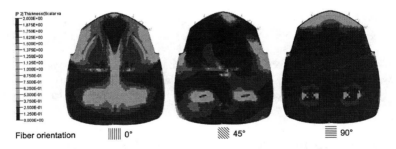

Fig. 8. Calculated distribution of lamina thickness in three different fibre orientations

Two loading cases dominate the optimization: the pressure to the roll-over bar and the loading from the anchor of the belts. The first case produces the thick areas on the left and right top side in 0° orientation and the thick region in 90° direction at the top. Also in 45° orientation, the influence is observable (the orientation of -45° is symmetric). The resulting amassments of the bolts are seen as two separate places at all material orientations.

Theoretically, there are a lot of regions without material on this side, but in practice it is not possible and not desired to build a composite laminate with holes.

Structural Optimization of Fibre-Reinforced Composites   153

As for the topology optimization, the result of the size optimization is regarded as information for the fabrication. This component was manufactured with two respective layers in ±45° with KGBX 2508 and locally reinforced with KGBX 2707 in 0° direction around the anchor of the belt and the left and right side of the top, as well as the top side of the wall with KGBX 2707 in 90° direction.

The result of the optimization of the side panel is explained in Figure 9. The main orientation of the laminate (0°) is directed along the trusses of the topology optimization. The reason is that these trusses basically are loaded in this direction. The upper figure shows these directions. Different laminate configurations were calculated for both sides of the sandwich. In practice, it is recommended to use

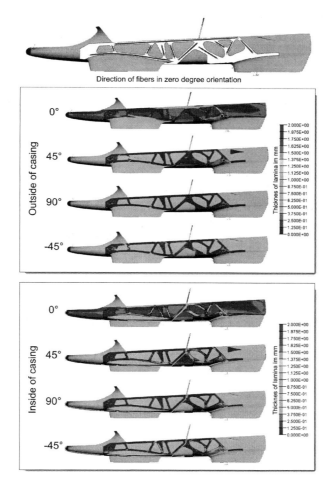

Fig. 9. Calculated distribution of lamina thickness on the sidewall sandwich. The 0° lamina orientation is shown in the above picture

equal configurations. In the used version of Hyperworks (8.0 sr1), it is not possible to abridge both sides so that the optimizer can create the same results. The inner and outer sides are displayed in two separate fields. Generally, it can be attributed to the fact that more material is needed on the inner side because, in addition, there is the aerodynamic cassis on the outer side. The highest lamina thickness is needed in the fibre orientation of $0°$. Especially in the connecting areas between the trusses, the laminae of $45°$ or $-45°$ become more important because of the influence between the trusses.

Generally, the weight of the chassis of Sax 3 could be reduced by one third as opposed to that of Sax 2.

6 Conclusion

With the aid of numerical optimization tools, the difficult task of optimization can be done more effectively, especially in the case of very different loadings and complex structures. Four different types of optimization were developed in the past for the different applications, but an optimization only makes proposals. It does not create a finished design.

For the vehicle Sax 3, topology and size optimization were used to reduce the weight of the chassis dramatically. This is a great advantage for the next races. Furthermore, a saving of the expensive materials was achieved. A shape optimization can be done optionally, but in the current state it was not necessary.

The full potential of optimization of the vehicle has not been utilized yet. Many safety factors were implicated caused by absent material properties or methods to use these properties in the simulation tool. Strain measurement on different places in the vehicle accomplishing at time, do check simulation results.

References

1. Information on http://www.fortis-saxonia.de
2. Bendsøe, M.P.: Optimization of structural topology, shape and material. Springer, Heidelberg (1995)
3. Schramm, U.: Automobiltechnische Zeitschrift, vol. 100, p. 456 (1998)
4. Schramm, U., Thomas, H.: 7th Symposium on Multidisciplinary Analysis and Optimization, St. Louis, MO, September 2-4 (1998)
5. Fredriksson, L.A., Schramm, U.: Shock & Vibration, vol. 8, p. 21 (2001)
6. Haftka, R.T., Gürdal, Z.: Elements of Structural Optimization. Kluwer Academic Publishers, Dordrecht (1991)
7. Bendsøe, M.P., Sigmund, O.: Topologie Optimization. Springer, Heidelberg (2004)

Optimisation and Characterisation of Magnetoelastic Microsensors

Bernhard Wielage, Thomas Mäder, and Daisy Nestler

Institute of Materials Science and Engineering, Chemnitz University of Technology,
09107 Chemnitz, Germany
bernhard.wielage@mb.tu-chemnitz.de,
thomas.mäder@mb.tu-chemnitz.de,
daisy.nestler@mb.tu-chemnitz.de

Abstract. The use of embedded sensors and actuators for condition assessment of machines and structures is on rising demand [1]. The load monitoring of parts made of composite materials, known as structural health monitoring, is of growing interest. Certainly, composite parts with high demands on safety and a long economic life-time, e. g. parts in airplanes and blades of wind turbines, will be equipped with strain- and impact-sensitive sensors in future. The monitoring will lead to a better understanding and a safer application of composite materials. Different types of sensors based on different physical effects are available.

The diameter or size of all of these available strain sensors is bigger than fibres in composite materials, and thus, these sensors do influence the structure and adulterate the strain measurements. Therefore, the aim of this work is to develop a strain sensor with a diameter equal to reinforcing fibres. Carbon fibre surfaces coated with magnetoelastic materials and structured by methods of micromanufacturing can serve as strain- and tension-sensitive microsensor systems capable of supporting safety and health monitoring functions in parts made of composite materials. The formerly developed fabrication method [3] based on thin-film deposition technology (cathode sputtering, plasma-enhanced chemical vapour deposition) [4] and microstructuring by means of focused ion beam (FIB) machining has been optimised. A novel PVD technique has been developed to deposit homogenous layers on single fibres and to improve the layer quality. Certainly, the deposition of magnetoelastic materials has been optimised. First sensor prototypes have been integrated into carbon-fibre-reinforced plastics. The results of all examinations will be presented.

Keywords: Structural Health Monitoring, Magnetoelastic Strain Microsensor, Coating of Carbon Fibres, Thin-Film Deposition, Single-Fibre Tensile Test.

1 Introduction

1.1 Structural Health Monitoring Technologies for the Monitoring of Polymer Matrix Composites

The non-destructive in-situ structural health monitoring (SHM) of highly loaded composite parts, in particular with a spatial resolution of strain down to fibre size, is still an unfulfilled quest. Market-available sensor solutions cannot be found. Different physical effects are used to realise embedded sensors for the macroscopic strain measurement in composite materials. Embedded sensors guarantee an artefact-free, lasting and, related to the cross-section of the part, representative strain measurement. Furthermore, these sensors are protected against external mechanical impacts [5]. Therefore, embedded sensors are more reliable than resistance strain gauges and should be favoured. In addition, strain gauges do not allow a space-resolved strain measurement down to fibre diameter. There are different strain sensor principles allowing an embedded application, based on different energy transformation effects. Piezoresistive [7; 8], piezoelectric [9] and magnetostrictive materials [10–13] are commonly used. Furthermore, optical fibres with and without engraved gratings are used to detect strain [5; 14; 15]. By means of optical fibres without gratings, a change in brightness or light intensity is used as an indicator for changing strain. In the fibres with so-called Bragg gratings changes in the running time of the light or the shift in the Bragg wavelength induced by strain in the fibre are used to directly measure changes in the fibre length. Finally, it is also possible to detect defects in structures using ultrasound-based sensors. Many sensor types have been refined and smaller sensors have been developed in recent years. The minimization of the sensor diameter is a goal for fibre-based sensors. A significant reduction of the diameter of optical sensors has been achieved in Japan [15]. The Japanese FBG sensor has a diameter of only 52 µm and is the smallest of its kind. Embedding this sensor reduces the notch effect evoked in the surrounding compared to common FBG systems. By directly using reinforcing fibres made of glass as a sensor, the notch effect can be completely omitted. The results of a research project have shown that the change in intensity of the light passing the glass fibres is a valuable measure for strain [14]. A direct correlation between intensity and load has been found. Due to a small diameter of 12 µm, these fibres induce no notch effect. A space-resolved monitoring by means of single-fibre sensors nevertheless has not yet been achieved. The change of the electric resistance of carbon-fibre-reinforced plastics can also be used as a measure for strain shifts. Carbon fibres are electrically conductive and comprise a graphitic microstructure of high order. Loading and changes in strain therefore induce a shift in the electric resistance within the fibres [8]. Carbon fibres show piezoresistive properties. Single carbon fibres are currently not used as strain sensors. Embedding of strain sensitive wires by stitching (e. g. constantan) is another possibility to realise piezoresistive strain sensors [7]. Wires with a diameter of less than 25 µm are commonly not available. An artefact-free strain measurement therefore cannot be guaranteed. Certainly, it is also not possible to realise a space-resolved measurement of strain with both of the above-described piezoresistive sensors.

Summarizing literature studies, the development of a strain microsensor is necessary to fulfil the above-mentioned demands – in particular artefact-free and space-resolved strain measurement. The goal of the present project therefore is the development of a strain-sensitive sensor which can be integrated into composite parts with a minimised influence on the composite structure. A minimum influence can be primarily realised by minimising the sensor dimensions. At a diameter of less than 10 µm, stress peaks, notch effects and damaging influences of the sensor on the composite structure can be omitted.

Fig. 1. Virtual model of a magnetoelastic microsensor

1.2 Principle of Magnetoelastic Microsensors

In the present research, the described sensor will be realised on the basis of the Villari effect. Exposing a magnetic material sample to strain, the properties of the material will change; in particular the magnetic permeability. This behaviour is called the Villari effect. It is inverse to Joule's magnetostriction which is the change in shape (elongation or contraction) of a magnetic sample in a magnetising field [16]. An alloy of 45 % Ni and 55 % Fe (NiFe 45/55) was chosen as the magnetoelastic material. It was deposited on a carbon fibre as a first layer. The measurement of changes in the magnetic permeability of the composite consisting of carbon fibre and the NiFe 45/55 layer under load can be realised applying two different principles. Using a transformer configuration, the NiFe 45/55 layer is surrounded by two separately deposited copper coils which are electrically non-conductive with each other as well as electrically isolated towards the NiFe 45/55 layer. By means of one of these coils, a voltage is induced to the other coil. If the magnetic permeability of the NiFe 45/55 layer is changed due to tension of the sensor, the electromagnetic induction changes as well. This can be measured in an altered induced voltage. Using the second configuration, only one coil is needed. This coil is connected in parallel with a capacity to build an oscillating circuit. Changes in the magnetic permeability can be detected in changes in the resonance frequency of the oscillating circuit to determine the applied tension. Decoupling of the layers is realised by non-conductive interlayers between NiFe 45/55 and Cu, e. g. using Si_3N_4 or Al_2O_3 layers [18]. A schematic of the one-coil sensor comprising the carbon fibre and different layers is shown in Fig 1.

To prove the feasibility of the concept, a 100:1 model of the instrumented fibre was built. In this prototype, a voltage of 100 mV was induced by applying a mechanical stress of 3.5 MPa, indicating the feasibility of the concept [10].

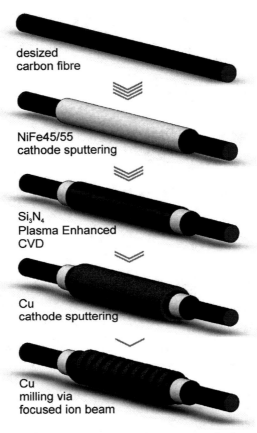

Fig. 2. Process steps for the manufacturing of the magnetoelastic strain microsensor

1.3 Manufacture of Magnetoelastic Micro-sensors

Utilising thin-film technologies, it is possible to manufacture a three-layer system of NiFe 45/55, Si_3N_4 and Cu, deposited on the circumference of a carbon fibre. Carbon fibres are commonly sized. Before the NiFe 45/55 layer is deposited, the polymer size must be removed completely [19]. A thermal treatment or a chemical solution by acetone is currently used for desizing. Results of former investigations have described the use of methyl ethyl ketone for desizing. The chemical state of carbon at the fibre surface depends on this process. As a result of the desizing process, oxidised forms of carbon or an amorphous pyrolytic phase are possible [19].

The small fibre size and its round shape pose a great challenge to find proper deposition techniques. Cathode sputtering [4] seems to be an adequate process to realise the deposition of NiFe 45/55 and Cu. To obtain homogeneous layers with a constant thickness at the whole circumference, the carbon fibres follow the original process route, manually rotated twice (120°) while fixed in a special holder to be coated from three sides. First deposition tests have shown a low adhesion of NiFe 45/55 on the carbon fibre surface. The use of Cr as a coupling agent was tested. Coupling of carbon and NiFe 45/55 has been significantly improved by Cr. A 50-nm Cr layer is generally applied as coupling agent [18]. By means of a PECVD process, the non-conducting layer of Si_3N_4 is deposited. Because of the PECVD process characteristics, rotation of the fibres is not necessary to form homogeneous layers. The combination of cathode sputtering and PECVD allows the manufacture of a three-layer structure on carbon fibres (see Fig. 2).

Using a focused ion beam (FIB), the electromagnetic coil is fabricated, structuring the previously deposited Cu layer [18]. A FIB system cuts the surface of specimens by scanning ion beams, in particular a Ga^+ ion beam [20]. To machine carbon fibres, a certain fibre-turning lathe has been developed. While macroscopic lathes feature stationary chucks for rotating the work piece, and a travelling support holding the cutting tool, the kinematics have been reversed for the fibre-turning lathe. This is necessary since the point of impact of the ion beam is fixed. Therefore, the portion of holding and rotating the carbon fibre resides in a traveller. A fibre support slightly pre-tensions the fibre and holds it in place at the location of the impact of the focus ion beam [22]. The last step of the sensor manufacture includes contacting the coil.

2 Methods and Experimental Procedures for the Sensor Optimisation and Characterisation

2.1 Sensor Embedment into Carbon-Fibre-Reinforced Plastics (CFRP)

To investigate the influence of the composite material and the production process to embedded sensors CFRP laminates with integrated sensors were manufactured. The manufacture has been carried out by means of a single side aluminium mould. Prior to the lamination the mould has been polished and treated with a mould release agent on the surface. Woven carbon-fibre fabric has been used as the reinforcing material. To be more precisely it has been made use of a pre-impregnated plain-weave-fabric material (prepreg), impregnated with epoxy resin. The lamination has been carried out via hand lay-up. A simple symmetric laminate construction having two plies 0°/90°, then the sensor prototypes either in 0° and 90° or in ±45° direction and another two plies of 0°/90° fabric on the top was chosen. The sensor prototypes have been manually integrated – in 0°, 90° and ± 45° orientation related to the woven fabric. After the last ply a perforated foil, and a suction fleece have been wrapped around the mould and above the sample surface in order to allow a barrier-free flow of excess resin. Everything has been put into a vacuum bag afterwards. Curing has been carried in an autoclave applying either 5 bar gas pressure or just atmospheric pressure and with curing temperature of 120 °C for 3 hours. After curing the CFRP plates have been removed from the mould and have been carried over for materialographic characterisations.

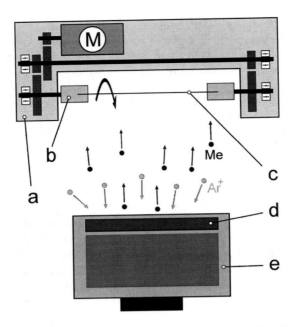

Fig. 3. PVD setup with rotational single-fibre holder – a: rotational holder, b: single-fibre clamp, c: single carbon fibre, d: sputter target, e: sputter magnetron, Ar⁺: argon ions, Me: metal atoms

2.2 Optimisation of Deposition Processes for Metallic Layers

The first optimisation step contained the reduction of layer thicknesses of NiFe 45/55 compared to early processes. The characterisation of thick NiFe 45/55 layers showed many voids and defects [23]. Furthermore these layers comprised residual stresses and tensile tests proved an early plastic deformation [25]. To avoid these characteristics, the thickness of the NiFe 45/55 layer was reduced to about 1 µm. So coated carbon fibres have been mechanically tested to determine the impact of this optimization.

Further process optimisation includes the development and application of a rotational holder for the PVD of metallic layers on single-fibre substrates. By means of the conventional PVD process, where the substrate is rotated twice for 120 ° in air, an interface between the single NiFe45/55 layers was found. Oxidation processes occurring while the sample is emerged to air during manual rotation are probably responsible. To avoid this effect, the fibre rotation should only be carried out within vacuum. Together with the company Malz & Schmidt Meißen a rotational holder was developed (see Fig 3). Driven by an electric motor the carbon fibre can be continually rotated during cathode sputtering with a varied rotational speed. The fibre is non-positively fixed between two polished metal plates which are kept in place by a clamp. A fixation with glue is not necessary and the clamping force does not crush the fibre. The holder is used in combination with a renewed B30-PVD system for DC-sputtering. The first experiments have been carried out depositing aluminium on single carbon (HTA) and glass fibres. The so produced specimens have been microscopically investigated.

2.3 Microscopic Characterisation

The objective of the characterisation of layers and their interfaces is to find voids, to determine layer coupling and to identify influences leading to certain characteristics. Using materialographic preparation techniques, coated carbon fibres are fixed in epoxy resin, then grinded and polished. High-grade materialographic specimens are produced. Uncoated fibres are fixed within Cu instead of epoxy resin by means of electrochemical or PVD processes. This leads to a better contrast in SEM observations and enables an easier detection of fibres. The specimen quality is checked by means of light-optical microscopy. This allows first conclusions about the layer quality. The following SEM studies were carried out using the LEO 1455VP and the NEON 40EsB. A detailed investigation of the layer microstructure, the observation of voids and the conclusion of reasons for their appearance were possible. Additionally, X-ray spectroscopy (EDXS) was used to identify the element formation of layers and fibres to investigate diffusion.

The investigations also include STEM observations for better interface and layer analyses. Special FIB-TEM preparation techniques have been carried out for the preparation of 100-nm-thick, electron-transparent thin films of single fibres, multilayer-sensor fibres and fibres coated with single layers. These techniques have been developed at the Institute of Materials Science and Engineering. They incorporate the options of a high preparation speed and excellent thin-film quality [26].

Additionally laminated plates with integrated sensor prototypes have been undergone a materialographic preparation as well. Fixed between plastic clamps the plates have been cut and pre-polished to find an evidence of the integrated sensors. Having clear evidence by seeing the cross-section of the sensor for instance the plates have been embedded in epoxy resin and have been grinded as well as polished after to prepare the cross-sectional view (Figure 4 d). The preparation of a longitudinal section started with a parallel cut a few micrometers aside the cross-sectional sensor evidence, followed by the epoxy embedment, the grinding and polishing. Grinding and polishing have been continued until the sensors longitudinal section could be seen in a whole under the light microscope (compare Figure 4), which was obviously a time consuming process.

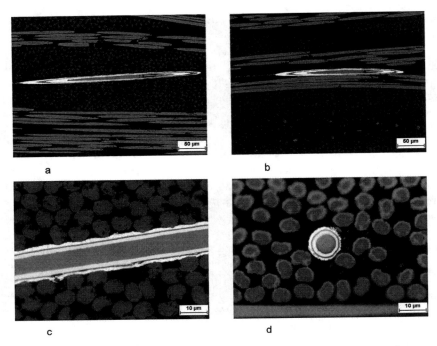

Fig. 4. Light-microscopic images of embedded microsensor prototypes – a and c cross-sectional embed-ment, b and c longitudinal embedment

2.4 Mechanical Characterisation

Mechanical tests of single fibres complete the characterisation of microsensors. These tests are carried out using the single-fibre tension test developed by the company Kammrath & Weiß. The tension tests include testing of carbon fibres in as-received state and testing of optimised layer structures deposited on carbon fibres. The influence of each single layer is tested by comparing test results of fibres with and without this particular layer. Tension tests of the layers deposited in the beginning of this project are compared with results of tests using optimised layers. The objective of the mechanical tension tests is the determination of influences on the mechanical properties induced by deposition processes.

3 Results

In the beginning of the research project and out of the first investigations, a variation of the layer quality was concluded [25]. Apart from some bad specimens, many samples already showed a really high layer quality. In the recent investigations numerous improvements have been achieved. The layer quality raised, keeps a constant level and a higher functionality of the microsensor can be expected. The following section specifies and discusses the determined results.

3.1 Results of the Characterisation of CFRP Containing Embedded Microsensors

The different light-microscopic images of cross-sections of CFRP with embedded microsensors document a perfect embedment of the sensors (compare Figure 4). In both orientations toward the reinforcing fibres, crosswise and along, the sensors are in direct contact with the fibres or only divided by a nanometre-thick resin layer. Fibres and sensors are enclosed with a resin film and completely wetted. An excellent load transfer from the composite material into the microsensor can be expected. Resin clusters, pores, cavities or other disturbing anomalies in the structural construction of the CFRP could not be found. Therefore, a formation of notch effects seems to be impossible. The embedment itself is accompanied by a self-alignment of fibres and sensors. This is a true proof for practicability of the objected dimensions of the microsensor.

Unfortunately there are some influences of the lamination process to the embedded sensor. In case of the autoclave curing with 5 bar gas pressure, which is necessary to achieve a high compaction rate for the composite structure, the reinforcing fibres deformed the outer cupper layer of the sensor (Figure 4 c). This deformation only occurred when the sensor was oriented crosswise towards the surrounding fibres. The contact pressure between sensor and crossing fibres is higher than compared to pressure in contact with parallel oriented fibres. It can be deduced, that the observed deformation might lead to an interruption of the electric connection within the electromagnetic coil and to the destruction of the sensor function. Further tests are necessary to investigate the deformational influence of the lamination process using other conducting materials with a higher stiffness and higher strength. Certainly aluminium will be tested in future investigations. Besides higher stiffness and strength aluminium would additionally guarantee a better elastic deformation characteristic compared to copper. This would increase the expectable life-time of the sensor. On the other hand it is questionable if a crosswise orientation of the sensor is useful. A structural health monitoring of composites in direction of the reinforcement orientation is of higher sense. Concerning research aspects the measurement of crosswise mechanics is nevertheless not useless. Therefore crosswise sensor functionality is still of interest and the investigation of alternative conducting materials will be done. Concerning CFRP cured under atmospheric pressure, no deformation of the conducting layer of the sensor was visible. The vacuum bag curing can be used without concerns.

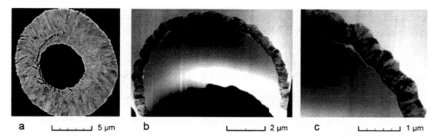

Fig. 5. Comparison of cross-sections of PVD-coated carbon fibres coated with the conventional PVD process (a) and coated using the rotational holder (b,c)

3.2 Results of the Microscopic Characterisation of Optimised Layers

By means of the conventional deposition route some specimens showed pores, oriented in radial direction (compare Figure 5 a). Shadow effects were identified as being responsible for an irregular layer growth on the fibre side not facing the sputter target. On fibre surfaces with a strongly grooved structure, this effect is intense. During the sputtering of NiFe 45/55 from one side, voids formed in the first deposition step provoke the formation of even bigger voids in the following deposition steps after manually rotating the fibre. An interface between the single NiFe 45/55 layers was found as well. Oxidation processes occurring while the sample is emerged to air during manual rotation were found to be responsible. Therefore the development of equipment to realise an automatic rotation within the closed and evacuated recipient was proposed to achieve better layer qualities. As already described a rotational holder for the coating of single-fibres was built. The first coating experiments, depositing aluminium on a carbon fibre surface were really successful. STEM investigations of so produced layers demonstrate a homogenous distribution of the layer microstructure (Figure 5 c). The layer comprises an even thickness around the whole circumference of the surface (Figure 5 b). Layers a compact and no voids or defects are visible (Figure 5 c). Coating experiments using other target materials, certainly NiFe 45/55, are necessary to allow a direct comparison between the conventional and the novel PVD process. But out of the first results with aluminium a further increase of the layer quality can be expected.

3.3 Results of the Mechanical Characterisation

Single-fibre tensile tests have been carried out to investigate the influence of the layer thickness and the use of chromium as a bonding agent on the mechanical properties of the microsensors, certainly on the mechanical properties of the NiFe 45/55 layer. In Figure 6 four representative stress-strain curves of tensile tests with different layer configurations, having different layer thicknesses and deposited with or without chromium, can be seen. It is obvious, that the coated fibre with a thick NiFe 45/55 layer (103 μm^2 cross-sectional layer area) and without the chromium layer show a plastic deformation for a strain above 0.5 %. The weaker bonding and residual stress, which was observed in thick NiFe 45/55 layer during earlier XRD analyses [25], are possible reasons for the plastic characteristic. The stress-strain curves of tensile tests of the coated carbon fibres with thin NiFe 45/55 layers comprising the chromium layer (HTA+Cr+NiFe – 24 μm^2 cross-sectional NiFe layer area) document a solely elastic deformation for this material. Additionally this type of a coated fibre reaches much higher strength compared to thick-layer fibres. The optimisation of the layer thickness shows successful results. Furthermore a single-fibre tensile test of a microsensor exhibits the same mechanical characteristic. Until a tensile load of 150 mN the sensor could be elastically deformed. After loading the load cycle was reversed. The sensor deformation returned to zero on almost the same linear route without hysteresis.

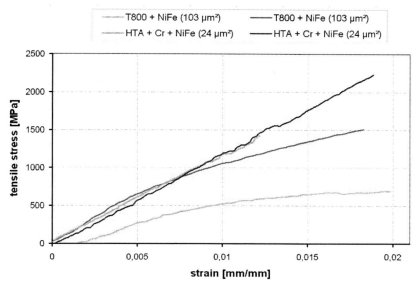

Fig. 6. Comparison of representative stress-strain curves taken from single-fibre tensile tests of NiFe 45/55-coated carbon fibres with different layer thicknesses (layer cross-sectional area in brakets)

4 Conclusion and Outlook

The formerly developed fabrication method based on thin-film deposition technology (cathode sputtering, plasma-enhanced chemical vapour deposition) and microstructuring by means of focused ion beam (FIB) machining has been successfully optimised. The optimised, conventionally manufactured, magnetoelastic NiFe 45/55 layer is elastic and was proved to have a perfect substrate bonding. The newly developed PVD technique furthermore allows the deposition of homogenous layers on single fibres and the further improvement of the layer quality. First sensor prototypes have been successfully integrated into carbon-fibre-reinforced plastics. This embedment guarantees an artefact-free strain monitoring due to the fact that no voids were found in the composite structure. Compaction of the composite material by high pressure autoclave curing leads to a deformation of copper layers on the microsensor. Further investigations are necessary to identify stiffer conducting materials to realise a lasting coil for compacted composite materials. Subsequent investigations will additionally deal with the sensor calibration and its characterisation. The electric contacting of the coil is also objective of future investigations.

Acknowledgments

The authors gratefully acknowledge the financial support of the Deutsche Forschungsgemeinschaft (German Research Foundation) and of the European Social

Fund allowing the experimental and theoretical work presented in this paper. The authors also thank the company Malz & Schmidt GbR, Meißen for their support in the development of the rotational holder in combination with the B30.TSP PVD system.

References

1. Karthik, T., Singh, R.: Micro electro mechanical systems based sensors for non destructive evaluation. In: Proc. Int. Struct. Eng. Constr. Conf., ISEC-4 - Innov. Struct. Eng. Constr., vol. 2, pp. S1121–S1127 (2008)
2. The British Composites Society: 17th International Conference on Composite Materials 2009 (2009)
3. Wielage, B., Mäder, T., Weber, D., Gatzen, H.H., Belski, A.: Analyses of Sensor Films on a Carbon Fibre. In: Banks, W.M. (hrsg.) 17th International Conference on Composite Materials (2009)
4. Seshan, K.: Handbook of thin-film deposition processes and techniques: Principles, methods, equipment and applications, 2nd edn. Materials science and process technology series Electronic materials and process technology. Noyes Publications/William Andrew Pub., Norwich/N.Y (2002)
5. Skontorp, A.: Composites with embedded optical fibers at structural details with inherent stress concentrations. J. Compos. Mater. 36 22, S2501–S2515 (2002)
6. Institut für Allgemeinen Maschinenbau und Kunststofftechnik: 11. Chemnitzer Textiltechnik-Tagung. Chemnitz (2007)
7. Kroll, L., Elsner, H., Tröltzsch, J.: Integration textiltechnisch hergestellter Drahtsensoren in Hochleistungsverbundbauteile für die Großserie. In: 11. Chemnitzer Textiltechnik-Tagung. Chemnitz (2007)
8. Matsuzaki, R.: Damage sensing carbon fibre composites based on electrical resistance. In: Banks, W. M (Hrsg.) 17th International Conference on Composite Materials 2009 (2009)
9. Schönecker, A., Gebhardt, S.: Microsystems technologies for use in structures and integrated systems. In: CIMTEC - Proc. Int. Conf. Smart Mater., Struct. Syst. - Emboding Intell. Struct. Integr. Syst., vol. 56, pp. S76–S83 (2008)
10. Dinulovic, D., Gerdes, H., Mucha, H., Wielage, B., Gatzen, H.H.: Carbon fiber with magnetoelastic sensing capability for composite materials. Sensor Letters 5,1, S218–S221 (2007)
11. Takahashi, S., Echigoya, J., Ueda, T., Li, X., Hatafuku, H.: Martensitic transformation due to plastic deformation and magnetic properties in SUS 304 stainless steel. Journal of Materials Processing Technology 108, 2, S213–S216 (2001)
12. Hecker, S.S., Stout, M.G., Staudhammer, K.P., Smith, J.L.: Effects of Strain State and Strain Rate on Deformation-Induced Transformation in 304 Stainless Steel: Part I. Magnetic Measurements and Mechanical Behavior. MTA 13 4, S619–S626 (1982)
13. Verijenko, B., Verijenko, V.: The use of strain memory alloys in structural health monitoring systems. Compos. Struct. 76 1-2, S190–S196 (2006)
14. Malik, S.A., Ojo, S.O., Harris, D., Fernando, G.F.: In-Situ Damage Detection in Glass Fibre Composites. In: Banks, W. M (hrsg.). 17th International Conference on Composite Materials 2009 (2009)

Optimisation and Characterisation of Magnetoelastic Microsensors

15. Ishikawa, T.: Overview of trends in advanced composite research and applications in Japan. Adv. Compos. Mater Off. J. Jpn. Soc. Compos Mater 15 1, S3–S37 (2006)
16. Ben Amor, A., Ruffert, C., Gatzen, H.H.: NiFe 45/55 and its application in a strain gauge sensor. In: Proc. Electrochem. Soc. PV 2004-23, pp. S481–S492 (2006)
17. TU Chemnitz, Lehrstuhl für Verbundwerkstoffe: Beschichtungstechnik: Tagungsband zum 11. Werkstofftechnischen Kolloquium. 01. und 02. Oktober, in Chemnitz, Chemnitz: Eigenverlag (Werkstoffe und werkstofftechnische Anwendungen 31) (2008); ISBN 978-3-00-025648-6
18. Belski, A., Gatzen, H.H.: Einsatz dünnfilmtechnischer Fertigungsverfahren zur Herstellung eines magnetoelastischen Sensors auf Kohlenfaserbasis. In: Wielage, Bernhard (Hrsg.). Beschichtungstechnik: Tagungsband zum 11. Werkstofftechnischen Kolloquium. 01. und 02. in Chemnitz. Chemnitz: Eigenverlag (Werkstoffe und werkstofftechnische Anwendungen, 31), S269–S274 (Oktober 2008) ISBN 978-3-00-025648-6
19. Than, E., Hofmann, A., Leonhardt, G.: Surface composition and structure of fibres for composite materials studied by means of XPS and AES. Vacuum 43(5-7), S485–S487 (1992)
20. Fujii, T., Iwasaki, K., Munekane, M., Takeuchi, T., Hasuda, M., Asahata, T., Kiyohara, M., Kogure, T., Kijima, Y., Kaito, T.: A nanofactory by focused ion beam. J Micromech Microengineering 15(10), S286–S291 (2005)
21. Proceedings of the 10th anniversary international conference of the European Society for Precision Engineering and Nanotechnology, May 18-May 22. Euspen, Zürich (2008)
22. Belski, A., Kammrath, W., Gatzen, H.H.: Focused Ion Beam Fiber Turning Lathe for Fabricating a Strain Gauge for Carbon Fibers. In: van Brussel, H. (ed.) Proceedings of the 10th anniversary international conference of the European Society for Precision Engineering and Nanotechnology, Zürich, Switzerland, May 18-May 22, pp. S129–S133. Bedford, Euspen (2008)
23. Wielage, B., Mäder, T., Weber, D., Mucha, H.: Charakterisierung der Verbundschichten von magnetoelastischen Sensoren. Materialwissenschaft und Werkstofftechnik 39 12, S947–S950 (2008)
24. Georgia Institute of Technology: 3rd International Joint Conference on Advanced Materials and Design (2009)
25. Wielage, B., Weber, D., Mäder, T.: Characterisation of composite layers and single-fibre composites for the development of magnetoelastic sensors. In: Garmestani, H. (hrsg.) 3rd International Joint Conference on Advanced Materials and Design, pp. S1–S7 (2009)
26. Mucha, H., Kato, T., Arai, S., Saka, H., Kuroda, K., Wielage, B.: Focused ion beam preparation techniques dedicated for the fabrication of TEM lamellae of fibre-reinforced composites. J. Electron Microsc. 54(1), S43–S49 (2005)

Advanced Energy Utilization – Fuel Cell

Challenges and Opportunities in PEM Fuel Cell Systems

Amir M. Niroumand and Mehrdad Saif

Simon Fraser University, Canada
amniroum@sfu.ca, saif@sfu.ca

Abstract. In this work, we discuss challenges and opportunities facing PEM fuel cell systems in three levels: system architecture, control, and diagnostics. In the system architecture context, we analyze various realizations for the fuel cell system as well as the hybrid system. In the control context, we discuss the optimal control strategy for flow, humidity, temperature, and power subsystems. In the diagnostic context, we discuss diagnostic tools based on the stage in the product lifecycle where they are required, i.e., research, development, manufacturing, and operation.

Keywords: Fuel cell, System, Control, Diagnostics.

1 Introduction

Polymer Electrolyte Membrane/Proton Exchange Membrane (PEM) fuel cells are electrochemical cells that combine hydrogen and oxygen to produce electricity, heat, and water. They consist of a Membrane Electrode Assembly (MEA), sandwiched between anode and cathode Flow Field (FF) plates, typically made of graphite (Fig. 1). The MEA consists of a polymer membrane, sandwiched between anode and cathode electrodes. Each electrode consists of a Catalyst Layer (CL), made from agglomerates of platinum particles, carbon grains, and ionomer, and a Gas Diffusion Layer (GDL), which is typically made of carbon paper (Fig. 1).

The membrane becomes proton conductive when it is well hydrated, while remaining insulated to electron and gas transport. The electrochemical reaction takes place at the three phase boundary in CL agglomerates at both sides of the MEA where carbon, catalyst, and ionomer meet. In the anode, hydrogen flows through the anode FF by convection and diffuses through the anode GDL to reach the anode CL sites. At the anode CL, hydrogen molecules dissolve at the ionomer surface and diffuse through the ionomer to reach the three phase boundary. At this point, hydrogen breaks into electrons and proton. Electrons are conducted through the anode CL carbon grains to the anode GDL carbon fibres, anode FF landings, and the load. They then take a similar path in the opposite direction to reach the three phase boundary at the cathode CL, where they combine with protons conducted through the membrane and the oxygen molecules diffused through the cathode CL ionomer to produce water.

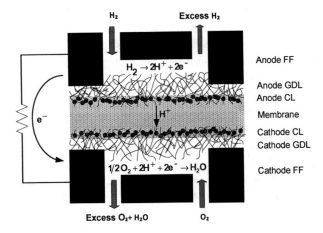

Fig. 1. PEM fuel cell schematic diagram

The output power of a fuel cell is proportional to its electrochemical active surface area, while its output energy depends on the size of its fuel tank. This allows fuel cells to scale power and energy independently, making them feasible for a wide range of applications, from mobile electronic devices to power plants. Their commercial viability, however, has been primarily constrained by their cost. In this paper, our emphasis is on cost reduction strategies for PEM fuel cell system architecture, control, and diagnostics contexts.

2 System Architecture

PEM fuel cell system architecture can be viewed in two levels: fuel cell system and hybrid system. Below we discuss challenges and opportunities at each level.

2.1 Fuel Cell System

A PEM fuel cell system consists of a PEM fuel cell stack and a control system. The fuel cell stack is where the electrochemical reaction takes place; while the control system is responsible for providing oxygen/air and hydrogen to the fuel cell stack at the desired temperature, pressure, flow rate, and humidity, as well as controlling the stack temperature and load. The control system requirements are dictated by the stack design, which in turn depends on the load power requirements. Fig. 2 shows rough estimates for the fuel cell stack and control system cost as a function of fuel cell output power for fuel cells at low volumes.

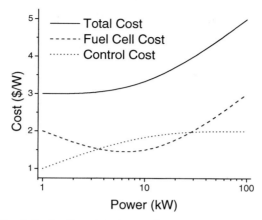

Fig. 2. Fuel cell stack and control system cost vs. power

When the fuel cell output power is up to $O(1\text{kW})$, the reaction heat can be dissipated to the environment by evaporating the water produced in the cathode, i.e., an open cathode stack architecture. Since cathode is the main source of losses, this is a very efficient method for cooling PEM fuel cells. For such stack, the control system requirements are minimal: no external cooling is required. In addition, the low pressure atmospheric operation of these stacks allows using fans instead of compressors, which are simple and reliable. These factors result in a low cost control system for the air cooled fuel cell architecture.

As the fuel cell output power increases, so does the heat produced due to losses. When the fuel cell output power reaches $O(10\text{kW})$, heat exchanger plates need to be integrated into the fuel cell stack for cooling. At these higher power outputs, the stack is also operated at higher pressures to improve reaction kinetics. Therefore, the cost of the fuel cell stack per power drops due to improved efficiency. However, the cost of the control system increases due to higher cathode pressure as well as cooling requirements. Also note that when a liquid agent, such as water, is used to cool the fuel cell stack, it ionizes as it flows in the channels, tubing, and heat exchangers. As a result, its resistance drops and it conducts electricity between the flow field plates, resulting in voltage loss. To avoid this, a deionizer is required to ensure the cooling agent resistance remains above a certain level. This contributes further to the control system cost of liquid cooled stacks.

When the fuel cell output power reaches $O(100\text{kW})$, in addition to the cooling and pressurized air requirements, even distribution of hydrogen becomes a challenge, requiring large manifolds or multiple sources, which add to the system cost. Furthermore, as the scale of the stack increases, implementing mass manufacturing practices becomes more challenging, adding to the stack cost per output power. The control system requirement is also more demanding in this class. In the lower classes, simple passive diffusive humidifiers can be used. The higher output power class however requires active humidification as well as ejectors to separate water from the output stream, all adding to the system cost.

While configurations other than the ones described here are also feasible and result in different system cost dynamics, the stack and control system architecture in general limits the control system design. This is because the requirements are primarily dictated by the stack design an little freedom for various architecture exist. An alternative approach is to breakdown the fuel cell system into MEA and control system. In fact, the flow fields are part of the reactant delivery as well as heat and electricity removal systems, hence art of the control system function. Such architecture allows additional degrees of freedom in the control system design, providing cost reduction opportunities.

Printed Circuit Board (PCB) fuel cell with MEMS based integrated control system is one realization of such system architecture. PCBs can be used as flow field separator plates to sandwich the MEAs to build a stack [1]. With PCB based fuel cells, MEMS based control components such as valves and sensors can be integrated into the fuel cell stack using electronic manufacturing techniques [2]. This results in smaller and lighter fuel cell systems with a cost dynamics that follows the economies of scale due to the mature manufacturing processes that has been developed for the semiconductor industry. More specifically:

- Graphite plates used to build FFs are heavy, brittle, and not suitable for mass manufacturing schemes. PCBs however are light and can be produced using batch manufacturing processes at a low cost.
- With conventional stacks, sealing gaskets need to be placed between the graphite plate FFs and the MEA, which add as a component during stack assembly. With PCB FF plates, it is possible to deposit the sealing on the plates, eliminating a component during the assembly process.
- With conventional fuel cell stacks, heavy end plates are required to hold the stack together. With PCB based fuel cell stacks, soldering can be used to maintain the stack structure, reducing stack weight and manufacturing complexity.
- Conventional control systems use machined components such as control valves; therefore their cost does not scale very well with volume. On the other hand, MEMS based components are produced using batch processes, hence following the economies of scale.
- The assembly process of a semiconductor based integrated control system is less than a distributed one, thanks to the mature manufacturing processes developed for the semiconductor industry.

PCB based fuel cells have been developed for micro fuel cells with an output power up to $O(100W)$. As the fuel cell output power increases, heat exchange becomes a more important issue. We have discussed PCB based fuel cell heat exchangers in [3].

2.2 Hybrid System

Fuel cells are used in a hybrid system with batteries and/or capacitors as a buffer to deliver power to the load. The size and configuration of the hybrid elements as well as the control strategy used affects the system cost, performance, and lifetime, as discussed below.

The two basic hybrid configurations are series and parallel. In the series configuration, the fuel cell essentially acts as a battery/capacitor charger, where in the parallel configuration, the fuel cell and the battery/capacitor can both drive the load. The parallel configuration can deliver higher power to the load with hybrid components similar to the series configuration, or alternatively similar power with smaller components. The serial configuration however results in more separation between the load dynamics and the fuel cell, resulting in a more relaxed operating regime for the fuel cell and hence improved lifetime. Combinations of serial/parallel configurations can be used to achieve various output power profiles.

The hybrid control objective depends on the architecture used. In a serial configuration, the fuel cell has to be operated such that it maintains the State of Charge (SoC) of the battery/capacitor in a certain boundary, while delivering power to the load from the hybrid pack (battery/capacitor). In the parallel configuration, the control system objective is to deliver power to the load from the fuel cell, given the slower dynamics of the fuel cell. In this configuration, the hybrid elements are essentially used to compensate for the fuel cell slow response compared to load changes. In both architectures, hybrid elements are also used for regenerative purposes. Fluctuations in DC bus voltage as well as boundaries between high load transients such as generative/regenerative cycles are some challenges that these controllers confront.

The choice of power operating point for the fuel cell affects the fuel cell efficiency and degradation rate, as well as the hybrid component requirements. These parameters can be used to find an optimized operating point for a given architecture and load profile that results in the minimum cost for the system, or preferably maximizes the economic value of the system (time included). An alternative formulation is to operate the fuel cell at an operating point at which it maximizes the energy delivered by the fuel cell over its lifetime. One obstacle in such analysis is that the relationship between operating point and degradation rates is not well understood and characterized. Characterization of the degradation rate as a function of the operating point provides key information for such optimization. Note that in addition to the steady state load, load transients also affect the fuel cell lifetime. A simplistic approach to mitigate from such degradation mechanisms is to heavily hybridize the fuel cell system in order to reduce load transients for the expense of the additional buffer components.

3 Control

The fuel cell control system can be divided into the following subsystems: flow, humidity, stack temperature, and load, as shown schematically in Fig. 3. The dynamical response of these systems has a wide range of time constants; from very fast electric response to slow temperature variations, and between. An alternative challenge for model based control is that while the dynamical response of the control subsystems can be well captured by first principle and/or empirical models, the dynamical response of the fuel cell stack depends on many parameters that are not well understood and characterized. The fuel cell polarization curve that is used to relate the load and voltage is a steady state representation of the fuel cell and does not capture fuel cell dynamics. This limits the performance of model based fuel cell control algorithms, specifically when their effectiveness is tested using simulations with the same model for the fuel cell [4].

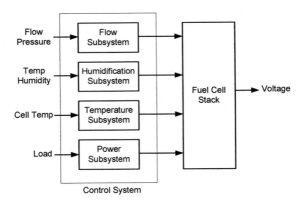

Fig. 3. Fuel Cell Control Schematic Diagram

An alternative problem is finding the optimal set points for the fuel cell under different operating conditions. Furthermore, as the fuel cell ages, the transport properties in the fuel cell components change, and so does the optimal operating point. An adaptive scheme that can find the optimal operating point during operation can improve the control system performance. These have been further discussed for each subsystem below.

3.1 Flow Subsystem

The flow subsystem is responsible for maintaining the anode and cathode streams pressure and flow at the optimal level. Fig. 4 shows the instrumentation diagram for a typical anode flow and humidification subsystems. In hydrogen fuel cells, the anode flow subsystem primarily consists of an input and an output control valve, along with an input flow sensor and an output pressure sensor. One control methodology is to use the input flow sensor with the input valve and the output pressure sensor with the output valve in two different control loops to control the input flow rate and output pressure at the desired set points. An alternative approach is to use a dynamical model of the anode flow field with a two input two output state space representation of the system. This approach could result in a better dynamical response for the flow control system for the cost of a more complicated control scheme. In this method, changes in the flow field transport properties, e.g. due to liquid water build up, can be treated as disturbances to the system.

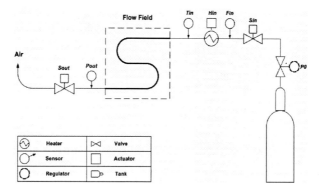

Fig. 4. Flow and humidification subsystems

The flow operating point of a hydrogen fuel cell anode is the point at which sufficient hydrogen is supplied to drive the reaction. This corresponds to a hydrogen stoichiometry (ratio of supplied to used reactants) close to one. Supplying a higher stoichiometry to the anode side does not have a significant effect on the cell voltage, since losses are dominated by the cathode side. The lowest anode stoichiometry at which the highest cell voltage is achieved can be obtained a priori of fuel cell operation, and used as the reference point during operation [5].

On the cathode side, the flow subsystem depends on the fuel cell system architecture, which in turn determines the source of gas supply, i.e., fans, air compressors, or pressurized oxygen/air tanks. Fans are used for open cathode air cooled architectures. In such systems, the output pressure does not need to be controlled (atmospheric). Only the input air flow rate is controlled using the fan voltage, resulting in a relatively simple control system. For higher power fuel cell systems that are pressurized, compressors or pressurized tanks are used along with the instrumentation, similar to that shown in Fig. 4.

In the cathode, higher air flow rates result in higher concentration for oxygen as well as improved water removal, hence improving the cell voltage. On the other hand, higher air flow rates translate into higher parasitic loss for the fans or air compressors used, which in turn reduces the overall system output power. Therefore, the control system needs a scheme for finding the optimal stoichiometry at each load that result in the highest output power for the overall system. One approach is to obtain the peak power point using *a priori* of real time operation. However, as the fuel cell and/or fan/compressor dynamics change, e.g., due to aging, an adaptive scheme is required to maintain the optimal operating point.

An adaptive scheme for the control problem can be formulated such that the $dP/dV = 0$ condition is satisfied, i.e., small changes in fan voltage does not affect the system output power, meaning that the peak power has achieved. To implement such scheme, a small signal low frequency oscillation can be imposed on the fan/compressor voltage, and changes in system output power can be measured. The sign and magnitude of the derivative can be used as a vector towards the peak power operating point.

During load transients, on both the anode and cathode sides, the flow should lead the current when the load is increasing and should lag the current when it's decreasing to ensure sufficient reactant is available to drive the reaction. Therefore, the dynamical response of the fuel cell to load changes is constrained by the dynamics of the flow control subsystem. This is of particular importance in parallel hybrid architectures where the fuel cell needs to follow the load.

The pressure operating point is determined by the stack design and remains constant during fuel cell operation. During start up and shutdown procedures, anode and cathode experience pressure transients. A constraint for the pressure control system is bound the difference between anode and cathode pressures to avoid the membrane to rupture. The pressure gradient limit is determined by the membrane specifications.

3.2 Humidification Subsystem

The fuel cell membrane and CL ionomer need to be well humidified in order to transport protons effectively. Therefore, reactants should be preferably humidified before entering the fuel cell to ensure proper humidification of the membrane and ionomer on both the anode and cathode electrodes is achieved. However, this imposes additional cost on the control system. Various humidification strategies are explained below, as they related to the stack output power and system architecture.

In micro fuel cells, $O(100W)$, where the current density is low, the water produced in the cathode can be managed to sufficiently humidify the membrane, eliminating the need for any external humidification. In these systems, the water produced in the cathode primarily humidifies the cathode. Furthermore, it diffuses through the membrane to the anode and humidifies the ionomer on the anode side, providing internal humidification.

When moving to higher power fuel cells, $O(1kW)$, where the open cathode architecture is still feasible, the water produced at the cathode can only be used to humidify the cathode. The higher current density regimes operated by these fuel cells compared to micro fuel cells result in a higher Electro-Osmotic Drag (EOD) of water molecules from the anode to the cathode side by protons that cross the membrane. This counter balances the water that diffuses from the cathode to the anode, resulting in a relatively dry anode if not externally humidified. Therefore, the input anode stream of these fuel cells need to be humidified before entering the stack.

As the fuel cell output power increases to $O(10kW)$ and higher, the cathode also needs to be humidified. This is because the higher flow rates required by these stacks will dry the beginning of the cathode, if a dry stream is used. Therefore, to ensure proper humidification at the beginning of the cathode, the cathode stream is also humidified.

The choice of humidifier also affects the system cost, efficiency, reliability and dynamical response. Bubbling humidifiers and diffusive humidifiers two main technologies developed. Bubbling humidifiers heat the water in a pressurized tank to a desired temperature. The reactants flow through the heated water, where they heat up and absorb water vapor, hence humidified to the desired dew point temperature. This method allows relatively accurate control over the reactant dew

Challenges and Opportunities in PEM Fuel Cell Systems

point temperature; however, the dynamical response for changing the dew point temperature is relatively slow due to the high heating capacity of water. This humidification technique can be scaled for use in large systems, for the expense of high heating parasitic losses.

Diffusive humidifiers are a cheap and passive option when compared to bubbling ones. In this technique, the wet gas stream exiting the fuel cell is passed on one side of a membrane, and the dry gas passes through the other side of the membrane. Water diffuses from the wet to the dry gas across the membrane, humidifying the fuel cell dry input streams. These humidifiers can not provide a humidification level as accurate as bubbling humidifiers. Controlling the humidification level with these humidifiers can be achieved by changing the flow rates on the wet and dry gas sides of the humidifier.

There are number of challenges in PEM fuel cells water management. The fuel cell needs to be well humidified before going under load, while water has not yet been produced due to the absence of load. This specifically causes complication for humidification schemes where they depend on the water produced in the cathode for internal humidification. Gradual increase in the fuel cell load, hence water production rate, is one approach for the start up of such systems.

An alternative complication in humidification control arises from the fact humidity sensors are not reliable for operation in high humid condition due to water condensation. As a result, assessment of the humidity conditions in the fuel cell need to be achieved by other measurements, which makes such assessment more complicated. Feed forward control is one approach to the problem.

Controlling the humidification level during load transients is another challenge, specifically given lack of measurement as well as the slow dynamics of the humidifier. However, the factor playing in favour of such control scheme is that changes in membrane humidity conditions is also slow, providing the control system with time to respond.

The reactant temperature entering the cell also affects its humidity level. Active heating of the reactants impose parasitic loss on the fuel cell, hence not desirable. Heat exchangers can be used to passively heat the reactants from the heat generated by the fuel cell reaction.

3.3 Temperature Subsystem

The temperature subsystem is responsible for maintaining the stack temperature at a fixed value as determined by the membrane specifications, typically around 80°C for PEM fuel cells. The cooling scheme used depends on the fuel cell system architecture, as explained below.

With the open cathode architecture, the cathode stream is responsible for providing oxygen, as well as removing the produced water and heat. Such architecture results in limited freedom for independent control of these variables, instead, it requires proper stack design to ensure all these requirements are met at the same time.

For pressurized systems, a liquid cooling agent such as water or oil is circulated in parallel FFs in the stack to remove the produced heat. The cooling agent is then circulated out of the stack and cooled in an external heat exchanger. As mentioned earlier, a major challenge with these systems is maintaining high resistance for the cooling agent, so that it does not conduct electricity between the various graphite plates with different potential. Therefore, de-ionizing the cooling agent is an additional requirement for these systems.

Temperature gradient in the fuel cell stack is another challenge that the stack temperature control subsystem need to address. Reaction rates are higher at the beginning of the stack due to higher reactant concentration, while losses are higher toward the end of the stack due to reduced reactant concentration. Therefore, there are higher heat removal requirements to maintain the desired operating temperature at the stack end compared to the beginning of the stack. The cooling agent can be circulated in the opposite direction of the reactants in the parallel heat exchange FFs, so that when the it has a higher heat removal capacity it removes heat from the stack end, and as it heat removal capacity reduces, it reaches the stack end where there are lower heat dissipation requirements. The cooling agent flow rate and temperature are the parameters that can be used to control the stack temperature.

3.4 Power Subsystem

The power subsystem is responsible for bounding the fuel cell current in an admissible region, and controlling the fuel cell output voltage at a fixed value.

The fuel cell voltage is a function of current, however, in most applications, a fixed DC or AC voltage is required to drive the load. The power subsystem controls the fuel cell output voltage at the desired level by means of a DC-DC or DC-AC converter. Maintaining a fixed output voltage given current and input voltage fluctuations is not a trivial task.

An alternative issue arises during load changes. When the load increases, so does the flow rate requirement. However, the dynamical response of the flow subsystem is slower than the load. Therefore, to ensure sufficient reactant is available for reaction, the fuel cell current has to be bounded to the maximum current that the existing anode and cathode flow rate can safely provide [6].

4 Diagnostics

Diagnostic tools can be used at different stages of fuel cell lifecycle to improve their reliability, performance, and lifetime. At the research phase, diagnostic tools are needed to provide better understanding of dynamical process in the fuel cell. Accuracy and repeatability are major requirements for such tools. During product development, diagnostic tools are used to characterize the fuel cell for design optimization.

At manufacturing, these tools are used for Quality Control (QC). Therefore, they need to be scalable and compatible with mass production schemes. During operation, diagnostic tools are used to monitor the state of health of the system and detect faults before they fully develop. Cost and reliability are two major requirements for diagnostic tools that are used at this stage. In this section, we discuss challenges and opportunities in fuel cell diagnostics as it relates to the stage of the fuel cell lifecycle.

4.1 Research

One of the major challenges in understanding the dynamical processes in fuel cells arises from the fact that there are two many parameters and relatively too little measurements, essentially the problem is ill-posed. Therefore, many different model realizations can reproduce the observation, which in turn affects the reliability of the models used to explain dynamical processes in PEM fuel cells. To address this problem, one approach is to use simplified models so that the number of parameters matches the measurements. The drawback is that such approach limits the analysis to the simplifying assumptions made. An alternative method is to perform *ex situ* measurements of the transport properties in order to reduce the number of fitting parameters in the model. While this approach can result in better stability and reliability for the solutions proposed, there is no guarantee that the *ex situ* and *in situ* values be similar. In addition, the values reported in the literature usually spread over a wide range. Therefore, rather than exact behaviour, trends can be better explained by this method.

Another approach is to increase the dimension of the observation space by introducing additional measurements. This, along with experiments that are designed to separate a specific phenomenon of interest can lead to further understanding of the dynamical processes in these systems.

Frequency based techniques, such as impedance spectroscopy, provide a wide spectrum of observations at different frequencies. These techniques use the difference in the time constant of various processes to separate their contribution and understand their effect. Although frequency based techniques have high information content, proper modelling and interpretation of the results remain a challenge in this field. We have recently proposed a new frequency based diagnostic tool, called pressure-voltage spectroscopy, in which the ratio between imposed pressure and measured voltage oscillations are used to separate dynamical processes in the fuel cell [7].

4.2 Development

While better understanding of dynamical processes is very useful for improving the fuel cell design, characterization is the essential tool for design optimization in the development stage. At this stage, diagnostic tools need to separate the contribution of individual components, rather than processes. This allows optimizing the components by trial and error, while fundamental understanding plays a key role in directing the trials.

A current challenge in this context is developing diagnostic tools that can characterize degradation mechanisms. Degradation experiments consume lots of resources, which makes trial and error practices for improving degradation rates very expensive. One approach for reducing the experiment duration is to expedite the rate of degradation by operating the fuel cell under stressful conditions. The challenge is that the mechanism by which the fuel cell degrades depends on the stress factor; therefore, it is not guaranteed that the same degradation mechanism occurs during rapid testing as it does during normal operation. Diagnostic tools that can characterize (classify and quantify) degradation mechanisms are required to address this challenge before rapid testing schemes can be reliably implemented.

4.3 Manufacturing

Diagnostic tools required for QC during manufacturing depends on the manufacturing technology used to build the fuel cell components and system. While some of the diagnostic tools developed for other stages of the fuel cell lifecycle can be adopted to be used for QC, not much has been done in this field by the academia. This is mainly because manufacturing processes used by the industry are treated as confidential, therefore, the requirements based on which these tools need to be developed/customized are not available and known to the research community. However, one would expect significant work done in this field by the industry.

4.4 Operation

Diagnostic tools at this stage need to detect and isolate fuel cell faults before they fully develop, in order to allow the control system to take compensatory actions. The fundamental challenge with fuel cell Fault Detection and Isolation (FDI) is that a mathematic model that captures the dynamics of the fuel cell under various operating conditions does not exist due to the complexity of the system. As a result, model based approaches have limited reliability for fuel cell FDI.

An alternative approach is to use a knowledge based technique for fuel cell FDI. Voltage drop is common to all fuel cell faults; therefore, it is possible to detect a fault in the fuel cell by comparing the measured voltage versus an expected reference value obtained *a priori* of operation. The expected reference value drops as fuel cell ages, an adaptive scheme is required to update the voltage reference value as the fuel cell degrades.

Since voltage drop is common to all fuel cell faults, other measures are required to isolate each fault, so that the control system can take compensatory actions. One approach is to use the fault signatures in order to isolate them. We have explained these signatures below based on their appearance in the measurement space.

The rate at which the fuel cell voltage drops could be used as an indication for isolating the fault occurred. Drying is a relatively slow process; therefore, it results in a slow decaying voltage [8]. On the other hand, flooding results in high fluctuations in the cell voltage. Starvation can cause sharp voltage drop to zero or negative values. CO poisoning could result in oscillations in the cell voltage, the frequency of which depends on the cell temperature.

Pressure measurements can provide information regarding the flow field condition. Flooding in the channels increases the pressure drop across the channel. Leakage in the stack sealing can also result in high pressure drop across the flow field. The flow rate drops further in the case of leakage compared to flooding, which can be used to isolate them.

Flow measurements can also be used along with current measurement to calculate the stoichiometry ratio used for isolating faults. Anode and cathode starvation caused by low reaction rates correspond with a low stoichiometry ratio. Flooding can also occur as a result of low stoichiometry, while drying can occur as a result of high stoichiometry.

As mentioned earlier, humidity measurements are not reliable due to water condensation at high humidity conditions, therefore, flooding can not be reliably isolated using humidity sensors; however, they can be effectively used to isolate drying.

Impedance measurement can also be used to isolate faults in the fuel cell, however, its high implementation cost as well as its sensitivity/unrepeatability makes it not suitable for many end use applications. Pressure-voltage spectroscopy however uses available control components, along with a more advanced algorithm to isolate fuel cell faults during operation [9].

5 Conclusion

We discussed challenges and opportunities in fuel cell systems, with an emphasis on cost reduction strategies. In system architecture, we proposed a new architecture that can result in smaller, lighter, and cheaper fuel cell systems. In the control context, we suggested developing adaptive schemes for finding the optimal operating point in real time. In the diagnostics context, we proposed using pressure-voltage spectroscopy as a new diagnostic tool that can be used at various stages of the fuel cell lifecycle.

References

1. O'Hayre, R., Braithwaite, D., Hermann, W., Lee, S.J., Fabian, T., Cha, S.W., Saito, Y., Prinz, F.B.: Development of portable fuel cell arrays with printed-circuit technology. J. Power Sources 124, 459–472 (2003)
2. McLean, J., Lindstorm, J.: Printed circuit board separator for an electrochemical fuel cell, US Patent No 6541147 (2003)
3. Niroumand, A.M.: Fuel cell separator plate with integrated heat exchanger, US patent application No 12/540309 (2009)
4. Pukrushpan, J.T., Stephanopoulou, A.G., Peng, H.: Control of fuel cell breathing. IEEE contr. Sys. Mag. 24, 30–46 (2004)
5. Niroumand, A.M.: PEM fuel cell low flow diagnostics, Ph.D. thesis, Simon Fraser University (2009)
6. Sun, J., Kolmanovsky, I.V.: Load governor for fuel cell oxygen starvation protection: A robust nonlinear reference governor approach. IEEE Trans. Contr. Syst. Techno. 13(6), 911–920 (2005)
7. Niroumand, A.M., Saif, M., Merida, W., Eikerling, M.: Pressure-voltage oscillations as a diagnostic tool for PEM fuel cell cathode. Elec. Comm. 12(1), 122–124 (2010)
8. Hissel, D., Pera, M.C., Kauffmann, J.M.: Diagnosis of automotive fuel cell power generators. J. Power sources 128(2), 239–246 (2004)
9. Niroumand, A.M.: Method and apparatus for evaluating the performance of a fuel cell, US patent application No 61/234185 (2009)

An Overview of Current Research Activities on PEM Fuel Cell Technology at Florida Atlantic University

Amir Abtahi and Ali Zilouchian

OME Dept., Florida Atlantic University, Boca Raton, FL 33431, USA
CEECS Dept., Florida Atlantic University, Boca Raton, FL 33431, USA
abtahi@fau.edu, zilouchi@fau.edu

Abstract. The Fuel Cell Laboratory at Florida Atlantic University (FAU) originally founded to support a research funded by the US Department of Energy under the leading author's supervision during 1996-1997 period [1]. In sequel, the laboratory has been upgraded with several industrial funding supports, including Teledyne Inc., Enerfuel Inc, in addition to the National Science Foundation (NSF) and FAU internal research funds. With the continuing support of local FC industries, the laboratory is considered a focal point for the research and educational activities pertaining to proven fuel cell testing technology. In this paper, an overview of several research and instructional activities   of FAU Fuel Cell Laboratory are briefly presented.

Keywords: Fuel Cell, Hybrid Golf Cart, Fuzzy Logic Controller, Nexa Modelling, Neural Network.

1 Introduction

The past ten years have seen an emphasis in advancing fuel cell technology to enable a decreasing reliance on fossil fuels [2]. Toward this end, a significant amount of governmental and industrial resources have been focused on developing Proton Exchange Membrane fuel cells (PEMFC) as a "cleaner" more efficient alternative to existing power generation technologies. PEMFCs convert chemical energy directly to electrical energy with conversion efficiencies approaching 60% [3]-[8]. In addition, the only bi-product of the reaction is water, which makes the PEM very attractive from an environmental standpoint. Foreseen applications of PEM fuel cells are widespread ranging from automotive, aviation, space, marine, military and stationary power production to name a few [9]-[11]. The technology is suited to provide for power demands from a few watts up to about 300kW [12],[13]. Some of the major advantages of PEM fuel cell over other fuel cell technologies are the low operating temperatures and quick startup.

Even though the flexibility of the PEMFC design makes this type of fuel cell an ideal candidate for use in a variety of applications, several areas within the technology need to be advanced to make them a practical and cost effective solution. Issues associated with water management, membrane longevity, and cost need to be addressed in order for the PEMFC to be seriously considered as a viable alternative to present day carnot-cycle/heat engines. As a result, a lot of activity within the fuel cell community is centered on improving membrane and electrode materials, developing methods for managing product water levels and implementing novel manufacturing techniques to reduce cost. In this presentation, projects developed in the College of Engineering and Computer Science at FAU are briefly discussed.

2 Fuel Cell Laboratory at FAU

The FC Laboratory at FAU features a fully instrumented PEM Fuel Cell fabricated and supplied by Teledyne System Energy Inc. under previous contracts with the Florida Energy Office, and the DOE. The single stack, expandable module is instrumented for temperature, moisture, and pressure detection. The novel thermocouple module and the nearly complete moisture detection module are designed to detect and record data in the x-y plane (parallel to the membrane), A novel design that incorporates AC Impedance analysis of a 4X4 grid determines moisture. A 4X4 corresponding grid is instrumented with a bank of thermocouples for temperature mapping of the FC. Along with pressure measurements, and humidification data for both the hydrogen, and oxygen flows, the test set-up can measure and plot a complete thermodynamic map of the FC. Since water generation in specific, and phase change in general are key factors in the optimization of FC performance, the FC set-up is an ideal diagnostic, analysis, and design tools.

The FC test station features another significant and unique capability. Multipressure supplies of hydrogen and air/oxygen can be connected through a series of high speed solenoid valves to create pressure and/or oxygen-to-air ratio pulses. This method of fluidic pulse generation is currently establishing benchmarking data that can be used not only for a better understanding of the impact of pressure and concentration on FC performance, but give us insight into the transient nature of the interaction of the mechanistic and electrochemical phenomena. In addition, the National Instrument's *Labview* which is our data acquisition platform, can implement various control strategies through active control of a number of input parameters. This capability will be used in implementing pulsing and fuzzy control strategies and signal analysis of the multi-input, multi-output model.

Table 1 and figures 1(a)-1(f) show different types of fuel cells stacks and their characteristics including active surface area utilized at the fuel cell laboratory for our research investigation. Fuel cells "a", "b" and "c" has been used to demonstrate power performance with respect to the effect of pulsing of reactants [18]. Fuel cell stack "d" has been utilized to provide a comprehensible picture of the pulsing effects [18]. The other fuel cells have been utilized in water management] and for the NN modeling investigations [14]-[17] as briefly described in more detail on sections 3 and 4.

Table 1. Utilized PEM FC Stacks

	Fuel Cell Type	Number of Stacks	Surface Area (cm^2)
a	RFC H2/O2/Air	1	5
b	Light FC - 2U	2	5
c	EcoFC-6AM	6	14.5
d	FC - 019K - 5	5	132.25
e	Teledyne	1	16
f	Nexa	42	14.5

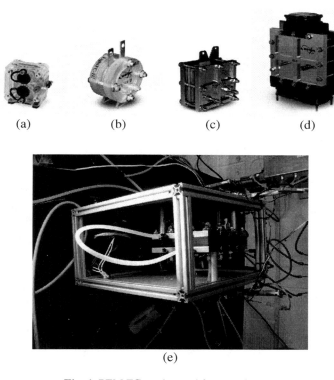

Fig. 1. PEM FC stacks used for experiments

(f)

Fig. 1. (*continued*)

3 Water Management of PEM

The membranes currently used in PEM fuel cells depend on the presence of water in electrolyte to facilitate proton transport. This includes perfluorinated membranes such as Nafion, Gore Select, etc. Since, the increase in the water quantity in the cell usually reduces the resistance to proton condition, there is a clear advantage to maintaining the membrane in a state of saturation. This can be achieved by either the gas that is in contact with the membrane be saturated with water vapor, or that the membrane is in contact with liquid water. However, the presence of excess water near reaction sites is not desirable. Liquid water can block free transport of reactants to electrocatalysts, and droplets in small channels cause pressure fluctuations, resulting in non-uniform reaction rates in the cell. Consequently, water management has been a major concern for PEM fuel cell developers for over two decades. There are several approaches for the control of the water quantity in a fuel cell stack, each with its own advantages and disadvantages.

3.1 Experimental Results-Single Stack FC

A number of pulsing techniques for PEMFCs have been successfully carried out at FAU's Fuel Cell Laboratory. Many have led to improved water management in the fuel cell channels. The overall system configuration for the Teledyne single fuel cell is shown in figure 2. During this experimental phase, only air was utilized due to potential hazards and additional costs of having instrumentation and equipment certified for the oxygen use. To control the frequency of pulsing an "in house" built circuit as well as external function generators were utilized.

Figure 3 shows polarization curves and power density vs. current density associated with different pulse frequencies for dual air source pulse results. It can be seen that the fuel cell's performance steadily increase with higher frequency. These performance curves were at an operating pressure of 172.4 kPa (25psig) and temperature of 25 °C.

An Overview of Current Research Activities on PEM Fuel Cell Technology at FAU   189

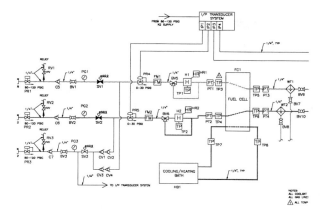

Fig. 2. Experimental set-up for Teledyne FC

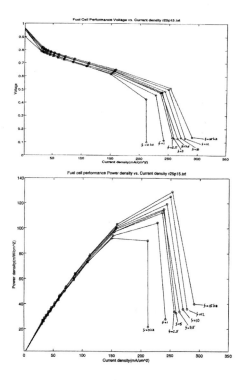

Fig. 3. Polarization curve and power density vs. current density for FC(e)

More details of the testing results for various pulsing frequencies are reported in [14]-[15].

3.2 Experimental Results: Multi Stack FCs

In order to determine the effectiveness of pulsing technique for muti-stack fuel cells, an extensive study of pulsing techniques was also conducted for muti- stack FC (b)-(d) as shown in Figure (1). For this phase of our experimental work, both air and hydrogen flows were pulsed. Detailed results can be found in [18]. For fuel cell (d), frequencies varied from 0.05Hz to 5 HZ. Flow rates and pressures of reactant inlets and outlets were constantly monitored and recorded by data acquisition equipment sampling at a rate of 20 samples per second.

As the experimental results show, any frequency below 3.3 Hz prevents over-hydration of the particular fuel cell under study. The variations of frequencies demonstrate a direct correlation between the pulsing rate and the power performance. Low frequency pulsing in PEMFC increases the differential potential pressure. In sequel, differential potential pressures while not considerable enough to damage the fuel cell membranes; permit the benefits of maintaining power performance. In addition, the larger the load resistance, the shorter time is required between pulses in order to maintain a desired power level.

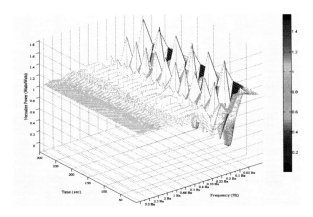

Fig. 4. Pulsing effect of fuel cell 1-(d)

3.3 Analysis of the Experimental Results

These studies demonstrate that by pulsing the reactant flows at selected frequencies, the development of pressure waves resulted in water purging of the fuel cell channels. Pulsing results in improved power performance of a PEMFC. By pulsing of fuel cell reactants, the water accumulation can be eliminated or delayed and consequently electrochemical activity at the electrode active surface area was kept within an optimal margin. Differential pressures inside of a fuel cell due to pulsing frequency of reactants avoid the over-imbibition of the water blockage within the channels when flow regimes are low. Appropriate drainage of the PEMFC has been obtained with pulsing techniques. Experimental results have shown that higher frequencies do not contribute significantly in improving power performance. Based on the results obtained, a relationship between frequency level and time for water accumulation has been determined. Activation polarization losses caused by the flow of intake reactants can be mitigated by pulsing of the flow of the reactants.

4 Modelling of Nexa Fuel Cell

In this section, we investigate how well back-propagation and RBF networks can predict the performance of a fuel cell system. The 1.2 KW Nexa™ fuel cell is shown in figures 1(f) and (5). It can produce the unregulated DC power from a supply of hydro-gen and air. The rated power at standard condition is 1200W with the voltage range 22V to 50 V and 26V at rated power [17]. The unit is self-operated with a capability to collect the data from computer via LabVIEW program.

The system configuration is shown in Figure 5. The system consists of 7 control signals and 9 sensors to monitor the significant variables. The hydrogen pressure is regulated by regulator valve. The system controls air compressor for air flow, and cooling fan for stack temperature.

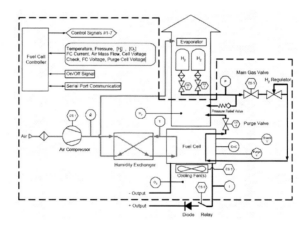

Fig. 5. Block diagram of Ballard™ fuel cell

4.1 Data Collection

We operated the 1.2 KW Ballard™ fuel cell with a load bank to increase the stack current up to 45A. There are 1072 data points collected. After removing some unwanted data at the beginning and at the transition time when we switched from resistive load to light bulb load, we have 883 data left to be utilized in neural network training.

The following data are collected; stack voltage(V), stack current(A), air flow(slpm), air temperature(°C), stack tempera ture(°C), fuel pressure(barg), fuel consumption(L), power(W), H_2 leak(%), O_2 concentration(%), purge cell voltage(V) and battery voltage(V).

4.2 Process Variable Selection

Even though there are many variables available to be utilized but from system knowledge, we know that the fuel cell will control air compressor and cooling fan and keep hydrogen pressure constant by using regulated valve. Therefore, we have selected only air flow and stack temperature as inputs and select stack voltage and current as outputs.

Inputs: mass air flow (slpm), stack temp. (°C)
Outputs: stack voltage (V), stack current (A)

We utilized the normalized data for both the back-propagation and RBF algorithms. The details of these algorithms can be found in [17].The results show that using the normalized data trends to give the better and faster results than using raw data or zero mean and unity stand deviation data. The error goals should be suitable, not too small and not too big. It is necessary to choose the right training algorithms in order to get a better and faster prediction.

Finally, we selected the best predictions from back-propagation and RBF networks and compared these results as shown in Figure 6 and Figure 7. The accuracies of both ANNs are almost the same for the prediction of stack current. For the prediction of stack voltage, there are few points of RBF prediction that lie a little bit far from the measured data.

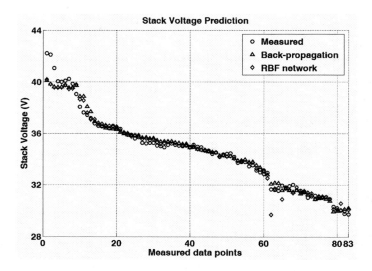

Fig. 6. Stack voltage prediction for various NNs

An Overview of Current Research Activities on PEM Fuel Cell Technology at FAU 193

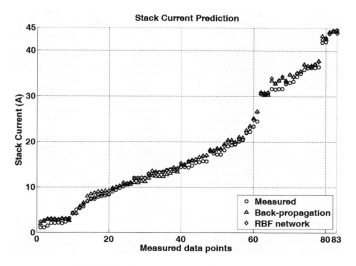

Fig. 7. Stack current predictions for NNs

5 Controller Design

In this section, the experimental set up and the equipment utilized for the controller implementation are briefly explained. The system control implementation is shown in figure 8. In order to control the air mass flow rates according to the control algorithms, the controller will send the control output signals via analog

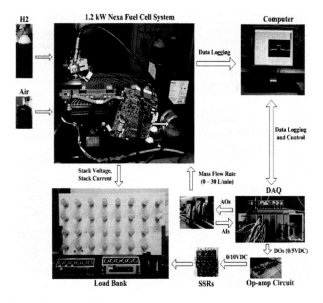

Fig. 8. Block diagram of FC experimental set-up and controller implementation

output module (0-5VDC) of data acquisition (DAQ) to the mass flow controllers. The mass flow controllers will also send the actual air mass flow rates to the computer via analog input module (0-5VDC). The load bank is varied and controlled by sending the digital outputs (DOs) to the op-amp circuit and the solid-state relays (SSRs). The nominal power of the Nexa system is 1.2 kW.

The membership functions of the input (stack power error) and output (air flow rate adjustment) are selected as shown in Figure 10.

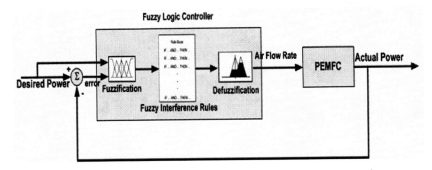

Fig. 9. Block diagram of fuzzy logic controller

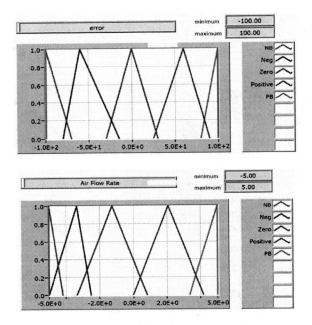

Fig. 10. I/O membership functions for the FL controller

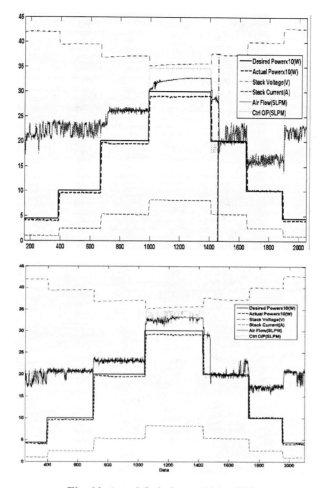

Fig. 11. Actual & desire tracking of FC

As it is shown in Figure 11, the actual power track the desired power very closely using the Neo-fuzzy controller. The comparison of the result with fuzzy controller as shown in {8} demonstrates the effectiveness of the proposed method. The controller action is much smoother than either a simple PID controller or a Fuzzy controller [17].

6 Golf Cart

In this section, the design and construction of a prototype golf cart with several renewable energy sources is presented. The first prototype was proven to be very successful using solar system. The combination of unique features such as the tracking of the sun resulted in significant increase of output power with respect to other solar carts. However, in certain regions where the average irradiance level is low, an extra source of energy is needed if total independence from the utility grid is required. The incorporation of an extra source of energy with PEM fuel cell requires a complex power management system. The proposed system manages the energy delivered to the batteries from either the solar platform or the fuel cell system. If the cart undergoes severe hours of driving, both sources are on in order to supply the energy needed to keep the batteries charged. The cart with its different energy sources becomes a small power plant that can supply energy to an emergency load in case of a natural disaster. The detailed work can be found in [19].

Fig. 12. Prototype of hybrid golf cart

6.1 Design Considerations

The existing prototype depends on the amount of sunlight and also on the electric grid in case the usage exceeded the energy production from the solar array. In order to look for alternative ways to make the cart totally independent of the utility grid, the solar cart has been be retrofitted with Nexa fuel cell system.

However, the storage of the hydrogen has been an issue. It was first thought of using carbon composite H2 tanks, but the issue of having a vessel at 4500 PSI encountered resistance among the university authorities. Liquefied hydrogen was out of the options due to its complexity and evaporation rate. The remaining solution was to use metal hydride canisters. One of the advantages of such selection is that the hydrogen is kept at a very low pressure (300 psi). The canisters are almost unbreakable. If there is a sudden leak of hydrogen, the rate of release will be slowed down since the canister cools down preventing a large amount of hydrogen release at once. Metal hydride can also hold a higher density of hydrogen than high pressure vessels in terms of volume.

6.2 Power Management System

The usage of three energy sources requires a complex system that maximizes the range of the cart, and the least amount of hydrogen used. Lead acid batteries lose their voltage output as the battery is discharged. However, the state of the charge of the batteries cannot be measured strictly from the battery voltage unless the batteries have not been used for 6 hours. This is the time the batteries required for the voltage to be stable. The measure of the state of the batteries will be determined by the current in and out of the battery. This procedure is not 100% reliable since at higher discharge rates, the batteries waste energy as heat due to its internal resistance. In order to overcome this problem, the state of charge of the batteries will be updated once the batteries haven't been used for a certain period of time. In addition, as the discharge rate increases, there will be a coefficient in which takes into account the higher discharge rate. This way, the state of charge of the batteries can be accurately measured.

7 Educational Activities at FC Laboratory

This NSF project is intended to substantially improve the capability of undergraduate instruction related to recent developments in PEM fuel cell technology. The laboratory has been used to supplement the introductory fuel cell course and support senior design projects.

The eductional activity of the laboratory consists of two test stations fully outfitted with extensive data acquisition systems, adapted from proven designs.

The authors has recently offered a new course entitled "Introduction to Fuel Cell Technology" in conjunction with the FC laboratory In addition, multidisciplinary student groups from Senior Design classes are utilizing the lab for FC design projects. It is anticipated that the current development of the new laboratory will have a direct impact on undergraduate education by creating a focal point for interdisciplinary learning, a balance between theoretical and hands-on experience in undergraduate teaching, and application of these educational tools in a vibrant technology sector. The evaluation plan for the course materials currently focuses on three general areas. The first focus is on the assessment of the course modules. The second focus is related to student-identified strengths/weaknesses of the course/modules. Finally, the third focus is to document the course/curricular refinements resulting from the evaluative data obtained.

Various topics such as effects of operating conditions on performance, effects of humidification, water management and heat management will be discussed in this course. The proposed laboratory experiments will be synchronized with the lectures.

The project has been supported by National Science Foundation under the grant # DUE-0341227.

8 Conclusions

In this paper, several research and educational activities at PEM fuel cell laboratory are presented. In addition to the above projects, new projects recently proposed and are under investigation. One of our future projects is the incorporation of fuel cells, solar panels and batteries to produce the next generation of Electrical Utility Vehicle (EUV). Such EUV with an extended range which does not require connection with the electrical grid is highly innovative. The successful completion of this project will have short term as well as long term economic benefits. In the short term this project will result in a seminal renewable energy technology demonstration. This demonstration will establish the market feasibility of the technology implementation strategies, support and generate new in-state developed intellectual property, support private and public organizations, and develop hydrogen infrastructure. In the long-term the economic impact could be enormous. It is hoped that the technology developed and demonstrated will result in a turn-key retrofit solution to transform gas powered or grid powered utility vehicles into renewably propelled vehicles. This solution could be used by golf clubs, planned communities, industrial complexes, commercial complexes, and university campuses to transform their utility and recreational vehicle fleet to renewably powered ones.

Acknowledgement

The authors would like to acknowledge the contributions of the following graduate students that have worked extensively during the implementation of the above described project: Michel Fuchs, Aucha. Saengrung, Max Saelzer and Aqeles Perez.

References

1. Abtahi, H., Das, P., Fuchs, M.: Final Report on Fuel Cell Flow Optimization, Contract # PRDA-DE-AC08-96NV1 (1986)
2. Barbir, F.: Recent Progress and Remaining Technical Issues in PEM Fuel Cell Development, World Hydrogen Energy Conference, WHEC XIII, Beijing, China (June 2000)
3. Wilson, M., Moeller-Holst, S.: C. Zawodzinski: Los Alamos National Laboratory: EfficientFuel Cell Systems, Annual Progress Report: Transportation Fuel Cell Power Systems, U.S. Department of Energy: Energy Efficiency and Renewable Energy Office of Transportation Technologies, 88-90 (2000)
4. Fuchs, M., Barbir, F., Husar, A., Neutzler, J., Nelson, D., Ogburn, M.: Performance of Automotive Fuel Cell Stack, 2000 Future Car Congress, Arlington, VA (2000)
5. Barbir, F., Fuchs, M., Husar, A., Neutzler, J.: Design and Operational Characteristics of Automotive PEM Fuel Cell Stacks, Fuel Cell Power for Transportation, SAE SP-1505. In: Proceedings of SAE International Congress & Exhibition, Detroit, SAE, Warrendale, PA, pp. 63–69 (2000)
6. Fuchs, M.F., Barbir, F., Nadal, M.: Performance of Third Generation Fuel Cell Powered UtilityVehicle #2 with Metal Hydride Fuel Storage, 2001 European Polymer Electrolyte Fuel Cell Forum, Lucerne, Switzerland (2001)
7. Fuchs, M., Barbir, F., Nadal, M.: Fuel Cell Powered Utility Vehicle with Metal Hydride Fuel Storage. In: Globe Ex 2000 Conference, Las Vegas, Nevada (2000)
8. Larminie, J., Dicks, A.: Fuel Cell Systems Explained. John Wiley & Sons, Ltd., West Sussex (2000)
9. Williams, M., Rastler, D., Krist, K.: Fuel Cells: Realizing the Potential, 2000 Fuel Cell Seminar, Portland, Oregon (2000)
10. Schmal, D., Bastianen, J., Barendregt, I.: Polymer Fuel Cell System Design for all Electric Naval Ships, 2000 Fuel Cell Seminar, Portland, Oregon (2000)
11. Velev, O., Hibbs, B., Parks, B., Boyer, C., Cisar, A., Andrews, G., Murphy, O.: Regenerative Fuel Cell System for Unmanned Solar Powered Aircraft, 2000 Fuel Cell Seminar, Portland (2000)
12. Washington, K.: Development of a 250kW Class Polymer Electrolyte Fuel Cell Stack, Fuel Cell Seminar, Portland, Oregon (2000)
13. Mathew, M.: Fuel Cell Engines. John Wiley and Sons, New York (2002)
14. Abtahi, H., Zilouchian, A., Fuchs, M.: Design and Implementation of an Intelligent Control Strategy for Fuel Cells. In: Proceedings of 1998 World Automation Congress, Anchorage, Alaska (1998)
15. Abtahi, H., Zilouchian, A., Fuchs, M.: Design and Implementation of a Hierarchical Control Strategy for Proton Exchange Membrane Fuel Cells. In: Proceedings of 37th IEEE Controls and Decision Conference, Tampa, FL, pp. 461–463 (1998)
16. Zilouchian, A., Abtahi, H.: Design and Implementation of a Fuel Cell Laboratory. In: 2008 National Science Foundation CCLI Conference, Washington, D.C. (2008)
17. Saengrung, A., Abtahi, H., Zilouchian, A.: Neural network model for a commercial PEM Fuel Cell System. Journal of Power Sources 172, 749–759 (2007)
18. Perez, A., Abtahi, H., Zilouchian, A.: Pulse-Width Modulation of Hydrogen Delivery for PEM Fuel Cells. In: Proceedings of PWR 2008, ASME Power, Orlando (2008)

19. Abtahi, H., Zilouchian, A.: Using Alternative Energy Sources and Appropriate Energy Utilization as Teaching Tools for Mechanical and Electrical Engineering Students. In: 4th Latin American and Caribbean Conf. on Eng. and Tech (LACCEI), Puerto Rico (2006)
20. Saelzer, M., Messenger, R., Zilouchian, A., Abtahi, H.: Solar-Powered Electric Cart. In: Proceedings of 19th Annual Florida Conference on Recent Advances in Robotics, Miami (2006)
21. Zilouchian, A., Abtahi, H., Saengrung, A.: Water Management of PEM Fuel Cells Using Fuzzy Logic System. IEEE Systems Man and Cybernetic Conference (2005)
22. Zilouchian, A., Abtahi, H., Fuchs, M.: Development of a prototype fuel cell laboratory. In: IEEE International Conference Systems, Man and Cybernetics, vol. 4, pp. 3398–3402 (2005)

Computational and Intelligent Systems Engineering

BELBIC and Its Industrial Applications: Towards Embedded Neuroemotional Control Codesign

Caro Lucas

Center of Excellence: Control and Intelligent Processing, Electrical and Computer Engineering Faculty, College of Engineering, University of Tehran

Abstract. The past few years have witnessed a proliferation of industrial and decision making applications of a novel neurocontroller designated as BELBIC: Brain Emotional Learning Based Intelligent Controller. Based on a proposed open loop descriptive model of the midbrain, where emotional processing is understood to mainly take place, its utilization has been motivated by the belief that most human decisions are made using bounded rationality. Successful control engineering and decision making applications are reviewed, where BEL has been used for satisficing action selection based on artificial emotions. Laboratory and Industrial scale applications are emphasized. Recent results on stability and performance guarantees are also examined. Finally flexible bioinspired SOC and other hardware/software implementations of BELBIC are investigated. It is argued that in order for VLSI implementations of neural networks to be commercially viable, it is crucial to minimize the redesign expenses for their optimal dedicated implementations for any given application on any desired platform. Furthermore, on line applications require recursive learning that need not precede the recall mode. High levels of adaptability, disturbance rejection, and fault tolerance are other important characteristics of the proposed IC.

Keywords: Decision and control, neural network, artificial emotion, model driven architecture, SOC implementation.

1 Introduction

The advent of the new wave of neural networks research in the late 80's marked a turning point in intelligent systems conceptualization. Connectionist and bioinspired approaches came to replace traditional symbolic and representational AI methodologies. However, despite the phenomenal growth in neural net research, which shows no sign of saturation after two decades, two of the most fundamental promises associated with that line of research remain largely unfulfilled. Firstly, the application of neural networks in control engineering, which was anticipated to mimic the central role that brain plays in human and animal decision making, has remained very underdeveloped. Neural nets have proven themselves as powerful

204   C. Lucas

tools in modelling and pattern analysis; but not in dynamical decision making. Specific neural networks like CMAC, however, have shown more promise in that respect. Secondly, hardware implementations of neural nets, so important for realizing the real parallel and massively distributed processing power of neural nets, have not registered notable commercial success. High design expenses, rapidly changing technologies that tend to shrink effective product life cycles, and the shift from mass production to custom design may be cited as three main reasons for the lack of success in hardware realizations of neural nets. As for control engineering applications, I wish to present two conjectures. The success of multilayer perceptron applications in static decision making has been gained at the price of diverting attention from development of other neuromorphic models mimicking specific parts of brain that are known to carry out dynamic feedback control or implementing human-like decision making algorithms that are not necessarily operating in the steady state conditions only. Also, despite the fact that neural network and adaptive system research agendas were initially one and the same, they have gone their separate ways later on. Although there is no compelling reason for the separation, learning and recall modes of operation tend to be strictly separated in neural net applications, whereas in adaptive algorithms, learning can take place while decisions are made dynamically, leading to the possibility of dual control and exploration- exploitation tradeoffs in machine intelligence and adaptive control applications. These considerations have been the main drivers of my research endeavours during the past few years. Two neuromorphic models: BELBIC and LOLIMOT have been used in many different application domains with great success. The first [1-35] is a neural net model mimicking the midbrain where emotions are known to be processed for quick decision making. The second [36-43], is a neurofuzzy net for which we have developed several modifications; reinforcement based emotional learning algorithms, and recursive techniques for incremental learning capability enabling it to be used as on-line feedback controller for plants or processes that are unknown or time varying. In this paper, I shall discuss BELBIC: the more direct physiologically motivated neural net configuration, but will try to present my arguments in general terms, so as to be applicable to other neural networks as well.

2 Emotional Decision Making

Emotion is often considered as interfering, non-rational element in decision making, often leading to bad decisions that are made almost reflectively and without proper deliberation. It is only in recent years that the real advantages of this faculty have increasingly come to be highly appreciated. Tools for assessing emotional intelligence, for example, have proven to be much better predictors for future success than traditional IQ tests. Some of the advantages of emotional action selection are as follows. They are quick because they do not need extensive deliberation. They are based on emotions formed through past experience. They, therefore, involve case-based reasoning and do not need assumptions necessary for predicting consequences of each possible decision. Being based on less assumptions means that emotional decisions are more robust. They are thus better suited

BELBIC and Its Industrial Applications

for environments involving uncertainties and/or cases where processing power poses limitation for decision making. Intelligent systems research has so far focused on high-level fully rational and optimization-based techniques for decision making. Neurologically, this means mimicking the operation of the upper brain that is supposed to be carrying out deliberative, conscious and intentional decision making accessible to volition and awareness. It is perhaps easier and more fruitful to capture lower mental and behavioural faculties first, even in artificially intelligent systems leading to satisficing action selection that, although may be suboptimal, but is good enough for a larger set of ambient conditions. Decisions based on emotions are not irrational, but are based on bounded rationality and have the above mentioned advantages in the majority of situations where prior experience exists and can inform the decision making process [44-46].

2.1 Brain Emotional Learning Based Intelligent Controller (BELBIC)

Our attempts on capturing emotional control have been based on assessment theory of emotions. A critic, according to this viewpoint, constructs the emotional cue based on goals, actual states, and other objective or subjective data. Emotional learning can thus be categorized as reinforcement learning. Decisions are made adaptively so as to decrease negative emotion levels or increase positive emotion levels. The success of our initial attempts to utilize functional models for emotional decision making [40-43] motivated us to try using structural models as well [1]. For this, we deliberately chose a computational model developed for descriptive purposes only. We could thus ascertain that no bias in model development had been introduced so as to assure better applicability. We used the open loop descriptive model without modifications or with slight changes in different feedback decision making applications and invariably it yielded very good results. The neural model is sketched in figure 1 and is composed of four subsystems: thalamus, sensory cortex, orbitofrontal cortex, and amygdala. The mathematical description of its STM and LTM variables is given by [1]

$$A_{th} = \max(S_i)$$
$$\Delta V_i = \alpha \max(0, S_i (stress - \sum A_i))$$
$$\Delta W_i = \beta S_i (\sum O_j - stress) \tag{1}$$
$$A_i = S_i V_i$$
$$O_i = S_i W_i$$
$$E = \sum A_i - \sum O_i$$

where stress is the negative emotion determined by the critic. To design the closed loop control circuit, one needs, in addition to determining the (negative) primary reward cue (the stress), to determine the sensory input(s) S. The determination of the former (the stress signal) depends on the control objectives (performance criteria), while the latter (sensory input) is determined via control engineering considerations (for example, by choosing error, error derivative, and/or error integral signals as feedback variables constituting the sensory inputs). The output E of the controller is fed to the plant as the actuating signal.

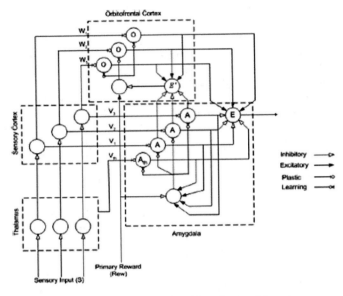

Fig. 1. Brain emotional learning model

2.2 Flexible SOC Implementation Tools

Realization of the full advantages of neural networks stemming from the biologically motivated design becomes possible only through dedicated hardware implementation and concurrent learning and recall. Two important aspects are parallel and distributed processing power and fault tolerance and error correction capabilities obtained from redundancy of the processing elements and connections and adaptability during the recall mode of operation. The latter reflects the self-healing property of living systems. Potential gains are enormous. However, the additional expenses can be prohibitive and killer applications for which dedicated embedded systems are absolutely required are not that many. In recent years, the advent of ubiquitous computing and the need to embed computational elements with sensors and actuators, as well as advances in VLSI technology and the decreasing costs of microcontrllers, DSPs, FPGAs, and ASICs have provided new impetus for hardware realization of neural nets as powerful processing tools that can become even more powerful if the enormously added speed and fault tolerance are also considered. The main remaining obstacles are the ever-increasing design costs and constantly shrinking effective life cycles. My research has been focused on overcoming these drawbacks through use of the method designated as "model driven architecture" (MDA) [2-4] in software engineering. The idea is to shift the bulk of design expenses towards the high-level, platform-independent, non-domain-specific phases of design and automate or semi-automate the development of detailed, domain-specific implementation on a given platform. The short-term advantage of MDA is to make the design itself reusable. When technology changes the new implementation is obtained with little extra cost. In the long-term, the implemented system can detect the required technological environment and transform itself so as to function in that environment. This means that the partition between software and hardware realization of the intended system (as well as other partitions like analog-digital, sever-side client-side) can be decided during the design process. A bioinspired method of evolvable hardware development categorizes system's changes in response to the environmental needs along three axes [5]. The phylogenesis axis captures evolution. The ontogenesis axis captures growth. And the epigenesis axis captures learning. Thus, for example, in a neural network implementation we can have adjustments of the adaptive parameters (e.g. connection weights) via some learning algorithm along the epigenesist axis. The changes in the net structure (e.g. connections) can be realized by using some evolutionary algorithm along phylogenesis axis. Finally, we can start with a minimal configuration and gradually let the network grow (e.g. by adding new neurons or connections) according to a given methodology (e.g. model tree in the case of LOLIMOT) along ontogenesis axis. The self healing property of the system can be due to any of the three processes. In every specific application one must decide which of the processes are to be realized, on-line or off-line, via hardware or software. If some components fail, the system will continue to function by growing new components and/or readapting its parameters and/or undergoing structural adjustments. The decision in any given case will be made considering the target platform; time, area, and energy consumption considerations; and total system requirements. For automatic model driven design we use converters for translating platform-independent general models into DSL descriptions (VHDL, C code, etc.). For semiautomatic design, we use design patterns (e.g. hardware versions of GOF patterns). The final design may not be as optimal as the hand-design for any given application. But will yield good near-optimal designs for a wide range of applications at little extra expense for any desired new application. In an instance of analyzing the performance of BELBC in a practical application involving overhead crane test bed [2-4] at different design stages from mathematical description and simulation of embedded controller applied to the target plant to empirical results obtained from actual lab experiments we have developed methodologies for representation of the mathematical description via actor oriented continuous simulation blocks, conversion of the blocks to VHDL models that are not synthesizable, utilization of design pattern for transforming this model into synthesizable modules, and finally, pin assignment and wrapping the core on targeted technology (Cyclone II FPGA). Results can be refactored via trading off accuracy with design objectives or use of design patterns for improvement of good design metrics. Our results show excellent correspondence between theoretical and experimental outcomes, and capability to carry out control objectives in the presence of disturbances and changes in plant characteristics.

Fig. 2. Laboratory experiment set-up.

3 Industry Applications

Design and implementation of intelligent drives was the first domain for which emotional control systems were developed. I was a member of a research team that was investigating the possibility of industrial-scale utilization of switched reluctance electromotors [47-50]. The research project included all the design aspects from electromechanical design of the motor and development of its intelligent control system [6-8] to implementation issues and business and economic aspects of the production line. BELBICs were subsequently utilized for controlling other electromotors [9-12], including interior permanent magnet synchronous motor drives, which implemented technology competing with that of the original project [9,10]. In a different domain of application, we used BELBIC for HVAC [13] and heat exchanger control [14]. The first task was chosen because parallel projects to design suitable controllers using advanced nonlinear, optimal, and intelligent control techniques was being conducted by my colleague at the University of Tehran. Also, the MIMO and non-minimum-phase plant characteristics posed serious difficulties especially for reinforcement-based control system development. It thus provided an excellent benchmark for examining the capabilities and advantages of BELBIC. Similarly, the micro-heat-exchanger control task was encountered by our colleagues at Karlsruhe Research Institute in their high energy physics monitoring efforts. Navigation and guidance constitute the third domain of application for BELBIC. Excellent results have been obtained for satisficing flight control [15-17]. A fourth decision making domain outside the scope of feedback control system design has been in prediction of monitored time-varying signals and design of warning or alerting systems [18-20]. My research efforts have been concentrated in crisis and risk management tasks especially naturally occurring calamities and medical engineering applications. Two important instances have been geomagnetic disturbance and epileptic seizure prediction problems. Business and industry benchmark problems like stock and security price prediction and load forecasting have also been attacked. An important motivation for choosing this domain of application has been the industrial importance of smart environments and ambient assisted living. Research in these areas have focused on spatial and web computing aspects; whereas detecting when a service is needed by the inhabitants of the environment is important for the intelligent system to proactively provide that service. The problem with most good predictors is that they show very poor performance during the prediction of peaks and sudden changes where alerting is usually required. The ability to define emotional cues other than least square error accumulation signals, and use of learning algorithms different than gradient-based corrections requiring smooth differentiable objectives are important advantages of brain emotional learning techniques. Our use of BEL-based predictors and alerting systems have shown their excellent ability to trade off accuracies in more significant regions with general accuracies. The natural next domain of application for BELBIC is to use it to control the provision of service process once the alert has been given [21-22]. I have begun with the problem of seizure control through electrical stimulators. The simple structure of the predictor and controller nets and the flexible embedded codesign methodology make the proposed techniques excellent candidates for implementation via wearables or implants. Robotics and multiagent systems constitute the next important application domains [22-25]. Use of BELBIC has registered phenomenal success in these applications. International competitions like Robocup and Trading Agent Contest (TAC) have become important testbeds where applicability of BELBIC has been extensively investigated. (robot movement, soccer, rescue simulation, travel package services provision, etc.) Unless sizable advantages are involved, participants in these competitions are extremely reluctant to apply new and untested ideas like utilization of artificial emotions and their embedded neural network implementations in decision making due to time limitations for design, implementation, testing, optimization, and generally, the preparation and maintenance efforts for their completed systems. Other application areas include automobile suspension, intelligent washing machines, AVR systems, power plants, servomotors, inverted pendulum, power systems, etc [26-35]. Besides my own research endeavours there are several other teams who have extensively used BELBIC in control engineering applications in the above mentioned as well as other domains and reported excellent results [50-54].

4 Conclusion

In a survey devoted to control based on human emotions [55], the authors point out that "this paper cannot be concluded without remarking the contributions of Caro Lucas". They go on to mention, "his prototypes are a good example of what to do with neuroscience applied to artificial intelligence. He and his co-workers, who included engineers from all around the world, have demonstrated that control of dynamical systems based on human emotions are possible". Indeed, the use of a neuromorphic model of emotion processing as a controller, furnishes an easy and efficient method for designing good, near-optimal action selection system for a wide range of control engineering and decision making application domains via bounded rationality philosophy. Emotional learning takes place concurrent with control action and the resultant adaptive capability endows the controller with very powerful capacity for disturbance rejection and adjustment to changes in plant characteristics One obvious advantage is the possibility of designing the controller not for a given plant but for a range of plants manufactured via a production line. Design of electromotor drives can be cited as an important example. Automation remains largely motion-centric and electromotors are used mainly as components in many home and industry appliances. The design of controllers in these applications should be robust with respect to manufacturing deviations from nominal characteristics occurring in production lines. Furthermore, utilization of the motor in position control or speed control applications means that it will not be necessarily operating in nominal speed or load conditions and it is very important that the controller can quickly adapt to load changes or changes in motor characteristics. The concurrent learning and control action is also very important for on line utilization of the neural net in dynamic applications. Although provision of stability and other performance guarantees remains largely an open problem in adaptive and intelligent control systems, far reaching results have already been obtained and reported by other researchers [51-54]. Alternatively, we have used BELBIC both as direct intelligent controller and as an indirect controller regulating the adaptable parameters of some other controller in our own research. In the latter mode of use [8,14], one can select a controller for which stability and other performance guarantees can be obtained for the whole range of adaptive parameter changes determined by BELBIC. The methodology for embedded realization of BELBIC further increases the usability of this neurocontroller. Considerable reduction in time to market, higher levels of reusability, interoperability, extendibility, and possibility of redesign involving addition of new higher level and non-functional concerns (e.g. in control engineering domain), and ability of integration of heritage systems are among main advantages. More generally, the methodology resolves the custom design and rapid technology change problems that has been main factors for lack of commercial success for hardware realizations of neural networks. Perhaps the most spectacular property of the proposed bioinspired approach is the auto-adjustment and self-healing capability of embedded BELBIC. The emotional approach also furnishes great potential for applications known to be regulated largely by emotions in humans [2-4,34-35,56-57]. Attention and contextualization are two important examples. We have conducted extensive research in use of BELBIC in attentional modulation, active perception, and context-aware systems applications. It is hard to underestimate the importance of attention and context-sensitivity in control engineering. Nevertheless, little progress has been reported in these areas. The role of self concept in undertaking of subtasks in multi-agent control and decision systems is another area we have begun to investigate. In the absence of full information and limitation of deliberation power, emotions are the best guide for deciding which monitored signals to process, which objective to follow, which method to use in any given context, and how to commit subsystems for conducting subtasks. Full rationality approaches, however, are often counterproductive, if not impossible due to existing uncertainties. Savings obtained by paying attention to the more significant processes, for example, can be lost if deciding which processes are significant involves complex deliberation requiring more computational power than that saved by neglecting the less significant processes. Our work shows that these are not very high-level cognitive capabilities useful only for the most advanced intelligent control systems, but are basic competencies that can considerably enhance the control performance in many industrial applications. On the other hand, the advent of ambient intelligence and prevalent computing will definitely put emotional learning based intelligent control high on the agenda of technological development. Moreover, human-like decision making, context-awareness, and sensitivity with respect to attention, enhances the believability of the intelligent srevosystem and contributes to the perception of being "domesticated" in smart environments and ambient assisted living systems [58].

References

1. Lucas, C., Shahmirzadi, D., Sheikholeslami, N.: Introducing BELBIC: Brain Emotional Learning Based Intelligent Controller. International Journal of Intelligent Automation and Soft Computing 10(1), 11–22 (2004)
2. Jamali, M.R., Dehyadegari, M., Arami, A., Lucas, C., Navabi, Z.: Real- Time Emotional Controller. Neural Computing and Applications 19(1), 13–19 (2010)
3. Jamali, M.R., Arami, A., Dehyadegari, M., Lucas, C., Navabi, Z.: Emotion on FPGA: Model Driven Approach. Expert Systems with Applications 36(4), 7369–7378 (2009)
4. Jamali, M.R., Arami, A., Hosseini, B., Moshiri, B., Lucas, C.: Real Time Emotional Control of Anti- Swing and Positioning Control of SIMO Overhead Traveling Crane. International Journal of Innovative Computing, Information and Control, IJICIC 4(9), 2333–2344 (2008)
5. Esmaeilzadeh, H., Jamali, M.R., Saeedi, P., Moghimi, A., Lucas, C., Fakhraie, S.M.: NnEP, Design Pattern for Neural Network-Based Embedded Systems. In: 15th IEEE International Conference: Mixed Design of Integrated Circuits and Systems, MIXDES 2007, Poznan, Poland, June 20-24, pp. 673–678 (2007)
6. Daryabeigi, E., Arab Markade, G.R., Lucas, C.: Switched Reluctance Motor (SRM) Control with Developed Brain Emotional Learning Based Intelligent Control (BELBIC). In: IEEE International Electric Machines and Drives Conference, IEMDC 2009, Miami, Florida, USA, May 3-6, pp. 979–986 (2009)
7. Parsapoor, A., Mirzaeian-Dehkordi, B., Lucas, C.: Speed Control of Switched Reluctance Motor by Brain Emotional Learning Based on Intelligent Controller. Electromotion 15(4), 246–252 (2008)
8. Rouhani, H., Sadekzadeh, A., Lucas, C., Araabi, B.N.: Emotional Learning Based Intelligent Speed and Position Control Applied to Neurofuzzy Model Switched Reluctance Motor. Control and Cybernetics 36(1), 75–95 (2007)
9. Azizur Rahman, M., Milasi, R.M., Lucas, C., Nadjar Araabi, B., Radwan, T.S.: Implementation of Emotional Controller for Interior Permanent-Magnet Synchronous Motor Drive. IEEE Transactions on Industry Applications 44(5), 1466–1477 (2008)
10. Mohammadi Milasi, R., Lucas, C., Araabi, B.N.: Speed Control of an Interior Permanent Magnet Synchronous Motor Using BELBIC (Brain Emotional Learning Based Intelligent Controller). In: Jamshidi, M., Foulloy, L., Elkamel, A., Jamshidi, J.S. (eds.) Intelligent Automations and Control-Trends, Principles, and Applications, vol. 16, pp. 280–286. TSI Press Series, Albuquerque (2004)

11. Daryabeigi, E., Lucas, C., Arab Markade, G.R.: Simultaneous Speed and Flux Control of Induction Motor with New Generation of Intelligent Controller Based on Emotional learning (BELBIC). In: IEEE International Electric Machines and Drives Conference, IEMDC 2009, Miami, Florida, USA, May 3- 6, pp. 894–901 (2009)
12. Mirzaeian Dehkordi, B., Parastegari, M., Lucas, C.: A New Method to Remove the Saturation Effect of Controllers in Vector Control of Induction Machines. Electromotion 27, 17–27 (2009)
13. Sheikholeslami, N., Shahmirzadi, D., Semsar, E., Lucas, C., Yazdanpanah, M.J.: Applying Brain Emotional Learning Algorithm for Multivariable Control of HVAC Systems. International Journal of Intelligent and Fuzzy Systems 17(1), 35–46 (2006)
14. Rouhani, H., Jalili Kharaajoo, M., Araabi, B.N., Eppler, W., Lucas, C.: Brain Emotional Learning Based Intelligent Controller Applied Neurofuzzy Model of Micro Heat Exchanger. Expert Systems with Applications 32(3), 911–918 (2007)
15. Mehrabian, A.R., Lucas, C., Roshanian, M.: Design of an Aerospace Launch Vehicle Autopilot Based on Optimized Emotional Learning Algorithm. Cybernetics and Systems 39(3), 284–303 (2009)
16. Mehrabian, A.R., Lucas, C.: Intelligent Adaptive Flight Control with a Physiologically Motivated Algorithm. International Journal of Modelling and Simulation 1(1), 205-4571-205-4617, 12–18 (2009)
17. Mehrabian, A.R., Lucas, C., Roshanian, J.: Aerospace Launch Vehicle Control: An Intelligent Adaptive Approach. Aerospace Science and technology 10(2), 149–155 (2006)
18. Babaie, T., Karimizandi, R., Lucas, C.: Learning Based Brain Emotional Intelligence as a New Aspect for Development of an Alarm System. Soft Computing- A Fusion of Foundations, Methodologies and Applications 12(9), 857–873 (2008)
19. Torkamani, M.A., Mahmoudzadeh, S., Asgari, J., Lucas, C.: Time Series Prediction Using Brain Emotional Learning Case Study: Iran's Pistachio Export. In: Third International Conference on Business, Turkey, June 16-17 (2007)
20. Gholipour, A., Lucas, C., Shahmirzadi, D.: Purposeful Prediction of Space Weather Phenomena by Simulated Emotional Learning. International Journal of Modeling and Simulation 24(2), 65–72 (2004)
21. Mostofi, N., Jahed Motlagh, M.R., Lucas, C., Farjadian, A.B.: Seizure Abatement in an In- Silico Model of Epilepsy Applying BELBIC. In: The 2010 International Conference of Electrical and Electronics Engineering ICEEE 2010, The World Congress on Engineering WCE 2010, London, UK, June 30-July 2 (2010)
22. Forouzan, S., Mirmomeni, M., Lucas, C., Moshiri, B.: Using a Multiobjective Controller Based on Brain Emotional Learning Method to Control Smart Environments. In: The International Conference on Knowledge Management for Composite Materials, Nanotechnology and Alternate Energy (Fuel Cell), KMCM 2007, Duesseldorf, Germany, July 3-6 (2007)
23. Niki Maleki, K., Valipour, M.H., Ashrafi, R.Y., Mokari, S., Jamali, M.R., Lucas, C.: A Simple Method for Decision Making in Robocup Soccer Simulation 3D Environment. Revista Avances en Sistemas e Informatica 5(3), 109–116 (2008)
24. Sharbafi, M.A., Lucas, C., Toroghi Haghighat, A., Ghiasvand, O.A., Aghazade, O.: Using Emotional Learning in Rescue Simulation Environment. Transactions on Engineering, Computing and Technology 13, 333–337 (2006)
25. Sharbafi, M.A., Lucas, C.: Designing a Football Team of Robots from Beginning to End. International Journal of Intelligent Technology (IJIT) 3(2), 101–108 (2006)

26. Pouladzadeh, P., Jamali, M.R., Lucas, C.: Active Automotive Suspensions Control with BELBIC. In: 7th France-Japan, and 5th Europe-Asia, Congress on Mechatronics, Mecatronics 2008, Le Grand-Bornand, Annecy, France, pp. 21–23 (2008)
27. Valizadeh, S., Jamali, M.R., Lucas, C.: A Particle-swarm-based Approach for Optimum Design of BELBIC Controller in AVR System. In: International Conference on Control, Automation and Systems, ICCAS 2008, Seoul, Korea, October 14-17, pp. 2679–2684 (2008)
28. Masoudinejad, M., Khorsandi, R., Fatemi, A., Lucas, C., Fakhimi, S., Jamali, M.R.: Real Time Level Plant Control Using Improved BELBIC. In: 17th IFAC World Congress, Seoul, Korea, July 6-11, pp. 4631–4635 (2008)
29. Mohammadi Milasi, R., Jamali, M.R., Lucas, C.: Intelligent Washing Mashine: A Bioinspired and Multiobjective Approach. International Journal of Control Automation and Systems 5(4), 436–443 (2007)
30. Lucas, C., Milasi, R.M., Araabi, B.N.: Intelligent Modeling and Control of Washing Machine Using Locally Linear Neuro- Fuzzy (LLNF) Modeling and Modified Brain Emotional Learning Based Intelligent Controller (BELBIC). Asian Journal of Control 8(4), 393–400 (2006)
31. Roshanaei, M., Vahedi, E., Lucas, C.: A Novel Method for Smart Antennas Application Using Brain Emotional Learning Based Intelligent Controller. In: Mosharaka International Conference on Communications, Propagation and Electronics MIC-CPE 2008, Amman, Jordan, March 6- 8 (2008)
32. Mohammadi Milasi, R., Lucas, C., Araabi, B.N.: A Novel Controller for a Power System Based BELBIC (Brain Emotional Learning Based Intelligent Controller). In: Jamshidi, M., Foulloy, L., Elkamel, A., Jamshidi, J.S. (eds.) Intelligent Automations and Control- Trends, Principles, and Applications. Proceedings of WAC, pp. 409–420. TSI Press Series, Albuquerque (2004)
33. Arami, A., Lucas, C., Nili Ahmadabadi, M.: Attention to Multiple Local Critics in Decision Making and Control. Expert Systems with Applications (forthcoming)
34. Roshtkhari, M.J., Arami, A., Lucas, C.: Imitative Learning Based Emotional Controller for Unknown Systems with Unstable Equilibrium. International Journal of Intelligent Computing and Cybernetics (forthcoming)
35. Pedram, A., Jamali, M.R., Pedram, T., Fakhraie, S.M., Lucas, C.: Local Linear Model Tree (LOLIMOT) Reconfigurable Parallel Hardware. Transactions on Engineering, Computing and Technology 13, 96–101 (2006)
36. Kalhor, A., Nadjar Araabi, B., Lucas, C.: A New Split and Merge Algorithm for Structure Identification in Takagi-Sugeno Fuzzy Model. In: 7th IEEE International Conference on Intelligent Systems, Design and Applications, ISDA 2007, Rio de Janeiro, Rio de Janeiro, Brazil, October 22-24, pp. 258–261 (2007)
37. Mirmomeni, M., Shafiee, M., Lucas, C., Araabi, B.N.: Introducing a New Learning Method for Fuzzy Descriptor Systems with the Aid of Spectral Analysis to Forecast Solar Activity. Journal of Atmospheric and Solar-Terrestrial Physics 68(18), 2061–2074 (2006)
38. Gholipour, A., Lucas, C., Najar Araabi, B.: Black Box Modelling of Magnetospheric Dynamics to Forecast Geomagnetic Activity. Space Weather 2(7) (2004)
39. Lucas, C., Jazbi, S.A.: Intelligent Motion Control of Electric Motors with Evaluative Feedback. In: International Conference on Large High Voltage Electric Systems, CIGRE (37th Session), Paris, 30 August-5 September8, vol. 104, pp. 11–104 (1998)

40. Lucas, C., Abbaspour, A., Gholipour, A., Araabi, N.B., Fatourechi, M.: Multi Objective Emotional Learning Based Fuzzy Inference System. WSEAS Transactions on Systems 2(4), 1094–1101 (2003)
41. Lucas, C., Abbaspour, A., Gholipour, A., Nadjar Araabi, B., Fatourechi, M.: Enhancing the Performance of Neurofuzzy Predictors by Emotional Learning Algorithm. Informatica 27(2), 165–174 (2003)
42. Seidi Khorramabadi, S., Boroushaki, M., Lucas, C.: Emotional Learning Based Intelligent Controller for a PWR Nuclear Reactor Core During Load Following Operation. Annals of Nuclear Energy 35(11), 2051–2058 (2008)
43. Damasio, A.R.: Descartes' Error: Emotion, Reason and the Human Brain. G.P. Putnam's Son, New York (1994)
44. Goleman, D.: Emotional intelligence. Bantam Books, New York (1995)
45. LeDoux, J.E.: The Emotional Brain. Simon and Schuster, New York (1996)
46. Rouhani, H., Nikkhah Bahrami, M., Araabi, B.N., Lucas, C.: Lumped Thermal Model for Switched Reluctance Motor Applied to Mechanical Design Optimization. Mathematical and Computer Modelling 45(5, 6), 625–638 (2007)
47. Farshad, M., Faiz, J., Lucas, C.: Development of Analitical Models of Switched Reluctance Motor in Two- Phase Excitation Mode: Extended Miller Model. IEEE Transactions on Magnetics 41(6), 2145–2155 (2005)
48. Lucas, C., Abbas Azimi, F., Moghani, J., Ghafoori Fard, H.: Design and Multiobjective Optimization of the Parameters of Switched Reluctance Motor. International Journal Computational Engineering Science 5(1), 225–233 (2004)
49. Lucas, C., Shahmirzadi, D., Bahrami, M.N., Ghafoorifard, H.: A FEM- Based Quasi-Static Neuro- Model for Acoustic Noise in Switch Reluctance Motors. International Journal of Computational Acoustics 12(1), 85–98 (2004)
50. Shahmirzadi, D.: Computational Modelling of the Brain Limbic System and Its Application in Engineering, MSc Thesis, Texas A&M University (2005)
51. Chandera, M.: Analytical Study of a Control Algorithm Based on Emotional Processing, MSc Thesis, Texas A&M University (2005)
52. Jain, A.: Computational Modelling of the Brain Limbic System and Its Application in Control Engineering, MSc Thesis, Thapar University (2009)
53. Jafarzadeh, S., Jahed Motlagh, M.R., Barkhordari, M., Mirheydary, R.: A New Lyapunov Based Algorithm for Tuning BELBIC Controllers for a Group of Linear Systems. In: 16th Mediterranean Conference on Control and Automation, Ajaccio, France (June 2008)
54. Antolines, J.D.R., Rivera, J.A.D.: Control based on human emotions: A Survey, http://www.udistrital.edu.co/comunidad/profesores/drairan/documents/doctorado/Survey07.pdf
55. Abdi, J., Rashidi, F., Lucas, C., Khaki Sedigh, A.: Combining Context and Emotional Temporal Difference Learning in Control Engineering. Sharif Journal of Science and Technology 30, 13–22 (2005)
56. Daneshvar, R., Lucas, C.: Introducing a New Approach for Defining Personality of Artificial Agents. WSEAS Transactions on Systems 3(1), 300–305 (2004)
57. Nercissians, E.: Social Anthropology of Convergence and Nomadic Computing. WASET 58, 684–687 (2009)

Intelligent 3D Programmable Surface

Michael Fielding, Samer Hanoun, and Saeid Nahavandi

Centre for intelligent Systems Research (CISR), Deakin University,
Victoria 3217, Australia
saeid.nahavandi@deakin.edu.au

Abstract. Creating a highly programmable surface operating at relatively high speed and in real time is an area of research with many challenges. Such a system has applications in the field of optical telescopes, product manufacturing, and giant 3D-screens and billboards for advertising and artwork. This paper covers certain aspects of a keynote presentation at ISDT 2010 including system design, modularity, programmability and the system control intelligence. An overview of the system architecture, actuator design, electronics and distributed control will provide an insight into how the system is controlled and self-tuned for a number of applications. A simulation environment that has been developed to streamline system reconfiguration will also be presented, demonstrating translation of complex mathematical functions into 3D shapes virtually before being displayed on the physical surface.

Keywords: Programmable 3-Dimensions Surface, Pneumatic Actuator Control, Audio Visualisation Techniques, Real-Time 3-Dimensitonal Mathematical Pattern Generation.

1 Introduction

To achieve the large number of individually controllable moving axes, many actuators are required. Large-scale control systems are used every day within automotive manufacturing facilities [1] and food processing plants, however few configurations require large numbers of actuators operating from a single controller at such high speeds. Even fewer systems [2] attempt to achieve this position control with pneumatic actuators in an open-loop fashion. The work presented herein details aspects of the design and control techniques proposed to achieve a high level of reliability and repeatability all while minimising cost. The paper is organised as follows: Section 2 give and overview of the mechanical design, illustrating aspects to be considered. Section 3 explains the electro-mechanical design of the system. The open loop control strategy is detailed in Section 4. Sections 5-11 present the system model and the control software implemented. Finally, the paper concludes in Section 12.

2 Mechanical Design

Four major considerations govern the mechanical design of the programmable surface: size, weight, reliability and cost. As the system was designed to be mobile and modular, each panel needed to be light and easily manoeuvrable. Each module weighs 70kg without a surface attached, making it suitable for a 2-person lift. High reliability is achieved through the use of simple double acting pneumatic cylinders. The cylinders are an inexpensive method of achieving linear motion with high mean time between failure (MTBF) and minimal service and maintenance. Each cylinder is controlled by an electro-pneumatic shuttle valve that can independently apply pressure to either inlet of the cylinder. The open-loop control strategy means costly input cards and additional processing are not required within the controller. Additionally, the complexity during assembly and wiring is significantly reduced.

Fig. 1. 5x6 Cylinder programmable surface backbone module

A. Actuator Selection

A number of pneumatic actuators [5] were considered. For a highly dynamic display surface in a typical configuration simple double-acting cylinders were used. For programmable-mould type applications a specialised brake system is mounted on each cylinder, allowing its position to be fixed at will. The brake can also be used for general display purposes, however lowers the maximum achievable refresh rate due to its switching time.

B. Valve Selection

A fast reacting valve was selected as this latency has direct impact on the reaction time of the system. Just as importantly, the reaction time needed to be highly repeatable. For a 700kPa line pressure a mere 50ms variance in reaction time could

Intelligent 3D Programmable Surface

mean a displacement error of ±150mm, or ~43% of FSD. A fast acting and economical valve supplied by Univer proved to have an acceptable repeatability. Figure 2 shows the delay observed for the cylinder rod to move, timed from when the control valve was first energised. Although this isn't strictly the 'reaction time' of the valve, it provides a useful figure that incorporates any consequential delays associated with the valve and pneumatic circuit for a given run of hose.

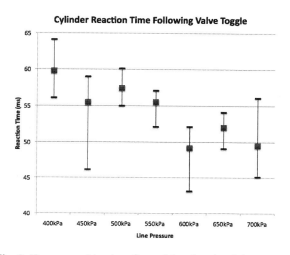

Fig. 2. Human machine interface of the virtual training system

As line-pressure is increased a reduced reaction delay is observed, likely due to mechanical resistance of the rod being overcome quicker. Each test was performed 10 times with average reaction time shown in red. The respective maximum and minimum reaction times are shown for each data set.

Fig. 3. Electro-pneumatic valves and valve manifolds

3 Electro-Mechanical Design

Figure 3 shows the pneumatic circuit for a single panel. Each manifold is fed by a primary air line and controls 6 cylinders and their brake units. Cylinders are held in place using rubber mounts at the base and 4 springs at the top. This allows the cylinder to move slightly and conform to the natural motion of the surface. Spring mounting the cylinders significantly reduces mechanical wear resulting from radial loading.

4 Open-Loop Control Strategy

An undesired characteristic of the double-acting cylinder is that it has different displacement profiles depending on its direction of travel. While extending, the cylinder achieves a velocity different to when retracting. This is due to the different cross sectional area on each side of the piston [6]. This difference in speed is of particular interest given that the cylinder is being positioned using open-loop control. The following equations give theoretical force applied when extending and retracting.

$$F_{Extend} = \rho_2 \pi (r_1^2) - \rho_1 \pi (r_1^2 - r_2^2)$$
$$F_{Retract} = \rho_1 \pi (r_1^2 - r_2^2) - \rho_2 \pi (r_1^2)$$

where:

ρ_1: Pressure applied to extend input

ρ_2: Pressure applied to retract input

r_1: Cylinder bore radius

r_2: Cylinder rod radius

In order to accurately control the cylinder position, characterisation of the displacement curve was recorded for varying line pressures from 400-700kPa in 50kPa increments, see Figure 6. The measured displacement curve was as predicted, retracting approximately 10% slower than extending.

5 Single-Cylinder Model

Ten samples were taken for each of the displacement curve measurements. Figure 4 shows the plotted results of the 10 tests conducted using the maximum line pressure of 700kPa while extending. A small oscillation is repeatedly observed as the piston approaches the hard-stop at the extent of its stroke. The selected cylinders have an in-built pneumatic cushion that slows the cylinder as it reaches the extent of the stroke to avoid mechanical damage as it contacts the end stop. Observed acceleration is almost instantaneous and the piston maintains a constant velocity of 2.45ms^{-1} for most of its move.

Intelligent 3D Programmable Surface 219

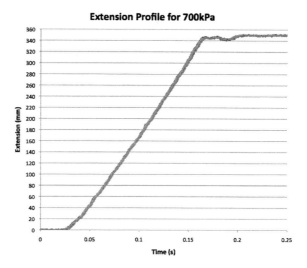

Fig. 4. Extension displacement curve, 700kPa

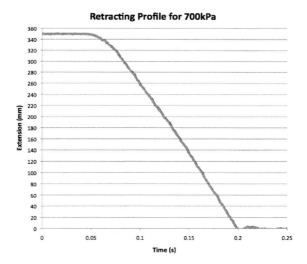

Fig. 5. Retraction displacement curve, 700kPa

The retraction displacement curve is noticeably slower to react initially due to the lower effective force applied by the pneumatic pressure. Once moving however, a constant velocity of 2.6 ms^{-1} is maintained. The full stroke consistently takes 155ms to retract, compared to 141ms when extending.

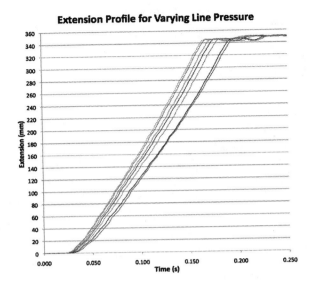

Fig. 6. 400-700kPa displacement curves

A. Repeatability

The results of the displacement measurements were used to build a look-up reference table that the main controller used to offset timing values. Pressure transducers located at each of the manifolds measured the instantaneous line pressure and communicated this back to the main controller. If a significant pressure drop was detected the controller could vary timing accordingly to ensure correct cylinder positions were achieved.

6 Surface Training

The programmable surface is capable of being set up in many different configurations, depending on the specific application. Different surfaces can be attached to achieve different visual effect. A learning algorithm is used to track the material properties and also to characterise the surface material performance. To achieve this a panel with high-resolution position feedback on each cylinder was used. To train the system, material is attached at the controller. A set of programmed moves are then run and the response of each cylinder is measured. Once the training is complete, results are stored in a surface properties file. At operation time, the controller loads saved data and adjusts the Safe Frame Check parameters to ensure the cylinders operate in a manner that won't damage the surface or itself.

7 Programmable Logic Controller Software

In order to achieve high-speed operations, a simple data structure is used to pass information between the supervisory computer and the PLC processor. Each frame of data is passed to the PLC as a block of 250 8-bit values.

Intelligent 3D Programmable Surface

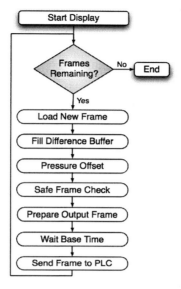

Fig. 7. Control flow diagram

The first 18 bytes are used to identify the data blocks and provide all specific configuration information. The following 232 bytes specify the desired state of each of the PLC outputs. The block can be quickly decoded by the PLC main loop, requiring no more than 50 lines of code to decode the frame and set the outputs. Figure 7 shows the basic logic contained within the PLC.

The following example shows a sample algorithm used to construct the data frames:

For all i = 0 to i<Rows * Cols in the frame

Begin

  if (FrameBuffer[i] == 0) && (OldFrameBuffer[i] == 0)) then DifferenceBuffer[i] = 0;

  if (FrameBuffer[i] > OldFrameBuffer[i]) then DifferenceBuffer[i] = 1;

  if (FrameBuffer[i] < OldFrameBuffer[i]) then DifferenceBuffer[i] = 2;

  if ((FrameBuffer[i] != 0) && (OldFrameBuffer[i] != 0) &&

  (FrameBuffer[i] == OldFrameBuffer[i])) then DifferenceBuffer[i] = 3;

END

The code compares the desired position to the current position and calculates the offset and direction. In the case of a null offset the cylinder is held in position, otherwise the valves are energised to allow movement in the required direction. Valves remain in this condition until the cylinder reaches the correct extension or is updated with a new desired position.

8 End-Point Exaggeration

To reduce the effect of accumulative error the control algorithm performs an operation that homes the cylinder each time the desired position is at the fully extended or fully retracted position. The "End-Point Exaggeration" operation holds respective valves in their on-state for longer than what would ordinarily be required to move the cylinder by the necessary distance. This ensured that the cylinder started from a known initial position for its next move.

9 Staggered Movement

Sometimes very dramatic movements are desired that require many cylinders to move simultaneously. As there is a finite air supply, a control is in place to minimise the impact these large movements have on the overall motion of the surface. As the *current position* and *desired position* frames are always known, the controller can quickly perform an operation to calculate the total displacement of the entire surface, essentially treating the system as a one large cylinder needing air. If the volume of air required to perform the move exceeds the available supply limit, the "Staggered Movement" control comes into effect. This groups neighbouring cylinders into smaller subsets, each of which requires lower total volume of air to perform its move. The control then triggers each of these sets to move slightly out of time reducing the instantaneous total air requirement. The operation is barely noticeable to the observer for most operations.

10 Run-Time Surface Protection

The run-time surface protection is an implementation of a 'neighbourhood watch' scheme that monitors the relative extension between any two neighbouring cylinders. This is performed by the Safe Frame Check, see Figure 7, to ensure that the controller is not attempting to drive adjacent cylinders to a position that may damage the surface material.

11 Interaction

The highly programmable surface is capable of responding to audio input cues and also sending triggers to an external lighting system. An Audio/Light Processor handles low-level signal processing and communicates with high-level commands to the Main Controller, shown below. Pattern cues can be triggered from multiple channels simultaneously and react to either frequency or amplitude.

In a typical configuration, patterns are not generated in real-time mathematically, but rather 1 or more pre-processed pattern files are accessed, mixed and displayed on the surface.

Intelligent 3D Programmable Surface 223

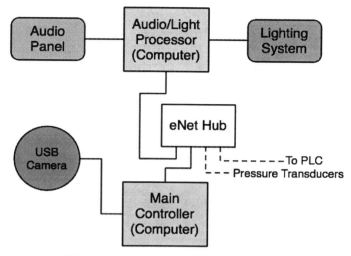

Fig. 8. Connection for interactive components

A. Pre-Processed Pattern Files

Each pattern file contains thousands of data points, all of which need to be accessed quickly when the pattern is being displayed. As such, patterns are stored as text files containing the desired position of cylinders for a series of frames. Files are quickly loaded and buffered as required. When displaying a pre-processed pattern the user loads the desired file, specifies a play rate and then starts the display. The software is capable of varying the speed of the pattern by updating the PLC frames at a user definable rate. A minimum update rate is maintained to ensure a fluid motion when displaying moving 3D shapes and contours. Figure 9 shows two examples of pre-processed pattern files being displayed.

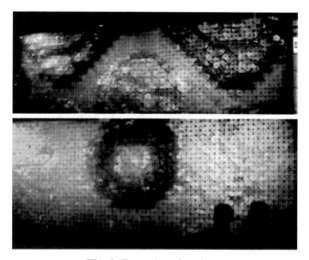

Fig. 9. Examples of surfaces

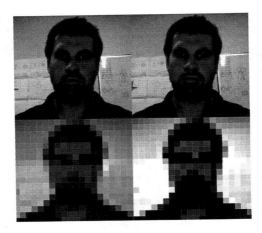

Fig. 10. Original image, greyscale conversion, pixel resize, covert to 4-bit

B. Visually Inspired Pattern Generation

The developed control software is capable of rendering captured video data in near real-time. Images are captured from a video source in full colour and processed into a 4-bit greyscale image with height and width size matching that of the display surface. The control software scales or crops the image to fit the surface if required.

C. Visualising Real-Time 3D Images

Pre-processed pattern files are essentially a table of x, y, z positions that are displayed in sequence. Just as 2-dimensional video can be displayed in real time based on pixel intensity, the programmable surface has the capability to display information from a 3D imaging system.

A number of 3D imaging systems are available that capture real-time 3D information which can then be fed into the controller to position the cylinders. A second standard video camera captures a normal video image which is then projected onto the surface. The projected colour video enhances the 3D surface, displaying a real time, full colour, 3D image.

12 Conclusion

There are many applications to which the highly programmable surface is suited, including multi-cavity mould generation for fibre glass part manufacturing to concrete pouring and boat making applications, large outdoor 3D display for advertising, art installation, stage backdrop, etc. High-powered cylinders could even be used for an adaptive floor skate park.

Acknowledgement

The contributions of Professors Mark Burry from RMIT University and Mark Goulthorpe of MIT School of Architecture and the Media Lab, on the façade design and especially the rubber squid attachment to the end of each actuator, is greatly appreciated and acknowledged.

References

1. Benner, P.: Solving Large-Scale Control Problems. IEEE Control Systems Magazine 24(1), 44–59 (2004)
2. Smaoui, M., Brun, X., Thomasset, D.: Systematic Control of an Electropneumatic System: Integrator Backstepping and Sliding Mode Control. IEEE Transactions On Control Systems Technology 14(5) (September 2006)
3. McLatchey, G.J., Billingsley, J.: Force and Position Control Using Pneumatic Cylinders in Walking Robots, Belgium (2006)
4. Ilchmann, A., Sawodny, O., Trenn, S.: Pneumatic Cylinders: Modeling and Feedback Force-Control. International Journal of Control 79, 650–661 (2006)
5. Wang, J.: A Practical Control Strategy for Servo-Pneumatic Actuator Systems. Control Engineering Practice 7, 1483–1488 (1999)
6. Ali, H.: A Review of Pneumatic Actuators (Modeling and Control). Australian Journal of Basic and Applied Sciences 3, 440 (2009)

Manage Competence Not Knowledge

A.G. Hessami and M. Moore

Vega Systems, London, UK
Freelance Competence Assessor, UK
Hessami@VegaGlobalSystems.com,
Mikemoore152@hotmail.com

Abstract. This paper develops a general framework for assessment and management of competence. It then illustrates a case study demonstrating how to pragmatically assist engineers and managers to confirm their competence, knowledge and understanding against occupational standards without placing undue pressure on their time. It proposes a form of continuous assessment over a 3-6 month period using electronic evidence provided by the candidate in response to a set of focussed emailed questions to build up a paperless portfolio. It also briefly looks how the process can be extended to maintain and update competence and possible future steps to quantify the assessed competence based on weighted performance measures.

Keywords: Knowledge Life-cycle, Competence Assessment, Competence Management, Competence Benchmarking.

1 Introduction

Traditionally knowledge and its creation/acquisition, development and application have been considered by the academic and management communities albeit with different perspectives. In this paradigm, knowledge is the key commodity and the focal point of all activities hence the term Knowledge Management (KM). We propose an alternative and utility based paradigm in which realisation of value through prudent application of knowledge is given prominence over mere acquisition, development, storage, use and ownership of concepts and facts. This is broadly referred to as competence which in a systems paradigm, involves a great deal more than knowledge alone. It is argued that in a real and pragmatic world, it is competence that matters rather than awareness, creation, appreciation and ownership of knowledge. Whilst knowledge is a key and fundamental component in this paradigm, many other factors come together in a systemic form to generate the key benefits from knowledge. This is a utilitarian perspective on knowledge and strives to establish a value system where knowledge is no longer the key commodity but its application in developing solutions to a tapestry of social, technical, global and political problems which are the transformational ability referred to as competence. This by necessity is a human attribute for the time being until cybernetic systems capable of emulating all facets of competence are developed and deployed.

2 Knowledge Life Cycle

In a similar fashion to any other product or service, knowledge undergoes a number of stages from creation and/or acquisition to disposal. This life cycle perspective is instructive in managing it prudently. The key Knowledge Life-cycle phases are:

1. creation, discovery, emulation or acquisition;
2. formalising and representation;
3. capture, encoding, storage and protection;
4. retrieval, dissemination and application;
5. review and enhancement;
6. adaptation and re-deployment;
7. release and disposal.

Each phase necessitates special skills and talents to ensure success. The first phase requires identification of a strategy for acquisition which may involve research, innovation, synthesis, emulation or mere procurement/licensing. These are quite rare capabilities. Formalisation and representation in text, mathematical or diagrammatic form likewise requires the mastery of selecting the most appropriate form or encoding for newly acquired or found knowledge. Once a representation style and form is chosen, the newly acquired knowledge can be captured or translated into this form, classified, encoded, stored and where appropriate, protected. The end users would subsequently retrieve, decipher and apply the captured and encoded knowledge. This is where a combination of other capabilities is called for to ensure the desired outcome at the requisite level of quality and to the satisfaction of the clients is achieved. Given knowledge can always be augmented and improved through usage, phase five involves incorporation of newly found aspects in the formalised knowledge hence enhancement. Knowledge is also often adapted for new environments and domains of deployment. This is where innovation is potentially the outcome since it involves synthesis and adaptation rather than creation of new knowledge in an attempt to realise new value. The phases 1-6 are necessarily iterative but, it is always possible that knowledge becomes out of date in view of discovery of new and more efficient methods and approaches to the same end. This eventually involves disposal or release of knowledge and its repository, paving the way for the new and more effective variations. In carrying out all these, some form of higher level knowledge (meta-knowledge) is required. The meta-knowledge required for successful application is called competence. It contextualises knowledge and deploys a portfolio of synergistic capabilities to realise the inherent value of knowledge in providing answers to real world issues and problems.

3 Competence

The European Guide to good practice in Knowledge Management (Euro Guide 2003) defines competence as an appropriate blend of knowledge, experience and motivational factors which enables a person to perform a task successfully. In this context, competence is the total ability to perform a task correctly, efficiently and consistently to a high quality, under varying conditions, to the satisfaction of the end client. This is a much more demanding portfolio of talents and capabilities than successful application of knowledge and portends that a competent person is much more than and knowledge worker. Competency may also be attributed to a group or a team when a task is performed by more than one person in view of the multi-disciplinary nature, complexity or the scale.

The recent debate about Competence dates back to early seventies when the validity of intelligence and aptitude tests as predictors of job success in the United States were challenged by a Harvard professor in an influential public lecture and later a paper (McClellan 1973). He questioned the undue confidence placed in the evidence that intelligence tests tapped the factors and abilities that are responsible for job success. McClellan also emphasised the need for a continual evaluation of achievement to give insight into the areas where growth in the desired characteristics takes place. In the same vein and based on an empirical "total" systems approach applied within a large-scale intensive study (2,000 managers holding 41 different jobs in 12 organizations), a search was instigated for the characteristics that enable managers to be effective in various management jobs (Boyatzis 1982). Apart fro the search for identifying the special characteristics, Boyatzis also focused on assessing and developing managerial talent whilst also introducing a model of individual competence. The impact of competencies on employee performance are now broadly recognised and adopted in progressive organisations as a cornerstone of the human resource development.

A competent person or team require a number of requisite qualities and capabilities namely;

1. The domain knowledge empirical, scientific or a blend of both;
2. The experience of application (knowing what works) in different contexts;
3. The drive and motivation to achieve the goals and strive for betterment/excellence;
4. The ability to adapt to changing circumstances and demands by creating new know-how;
5. The ability to perform the requisite tasks efficiently and minimise wastage of physical and virtual resources;
6. The ability to sense what is desired and consistently deliver that at a high quality to the satisfaction of the end client.

The right blend of these abilities renders a person or group of people (a team) competent in that they would achieve the desired outcomes consistently, efficiently, every-time or more often than not satisfying or exceeding the expectations of the clients over varying circumstances. Such persons/groups will be recognised for their mastery of the discipline and not just considered a font of relevant knowledge. In this spirit, competency is the ability to generate success, satisfaction, value and excellence from the application of knowledge and a blend of other attributes. This supports our axiom that competence matters more than knowledge alone.

4 Competence Assessment and Management, a Systems Approach

Given the six facets of competence elaborated earlier, the acquisition, assessment, development and management of competence poses a challenge beyond the traditional education and curriculum vitae. Whilst a blend of all six facets is a prerequisite for competency and mastery in a given discipline, the significance of each is highly dependent on the context and requirements of a given domain. Whilst theoretical knowledge plays a more significant role in abstract scenarios, experience of application, adaptability and creativity may become more prominent in other domains. Whatever the domain however, a systems framework for the evaluation, development and enhancement of competence is called for. This by necessity comprises two inter-dependent framework one focused on evaluation and assessment and the other on the management of competence.

4.1 Assessment of Competence

The competence assessment framework provides an integrated perspective on competence in a given context whilst additionally empowering the duty holders or the organisation to benchmark each aspect, measure, assess and where necessary take actions to enhance various elements in the framework. This is illustrated in the Weighted Factors Analysis (Hessami 1999), schema of Figure 1. The latter aspects of benchmarking, evaluating, assessing and potentially enhancing competence are inherent in the underpinning WeFA methodology (Hessami & Gray, 2002) and not elaborated here.

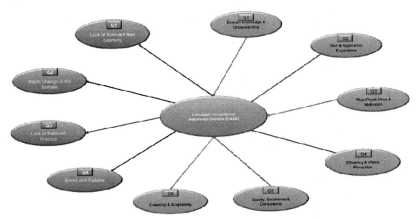

Fig. 1. The Systemic Competence Assessment Framework

Manage Competence Not Knowledge

The determination, benchmarking, evaluation and quantified performance assessment of six *driver* and four *inhibitor* Goals in the above WeFA schema is carried out as follows;

4.1.1 Driver Goals

The requisite domain knowledge in a given context as depicted in the driver Goal 1 (G1) is broadly supported by relevant industry's skill/competence frameworks. There are a number of such frameworks in use mainly within various engineering disciplines in the UK, for example OSCEng (2006), IRSE (2007) and IET (2007).

The composition and extent of relevant experience in a given context as depicted in the driver Goal 2 (G2) in the assessment framework is supported by subsequent decomposition of G2 into lower level WeFA structures, the so called Level 2 and Level 3 goals. This principally helps determine the driver and inhibitor goals for the higher level goal, the domain experience and the demonstrable skills.

The nature and degree of motivation and psychological/physical drive in a given context as depicted in the driver Goal 3 (G3) in the framework is supported by subsequent decomposition of G3 into lower level WeFA structures in WeFA. This principally helps determine the driver and inhibitor goals for motivational and drive aspects for a given role.

The essential determinants and degree of efficiency in carrying out tasks and avoidance of wastage of resource in a given context as depicted in the driver Goal 4 (G4) in the framework is supported by subsequent decomposition of G4 into lower level WeFA structures as appropriate within a given context and role.

The key determinants of quality and consistency in carrying out tasks in a given context as depicted in the driver Goal 5 (G5) in the framework is supported by subsequent decomposition of G5 into lower level WeFA structures, drivers and inhibitors respectively. These are specific to the environment and would differ even for the same role in varying settings depending on dominant cultures and value systems.

Finally, the extent to which a degree of adaptability, sensing the desired outcomes and creativity/innovation is instrumental in success in a given environment and role is depicted by the driver Goal 6 (G6) in the framework. This implies talents and capabilities beyond empirical and classical knowledge which is called for when rapid change or complexity poses challenges requiring novel solutions.

4.1.2 Inhibitor Goals

The key aspects and the extent of absence of relevant new learning in a given context of application as depicted in the inhibitor Goal 1 (G1) in the proposed framework is supported by subsequent decomposition of G1 into lower level WeFA structures, the so called Level 2 and Level 3 drivers and inhibitors in WeFA.

The key determinants and the extent of change in a given domain/context as depicted in the inhibitor Goal 2 (G2) in the proposed framework is supported by subsequent decomposition of G2 into lower level WeFA structures to aid clarity and presentation.

The key predictors and the extent of the currency of relevant practice in a given context as depicted in the inhibitor Goal 3 (G3) in the framework is supported by subsequent decomposition of G3 into lower level WeFA structures.

Finally, the existence of errors and failures are indicative of diminishing degrees of competency as depicted in the inhibitor Goal 4 (G4). Naturally, the types and consequences matter more than mere count of such occurrences and that is what the decomposition of this goal into lower levels would indicate in a given role and context.

A suitably developed, calibrated and validated WeFA schema for competence assessment in a given role, context/domain additionally requires a measurement scale for each goal (driver or inhibitor) as well the weights, i.e. the strengths of influence(s) from each goal on higher level goals. Once established, the weighted framework lends itself to application for assessment and management of individual's or groups' competence in fulfilling tasks in the particular context as depicted by the framework. This would render a number of advanced features and benefits namely:

- Up to 5 levels of competence comprising apprentice, technician, practitioner, expert and leader in a given role/domain;
- Identification of the gaps and training/experience requirements;
- A consistent and systematic regime for continual assessment and enhancement.

It should be noted that assessment here is devised and intended as a tool in the service of systematic approach to staff development and should not be misconstrued as an adversarial instrument for classification of people's contributions to the organisation.

4.2 Management of Competence

The deliverables of the engineering process applied to the creation and realization of parts, products, systems or processes often follow a life cycle from concept to decommissioning as popularized by engineering standards typically comprising;

1. Concept & Feasibility
2. Specification & Design
3. Development
4. Commissioning
5. Deployment
6. Maintenance & retrofit
7. Decommissioning

In this spirit, the human resource involvement/employment within an engineering environment, organisation or project likewise follows a life-cycle comprising seven key phases essential to the systematic and focused management of knowledge namely;

1. Proactivity: comprises corporate policy, leadership, mission, objectives, planning, quality assurance and commitments to competency and service delivery for the whole organisation;

2. Archtecting and Profiling: which comprises specification and development of a corporate structure aligned with the strategy and policy objectives together with the definition of roles and capabilities to fulfil these;

3. Placement: which essentially involves advertising and attracting candidates matching the role profiles/requirements involving search, selection and induction. Selection relates to deriving role focused criteria and relevant tests to assist with the systematic assessment, scoring and appointment tasks. Induction, involves a period of briefing, familiarisation and possibly training the extent of which is determined by the familiarity and competence of the individual concerned and the complexity and novelty of the role.

4. Deployment & Empowerment: which involves a holistic description depicting the scope of the responsibility, accountability and technical/managerial tasks associated with a specific role and empowering the individual to fulfil the demands of the role. This would include training, supervision, coaching, resourcing, delineation of requisite authority and accountabilities, mentoring and potential certification as means to empowerment for achievement and development;

5. Appraisal: which involves the planning and setting performance objectives, and identification of the performance indicators/predictors synergistic to the demands of a role and the individual's domain knowledge, aimed at ensuring all relevant and periphery aspects of the role are adequately addressed and the necessary provisions are made for learning where a need is identified. The evaluation and appraisal provides the necessary feedback on compliance with individual and organisational objectives and achievement, enabling the organisation to identify and reward good performance and develop remedial solutions where necessary;

6. Organisation and Culture: which involves clarification of role relationships and communications, support, reward and motivational aspects for competency development including requisite resources and learning processes for attaining the policy objectives. This is intended to develop and foster a caring and sensitive approach/culture nurturing talents and paving the way towards an innovating organisation.

7. Continual Development and Progression: this comprises identifying the synergistic aspects which may serve as a complementary and rewarding extension to individuals'/teams' specific roles. Development may involve managerial, technical, support functions or an appropriate blend of duties at the whole life-cycle level or extensions to the role specific activities and vision/ career

paths above an existing role into other parts of an organisation and even beyond. The review and assessment of success in all the principles inherent in the framework also fall within the Continual Development principle.

The seven focal areas/principles constitute a systematic competency management framework. It is worth noting however that employment and project/product lifecycles are orthogonal in that securing the requisite human resource and competence for any phase of an engineering production activity would potentially involve all the seven phases of the competence management.

The systematic framework for management of competence is depicted in the WeFA schema of Figure 2. Note that the two frameworks for assessment and management of competence are inter-related and complementary. Whilst assessment focuses on the individual and/or the team in terms of performance, the management framework addresses broader issues relating to the corporate's policy and a nurturing environment to foster talent and innovation as an embedded culture thus creating a sustainable business/service provision.

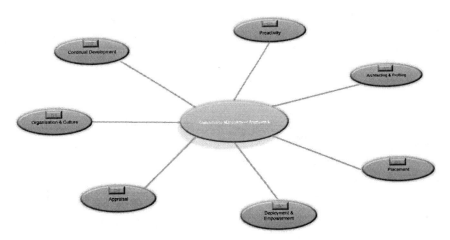

Fig. 2. The Systemic Competence Management Framework

A case study of an industry process for competence assessment and management is presented in section 5 to illustrate current practice and highlight the necessity and potency of systemic frameworks.

5 Case Study, Competence Assessment and Management in Industry

5.1 Making Time for the Assessment of Competence and Knowledge

One of the problems with getting commitment to a scheme that assesses knowledge and competence is to convince both the candidate and their line manager that the time spent is worthwhile and that there should be sufficient resources allocated to the process. In addition to the benefits of a quantifiable system for developing skills, there will be the need for an auditable record which shows that work has been carried out by those who have met the required occupational standard as part of a quality management process. Persuading people can be difficult particularly when individuals think they have been carrying out an activity satisfactorily for some considerable time. It is preferable to use the term "confirming competence" to define the assessment scheme, rather than the need to "demonstrate competence" which implies that, prior to the assessment, the candidate may be "not yet competent".

For tasks with a high practical element such as manufacturing, installing or maintaining, the most accurate assessment method is usually by observation of the candidate carrying out the task with an examination of their completed work. Any knowledge that could not be inferred from the observation but is required to meet the occupational standard would be covered in questioning or tests. The candidate is assessed in the work environment carrying out the activity they would normally be doing. The non–productive time for the candidate would then be limited to answering any questions raised by the assessor.

It is different however for many engineering and managerial roles including project engineers and designers whose work is usually more desk based. The activities tend to be spread over a longer time span covering discussions and analysis which are difficult to observe in action and the candidate may need to gather information, arrange meetings etc before an outcome can be demonstrated. In these instances the conventional way of assessing managers and engineers has been for them to write a personal report stating how they carry out their work and assemble a portfolio of documentary evidence cross-referencing to the occupational standard.

There tends to be some reticence against this additional work to "demonstrate I am doing my job properly" particularly in the climate where there is little "free" time for personal development unless there is some financial inducement or it is necessary to meet contractual or regulatory requirements. There is also the assessor's time that needs to be considered and the probable lack of assessors at a senior level. The assessor should be occupationally competent at the level of the candidate as well as being a qualified assessor. Many organisations may feel that an engineer or manager at this level is more productive carrying out engineering or managerial duties than assessments. Not having sufficient assessor resource also has a negative effect since if a candidate has put in considerable time to assemble a portfolio and then has to wait for an assessor to be available; the news soon gets around and has a detrimental effect by deterring other candidates from starting.

Therefore in order to gain acceptance of a scheme to assess competence and knowledge, it is necessary that the process does not become a burden on both the candidate's and assessor's time.

5.2 Automating the Process

Some processes for demonstrating competence rely on the candidate "raiding the filing cabinet" searching for historic evidence. However, a more staged approach based on short answers to email questions ensures currency and helps the candidate to compile a portfolio without setting aside large amounts of time and interfering with their day job. There are now many emails sent and received by managers and engineers as routine and the plan builds on their responses to short email questions which have been aligned to the occupational standard. Then over a period of time, the candidate will have effectively carried out a self-assessment against the occupational standard for their tasks and supplied sufficient evidence to the assessor for a decision on their competence to be made.

Each of the performance requirements needs to be converted into a format of a suitable email question, such that the answer is a short statement which, when accompanied by supporting documentary evidence, would be acceptable to the assessor. By asking the question in the form of "Please describe how you did.........", the candidate is encouraged to undertake a reflective review of how they carried out the work in their reply.

The questions are graded such that the first few cover daily or weekly routines for which the candidate will have little trouble in answering and finding the evidence, so giving them confidence to complete the program. As the candidate progresses, several related performance requirements can be grouped in the email questions. If the candidate has not recently carried out the specific activity but plans to in the near future, then a response indicating a later date is acceptable. Finally, to complete the process a professional discussion between the candidate and the assessor is held to the confirm the authenticity of the work submitted and resolve any outstanding issues.

5.3 Example from an Occupational Standard

A number of occupational standards have been used in the process including those from MCI (1997) for management standards and from OSCEng (2006) for engineering standards. As an example, consider the following performance and knowledge requirements taken from the MCI 1997 management standards:

Unit D6	"Use information to take critical decisions"
Element D6.1	"Obtain the information needed to take critical decisions"
D6.1e	The information you obtain is accurate, relevant, and sufficient to allow you to take decisions
D6.1f	Where information is inadequate, contradictory or ambiguous you take prompt and effective action to deal with this "Associated knowledge"

- How to judge the accuracy, relevance and sufficiency of information to support decision making in different contexts
- How to identify information which may be contradictory, ambiguous or inadequate and how to deal with these problems

Manage Competence Not Knowledge 237

5.3.1 The Question to the Candidate Is Emailed (On Say a Monday Morning) in a Format as Follows

Q1 - Please reply to this email by next Monday with a brief statement describing how you have obtained information to take a critical decision. Explain how you ensure that the information obtained was accurate and sufficient; where any information was suspect, describe how this was resolved.

You should attach to the reply some recent supporting documentary evidence such as:

- Examples of accurate information used
- Examples of information that is incorrect
- Correspondence requesting clarification of the information
- Documents you have returned where you have marked ambiguities or errors

Please use a unique file reference for each attached piece of evidence eg XX01 (where XX are your initials).

5.3.2 The Short Answer May Be in the Form of

I obtained performance statistics from the company's information management system and also data directly from my 3 supervisors to help me decide on resource planning for next year. However, the data received from area AAA was inconsistent with that on the system and I requested clarification from a second source. I also visited the site to establish the facts first hand.

Supporting evidence (attached)
AB01   Statistics for 3 areas
AB02   Data from AAA
AB03   Clarification request
AB04   Notes of site visit

5.4 The Candidate's Electronic Portfolio

Microsoft Access is used to create an electronic portfolio, which permits documentary evidence to be embedded as electronic files and therefore the creation of a separate paper portfolio is not required. Each electronic portfolio consists of a set of performance requirements, which define the standard that needs to be achieved with guidance and a list of suggested types of evidence that could be used to demonstrate competence. Where the candidate is completing the portfolio themselves, they will enter short statements describing how they meet the performance requirements as they carry out their engineering or management responsibilities. Each statement can be supported with a range of electronic evidence, such as reports, spreadsheets, emails, witness testimonies, digital pictures, short videos etc. The assessor reviews the statements and by clicking on the embedded evidence listed can view the supporting documents.

In the process described in this paper, the electronic portfolio is used by the assessor who enters the email responses and the attached supporting evidence on behalf of the candidate. The assessor then carries out an immediate assessment of the evidence submitted and provides feedback to the candidate.

5.5 Flow of Information

After briefing the candidate on the process, the assessor sends out an email in the format as shown in the example in 5.3.1; this should generate a reply similar to that shown in 5.3.2. On receipt, the assessor adds the text and attached files to the candidate's electronic portfolio candidate. Should more information be required or additional documentary evidence needed, then the assessor would reply with a second email request. If the initial reply shows that the candidate has not met the associated knowledge requirements, then the assessor would ask a direct question to cover a specific area of knowledge.

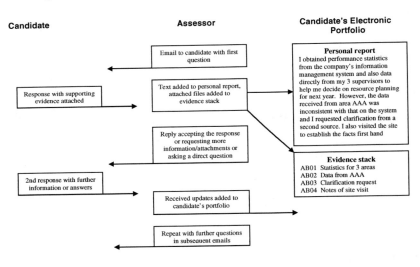

Fig. 3. Flow of information

One advantage of the process is that the assessor can provide guidance and feedback whilst carrying out the staged assessments remotely, thus reducing non-productive travelling time to meet the candidate and review the evidence. This can be a significant benefit particularly when there is a limited assessor resource which may not available locally.

The emailed questions could be sent out in conjunction with a "real-time" plan, i.e. questions that match dates when certain regular activities take place and actions are due. For example within operational management, decisions may be based on a 13 week cycle (roster planning meetings, ordering of materials etc); or tie in with know seasonal or climatic changes, then the email questions could coincide with these activities prompting responses from the candidates concurrently as they carry out the work.

5.6 Adding the Personal Statements and Evidence

An extract of the "Assessor's Page" of an electronic portfolio is shown in Figure 4; across the top are the unit and element titles followed by a brief summary of what needs to be provided to demonstrate competence. The email response from the candidate has been entered under "Candidate's personal report" together with the supporting evidence. The assessor then adds their judgement including any questions to ask the candidate and responds back accordingly. The program has the facility to export a report with the assessor's feedback and indicates those requirements which have yet to be met. The result is that over a period of between 3 to 6 months depending on the complexity of the standard, either the candidate is confirmed as competent, or a training and development need is identified. In the latter case, action can be taken immediately rather than waiting for the final assessment. The exercise should take no more than 15-30 minutes a week for both the candidate and the assessor. The final report includes an overall summary and any independent assessment or verification of the portfolio that may be required prior to submitting for an award.

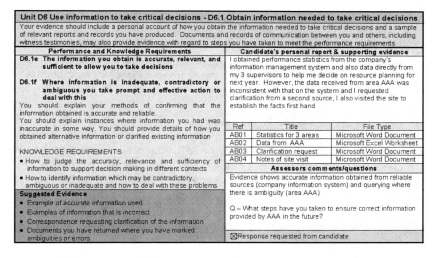

Fig. 4. Extract of the "Assessors Page" in the electronic portfolio

5.7 Maintaining and Updating Competence

Having achieved the target of confirming competence at the end of the exercise the process can be adapted to monitor and reassess the candidate to ensure that they maintain the level of skill, experience and knowledge required for satisfactory performance. For example a review could be planned after 12 months as part of the candidate's development. This can be initiated by an email from the assessor, but this time looking only at areas where an improvement is expected, perhaps in response to a development objective set by the candidate's line manager as a result of a performance review.

The focus and frequency of monitoring would depend on a number of factors taking into account development plans, the introduction of new processes or equipment, dealing with infrequent events and any risk associated with poor performance. A plan could be devised showing target levels to be achieved by certain dates following the completion of any training and development courses. The email questions could be triggered requesting evidence of the newly developed skills.

5.7.1 Certification Schemes

Many competence certification systems such as those accredited to the European Standard BS EN ISO/IEC 17024 (BSI 2003) will have an expiry date on the "certificate of competence". In addition they require that those assessed as competent need to demonstrate that they maintain their competences and continuously carry out tasks to the required standard. ISO17024 clause 6.4 "Surveillance" requires a "pro-active surveillance process to monitor certificants' compliance with the relevant provisions of the certification scheme" and that there is "impartial evaluation to confirm the continuous competence of the certified person". Regular email questions can be sent requesting a current "Reflective review" with supporting evidence that demonstrates continuous competence.

The IRSE (2007) Licensing Scheme, as part of its compliance with ISO17024, requires that there is evidence of the Licence Holder carrying out licensable work to the required standard on an annual basis. An email by the assessor could request evidence of this review, and if there was insufficient evidence of continuous competence then action can be taken immediately.

5.7.2 Competence below the Expected Level

The Office of Rail Regulator (ORR 2007) in their publication Railway Safety Publication No 1 "Developing and maintaining staff competence" discusses situations where competence may start to fall below what is expected. This may be due to the lack of opportunity to practice skills because of their low level of occurrence (infrequent event) but at the other extreme because of over familiarity when people "reach a level of almost automatic performance" and "regress into bad habits and lapses". A program of email questions with a number of case studies would assist in dealing with the lack of practice of those skills that are rarely used, whilst a program which focuses email questions on those areas where lapses are most likely to occur should pick up the possible onset of bad habits before they take hold.

5.8 Quantifying Levels of Competence

The process described is to confirm and maintain competence and knowledge at a specific level, but it can also be used to motivate the candidate's career development. There may be different levels of competence within an occupational area, but whereas for NVQs, the candidate is either "competent" or "not yet competent", a measure of competence could be incorporated within the program, with a rating between 1 to 5 such as that proposed by the IET in their "Competence Framework – Assessing Competence" (IET 2007). Since each of the performance requirements may also have a different weight in relation to its importance to the job role, the program could compute a score based on the weighting factor and a judgment of the level by the assessor. Indeed the candidate's job role may not require the same level throughout the related occupational standard, for example, they may need to be at level 4 for some performance requirements but at level 2 for others. The evidence supplied by the candidate would be judged against the target levels for their job role.

5.9 Enhancements

The process could be enhanced with the use of a central server, to which the candidate and assessor have access, enabling the candidate to view their progress on line. In addition, auto reminders could be used in order to reduce the build up of any backlog if replies were not received by the due dates.

It may also be possible to set up simple rules within the email programs such that when the candidate uses selected words or phrases in their emails as part of their normal work, then a copy of it with attachments is automatically sent to the assessor or routed directly to the candidate's electronic portfolio. Alternatively, an online and secure evaluation and assessment system can be devised enabling the candidates and assessors to access and evaluate competence against a defined role profile. This would obviate the need for email as a basis for solicitation of evidence and enhance the integrity and efficiency of the process.

6 Conclusions

With competence gaining pervasive prominence in preference to mere focus on knowledge, the adoption, deployment and continual enhancement of competency frameworks founded on systemic principles and a systematic approach provide an advanced basis for management of this strategic capability. We have illustrated candidate architecture for an advanced and systemic competence assessment and management framework which can fulfil the requirements and meet the challenges of this complex domain whilst illustrating the current practice.

References

1. Boyatzis, R.E.: The Competent Manager: A Model for Effective Performance. Wiley, Chichester (1982) ISBN: 978-0-471-09031-1.
2. McClellan, D.C.: Testing for Competence rather than for Intelligence. American Pshychologist Journal (1973)
3. European Guide to Good Practice in Knowledge Management, Work Item 5: Culture Working Draft 6.0, CEN-ISSS (July 2003)
4. Hessami, A.: Risk, a Missed Opportunity. Risk & Continuity-The International Journal for Best Practice Management 2(2) (1999)

5. Hessami, A., Gray, R.: Creativity, the Final Frontier? In: The 3rd. European Conference on Knowledge Management ECKM 2002, Trinity College Dublin, September 24-25 (2002)
6. MCI, The UK's National Occupational Standards for Management were originally developed by the Management Charter Initiative and have been taken over by The Management Standards Centre (1997), http://www.management-standards.org
7. OSCEng, The Occupational Standards Council for Engineering publishes occupational standards for engineering and manufacturing (2006), http://www.osceng.co.uk
8. IRSE, Institution of Railway Signal Engineers Licensing Scheme (2007), http://www.irselicences.co.uk
9. IET, Competence Framework – Assessing Competence, The institution of Engineering and Technology, UK (2007), http://www.theiet.org/careers/cpd/competences
10. ORR, The Office of Rail Regulator Railway Safety Publication No. 1 Developing and maintaining staff competence (2007)
11. BSI, European Standard BS EN ISO/IEC 17024 Conformity Assessment - General criteria for certification bodies operating certification of personnel (2003)

Entropy Measure and Energy Map in Machine Fault Diagnosis

R. Tafreshi, F. Sassani, H. Ahmadi, and G. Dumont

Mechanical Engineering Program Texas A&M University at Qatar, Doha, Qatar
The Department of Mechanical Engineering
The Department of Electrical and Computer Engineering
The University of British Columbia, Vancouver, BC, Canada,
Also with the Department of Electrical and Computer Engineering,
University of Tehran, Center of Excellence in Intelligent Signal Processing
rtafreshi@tamu.edu, sassani@mech.ubc.ca, noubari@ece.ubc.ca,
guyd@ece.ubc.ca

Abstract. This paper presents a novel wavelet-based methodology for fault diagnosis and classification. To compare the performance of the proposed approach with major existing methods, various sets of real-world machine data acquired by mounting accelerometer sensors on the cylinder head have been extensively tested. The developed method not only avoids the demerits of the previous techniques but also demonstrates superior performance.

Keywords: Fault diagnosis, pattern recognition, wavelet packet analysis.

1 Introduction

As machinery becomes more complex and costly to build and operate, preventive maintenance measures also become increasingly important. Implementing machinery performance analysis has been found to offer several benefits, including financial, operational, and even environmental advantages. A machine breakdown can lead to significant problems, including downtime cost, loss of functionality and productivity, catastrophic failure, which might be beyond repair, and loss of life.

There is currently a great need for equipment and software to automatically predict, detect, and diagnose faults in machines and components. Important examples of such systems include turbines, compressors, and engines as well as components such as bearings and gearboxes, and cutting and drilling tools.

Two different approaches have been used to diagnose faults in machinery such as internal combustion engines. In the first approach, a mathematical model of the specific engine or component under investigation is developed and a search for causes of changes in engine performance is conducted based on the observations made in the system output. In the second approach, the specific engine or component is considered a black box. Then, through observations of some sensory data, such as cylinder pressure, cylinder block vibrations, exhaust gas temperatures, and acoustic emissions, and analyzing them, fault(s) can be traced and detected. This paper focuses on the latter approach in which vibration data is used for the detection of malfunctions in reciprocating internal combustion engines.

The objective here is to develop effective data-driven methodologies for fault detection and diagnosis. The main application is the detection and characterization of combustion-related faults in reciprocating engines, faults such as knock, improper ignition timing, loose intake and exhaust valves, and improper valve clearances.

This paper first presents different researchers' use of vibration data for engine diagnosis and then reviews wavelets and data analysis by wavelet packet decomposition. A description of pattern recognition and classification and the concept of discriminant measure follow. Then the local discriminant bases search algorithm is critically reviewed and analyzed. The new cross-data entropy approach is introduced. Classification results of this scheme under real-world data are presented.

2 Use of Vibration Data

In engine monitoring applications, non-intrusive measurements, such as the use of the accelerometer on the cylinder head to acquire vibration data, offer clear advantages over other measurement types such as intrusive cylinder pressure measurement. One important benefit is that vibration measurement techniques can be used in existing engines without a major modification to retrofit sensors. The use of vibration data has become popular in a wide range of fault diagnosis applications, including the detection of knock [1, 2, 3, and 4], the detection of valve clearance and gas leakage in both intake and exhaust valves [5], as well as the detection of drift in ignition timings [6].

In recent years, several research attempts have been made to diagnose different faults relating to knock [7,8,9], loose or cracked roller bearings [10,11], ignition timing [6], operation under loose valve conditions [12], cracked teeth in gear trains [13,14,15], cylinder and ring wear, and injection system problems in engines [16]. In diagnostic applications, the problem becomes more complex when faults begin to develop concurrently.

Often, fault diagnosis includes two main stages: feature extraction and classification. In the feature extraction stage, we have used wavelets to analysis acceleration data acquired at the cylinder head location to capture features that contain information about the engine's state. Wavelets have proven to provide suitable signal processing tools for the analysis of transient data and noise reduction. Wavelet packet decomposition structure, as a generalization of standard discrete wavelet transform, offers a powerful data analysis structure for the extraction of features that are capable of identifying combustion malfunctions.

Wavelets, as a class of functions, have information localization ability in both time and frequency. They have been used as an efficient analytical tool in fault diagnosis. The basic concepts of wavelet analysis are introduced next.

3 Wavelets

Wavelets are classes of wave-like functions that are often irregular, non-symmetric, and not always with analytical expression. They have a finite number of oscillations, often with compact support [17,18]. Wavelets are used as basis functions for signal decomposition and signal reconstruction.

Wavelets in signal processing can be considered as windowing functions that extract signal information at localized regions of varying sizes. They allow the use of windows with long time intervals for low frequency information extraction and for short time high frequency oscillation to extract high frequency information. Wavelet transform projects a given signal onto a two-dimensional array of coefficients parameterized by scale (frequency) and translation (time), while Fourier transform maps a one-dimensional function into a sequence of single parameter coefficients. Two-dimensional signal representation allows localized extraction of signal information both in time and in frequency. In standard *discrete wavelet transform* (DWT) this representation is achieved using basis function ψ dilated with a scale parameter j, and translated by k:

$$\psi_{j,k}(t) = 2^{j/2} \psi(2^j t - k) \qquad\qquad j,k \in Z \qquad\qquad (1)$$

where Z is the set of all integers and the factor $2^{j/2}$ maintains a unity norm independent of scale j. Accordingly, any finite energy signal f in $L^2(R)$ can be decomposed using wavelet orthogonal basis $\{\psi_{j,k}\}$ [17]:

$$f(t) = \sum_{j,k} a_{j,k} 2^{j/2} \psi(2^j t - k) \qquad\qquad (2)$$

or

$$f(t) = \sum_{j,k} a_{j,k} \psi_{j,k}(t) \qquad\qquad (3)$$

where the two-dimensional set of coefficients $a_{j,k}$ is referred to as the wavelet coefficients of $f(t)$ and can be determined using inner products defined as

$$a_{j,k} = \langle \psi_{j,k}(t), f(t) \rangle = \int f(t)\, \psi_{j,k}^*(t) dt \qquad\qquad (4)$$

The decomposition of a signal can be carried out using a filter bank structure by breaking the signal into a set of low and high frequency components, as illustrated in Fig. 1, where $f = f_v^0$ is the original signal, h^j and g^j are low and high pass filters of stage j, $f_v^{\,j}$ and $f_w^{\,j}$, $j = 1,\ldots,J$ are called the approximation and detail at resolution level j, respectively. J is the number of the decomposition level considered for signal analysis. In standard wavelet transform, low pass and high pass filters h^j and g^j remain unchanged for all stages. At an arbitrary stage j, the original signal can be reconstructed from the sum of all the details up to stage j plus approximation at that stage i.e., $f = f_v^i + \sum_{i=1}^{j} f_w^i$ [19].

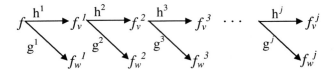

Fig. 1. Filter bank analysis and multi-resolution signal decomposition.

A wavelet transform of a given signal can be interpreted as a decomposition of the signal into a set of components described at different frequency channels. In standard wavelet decomposition, low frequency channels have a narrow bandwidth, and high frequencies have a wide bandwidth [42]. While this kind of signal decomposition is appropriate for many purposes, there are applications that need a more flexible frequency partitioning, which is the theme of the next section.

4 Wavelet Packets

Wavelet packet decomposition provides a more suitable structure for signal decomposition with a narrow frequency band data analysis. Wavelet packet can be considered a generalization of the standard discrete wavelet transform in which the outputs of high pass filters are further decomposed into high and low frequency signal components. Decomposition can be continued up to a level in which the last stage consists of a single sample only. A binary tree with a root as the original signal can describe wavelet packet signal decomposition. Each node is associated with a basis function spanning the signal component at that node.

While a Fourier basis provides a poor representation of functions for transient data analysis that are highly localized in time, standard discrete-time wavelet transform is also not well suited to represent functions whose Fourier transforms have a narrow high frequency bandwidth. To overcome this problem, wavelet packet, as a generalization of standard discrete wavelet decomposition, offers a richer range of possibilities for signal analysis. The standard wavelet technique decomposes the frequency axis in dyadic intervals where the size of bandwidth increases exponentially [21]. Wavelet packet, introduced by Coifman, Meyer and Wickerhauser [22], on the other hand, generalizes dyadic construction by decomposing the frequency axis in separate intervals of varying sizes.

In effect, wavelet packet is a *redundant* signal decomposition. The term "redundant" refers to the fact that there are more than one set of basis functions that can span a particular space of a given function. The non-orthogonality of wavelet basis functions at the parent/child nodes and the subsequent branch-off trees leads to a redundant signal representation. Redundancy in wavelet packet provides a wider collection of basis functions for selecting the projection directions that are "most suitable" for a given application, and as such can have a great influence on fault diagnosis results. Using wavelet packet, we can select bases from a library of basis functions that best meets the criteria chosen for information extraction in a given application. This approach may also include the selection of basis functions that best match with the signal components at different resolutions both in time and frequency.

5 Pattern Recognition

Fault detection and quantification problems may be analyzed within the scope of pattern recognition problems whose goal is to classify objects or patterns into a number of categories or types [8]. Pattern recognition is an integral part of most machine diagnosis systems built for decision-making applications. The major problem associated with pattern recognition is the so-called *curse of dimensionality* in reference to high dimensionality of the given raw data. Often a considerable amount of superfluous information in the original data leads to excessive computational time in the data analysis phase.

On the other hand, in classification problems, not only do we look for features that contain non-superfluous information but we also seek information that can separate classes from each other as distinctly as possible. This type of information is referred to as "discriminant." Usually, it is the superfluous information that turns the classification into a difficult task. The main objective in feature extraction and classification problems is to find a coordinate system that by projecting the signal on those directions we obtain high discriminatory information residing on a few axes while other axes contain insignificant information.

A linear projection from R^n to R^m is a linear map B represented as an $n \times m$ matrix:

$$Z = B^T X, \quad X \in R^{n \times l}, \quad Z \in R^{m \times l} \tag{5}$$

which transforms the n-dimensional data set X (consists of l data in each column) into an m-dimensional space; Z is the m-dimensional transformed data set and ^{T} is the matrix transpose operator. Suppose b_i is the n-dimensional column vectors of matrix $B=[b_1 \ b_2 \ ...b_m]$. If b_is are orthogonal to each other, the projection is called orthogonal, and if they have unit magnitudes, the projection is called orthonormal. If $m = 1$, then B is a one-dimensional projection, and Z is a scalar sometimes referred to as the *projection score*.

Fig. 2 shows the main stages of classification in which X is the input signal, Y, its corresponding class label (e.g., *faulty* or *healthy* conditions), and F, feature space, which is the discriminant subspace of the reduced dimension ($m > n$). The maps $f : X \to F$ and $g : F \to Y$ are called *feature extractor* and *classifier*, respectively. It is computationally more efficient to analyze the data in a discriminant subspace of the lower dimension. The classification goal is to determine which class a given data X belongs to by constructing a feature space F that provides the highest discriminant information among all classes.

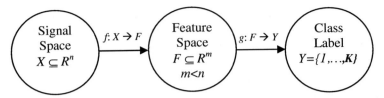

Fig. 2. Main stages in classification.

The vibration signals of a machine always carry information about the state of machine operation and the dynamic behavior of various machine events such as combustion. They can be used to identify faults in machine operation. Vibration signals in internal combustion engines are characterized by transient time behavior that are corrupted by large noise content. Wavelets are considered to be highly suitable for the analysis of transient signals to extract features that are used in fault detection problems.

The next section introduces a measure to evaluate the distance between two or more classes of data referred to as "discriminant measure."

6 Discriminant Measures

The principal objective in a classification problem is to develop measures that are capable of discriminating between different classes as much as possible. The accuracy of the classification results is highly influenced by the extent of class separation in feature space generated by the chosen discriminating measure. Discriminant measure, in general, is designed to evaluate the statistical distance between different classes. The choice of discriminant measure depends on the application on hand. Different authors have used different discriminant measures in various applications [16,23,24]. The approach used in this work is based on *relative entropy* as a measure for discriminating different classes that is now defined.

In a two-class case, suppose that $\mathbf{s}^{(l)} = \{s_i^{(l)}\}_{i=1}^{n}$ for $l=1,2$ are two non-negative sequences satisfying:

$$\sum_i s_i^{(1)} = \sum_i s_i^{(2)} = 1 \tag{6}$$

Symmetric relative entropy for the two classes is then defined as:

$$D(\mathbf{s}^{(1)},\mathbf{s}^{(2)}) = \sum_{i=1}^{n} (s_i^{(1)} \log \frac{s_i^{(1)}}{s_i^{(2)}} + s_i^{(2)} \log \frac{s_i^{(2)}}{s_i^{(1)}}) \tag{7}$$

assuming that

$$\log 0 = -\infty, \quad \log(s_i/0) = +\infty \text{ for } s_i > 0, \text{ and } 0.(\pm\infty) = 0 \tag{8}$$

If we use only the first term in the right-hand side of (2) then the *relative entropy* is defined as:

$$D(\mathbf{s}^{(1)}, \mathbf{s}^{(2)}) = \sum_{i=1}^{n} s_i^{(1)} \log \frac{s_i^{(1)}}{s_i^{(2)}} \tag{9}$$

Lemma: Equation (4) is always non-negative and will be zero if distributions $s^{(1)}$ and $s^{(2)}$ are the same.

Proof [24]: Recall the elementary inequality for real numbers

$$\log x \le x - 1 \tag{10}$$

with equality if and only if $x = 1$. Then, considering conditions (1) we have

$$\sum_i s_i^{(1)} \log \frac{s_i^{(2)}}{s_i^{(1)}} \le \sum_i s_i^{(1)} (\frac{s_i^{(2)}}{s_i^{(1)}} - 1) = \sum_i s_i^{(1)} - \sum_i s_i^{(2)} = 0 \tag{11}$$

Therefore, $\sum_i s_i^{(1)} \log \frac{s_i^{(1)}}{s_i^{(2)}} \ge 0$. Equality holds if and only if $s_i^{(2)} = s_i^{(1)}$, for all i.

Note: Conditions (1) are not necessary for a symmetric relative entropy to be non-negative because the right-hand side of Eq. (6) in a symmetric case results in the cancellation of the corresponding terms:

$$(\sum_i s_i^{(1)} - \sum_i s_i^{(2)}) + (\sum_i s_i^{(2)} - \sum_i s_i^{(1)}) = 0 \tag{12}$$

As can be seen from the above lemma, if two random variables have the same distributions, discriminant measure D will be zero. In classification applications, one is interested in those features that can separate the distribution associated with each class; therefore, one should look for those features that maximize D. The more separate these distributions are, the higher the discriminant measure D.

Now, the challenge is to find appropriate features that provide discriminant classification results. By projecting a set of data in different classes (vibration signals in our application) onto a set of basis (coordinates) and using the associated coefficients in different classes as $s^{(l)}$, we can find the corresponding discriminant measure D. In a classification problem, the objective is to locate those bases (from a dictionary of bases) that maximize the value of D. In this manner, the distribution of each class can be transformed into a sharp distribution with the least overlap with other distributions.

The discriminant function D measures how different the distributions of two classes are. Since $s^{(l)}$ are non-negative sequences assimilating pdf functions, we can successfully employ the normalized energy of the coefficients in each class as $s^{(1)}$ and $s^{(2)}$ in the classification problem.

In general, if there are L classes, one can use this simple approach:

$$D(\{\mathbf{s}^{(l)}\}_{l=1}^{L}) \equiv \sum_{i=1}^{L-1} \sum_{j=i+1}^{L} D(\mathbf{s}^{(i)}, \mathbf{s}^{(j)}) \qquad (13)$$

Equation (8) is not always the best choice when the number of classes is more than three, since a large value for D may be obtained because of the summation of many small terms, which is not a favorable outcome. In fact, the desirable situation for the classification is a large D due to a few significant terms, with most others having negligible values. To overcome this shortcoming, Watanabe and Kaminuma [25] split training data into class i and non-class i, form L sets of two-class problems, and finally construct a classifier for each of them.

7 Local Discriminant Bases

Coifman and Wickerhauser [26] established a mathematical foundation for a method that permits efficient compression of a variety of signals such as sound and images. Their method selects a set of basis that is best adapted to the global properties of signal. Such bases are appropriately selected from a dictionary of orthogonal bases such as local trigonometric functions or the wavelet packets family.

In wavelet packet transformation the selection of the best decomposition tree for signal representation is usually done by the entropy cost function introduced in [26]. Even though entropy is considered appropriate for signal compression, it may be unsuitable for signal classification [23, 27]. Saito and Coifman [23] used the concept of relative entropy as a cost function in classification applications, and called their algorithm the *local discriminant basis* (LDB). Like the process in [26], LDB selects an orthonormal basis from a dictionary, which tries to discriminate between different classes in a given set of data belonging to several classes. They later modified their algorithm by using the relative entropy of the empirical probability density estimate of each class in a wavelet packet domain [28].

Englehart *et al.* [29] applied LDB for myoelectric signals (MES) to clinical diagnosis in biomedical engineering. As in LDB, they used the time-frequency energy maps of each class as input to a symmetric relative entropy measure, in conjunction with principal component analysis (PCA) for dimensionality reduction. In another MES application, they proposed time-frequency methods such as short time Fourier transform (STFT), wavelet transform (WT), and wavelet packet transform (WPT) for feature extraction, along with PCA, and found WPT superior for classification purposes [27,30].

The time-frequency energy map of $\{\mathbf{x}_i^{(l)}\}_{i=1}^{N_l}$, a set of training signals belonging to class l, along the direction of a given basis $\mathbf{b}_{j,k,m}$, is a table defined by:

$$C_l(j,k,m) \equiv \sum_{i=1}^{N_l} (\mathbf{b}_{j,k,m}^T \cdot \mathbf{x}_i^{(l)})^2 / \sum_{i=1}^{N_l} \left\| \mathbf{x}_i^{(l)} \right\|^2 \qquad (14)$$

Entropy Measure and Energy Map in Machine Fault Diagnosis
251

for $j=0,1,...,J$, $k=0,1,...,2^j-1$, $m=0,1,..,2^{n_0-j}-1$, where triple index (j,k,m) are scale, translation (position), and oscillation indices of wavelet packet decomposition, respectively. $n_0=\log_2 n \geq J$ and N_l is the number of training sets in class l. Here "." denotes the standard inner (dot) product in $\mathbb{R}^n$. The energy map defined by (9) is computed by summing the squares of the expansion coefficients of the signals at each position and then normalizing them with respect to the total energy of the signals belonging to class l.

One can obtain expansion coefficients by decomposing a signal of length n into a tree-structured basis such as wavelet packet or local trigonometric dictionaries, in which the computational efficiency is $O(n \log n)$ and $O(n(\log n)^2)$, respectively. Here we have used the wavelet packet dictionary.

The discriminatory power associated with a given wavelet packet node indexed by j,k is the sum of the discriminatory power of its constituent basis $\mathbf{b}_{j,k,m}$ measured in the coefficient domain. The additive property of the discriminatory measure is used here as follows to guarantee high computational efficiency:

$$D(\{C_l(j,k,.)\}_{l=1}^L) \equiv \sum_{m=0}^{2^{n_0-j}-1} D(C_1(j,k,m),...,C_L(j,k,m)) \qquad (15)$$

As stated earlier, LDB selects a local orthogonal basis from a dictionary of bases in a wavelet packet, which properly categorizes the given classes, based on the discriminatory measures of their time-frequency maps. Suppose that $A_{j,k}$ represents the desired local discriminant basis restricted to the span of $B_{j,k}$, which is a set of basis vectors at (j,k) node, and $\Delta_{j,k}$ is the array containing the discriminant measure of the same node.

The algorithm first chooses a time-frequency decomposition method such as wavelet packet transform or local trigonometric transform. Then, for a given training dataset consisting of L classes of signals $\{\{\mathbf{x}_i^{(l)}\}_{i=1}^{N_l}\}_{l=1}^L$, the local best-basis can be found by induction on j as follows:

Step 1: Decompose the given signal $\mathbf{X}$ by expanding it into a dictionary of orthogonal bases to obtain coefficients and construct time-frequency energy maps C_l for $l=1,...,L$.

Step 2: For the start of the algorithm, suppose that $\mathbf{A}_{J,k} = \mathbf{B}_{J,k}$ and set $\Delta_{j,k} = D(\{C_l(J,k,.)\}_{l=1}^L)$ for $k=0,...,2^J-1$.

Step 3: Set $\Delta_{j,k} = D(\{C_l(j,k,.)\}_{l=1}^L)$ and search for the best subspace $\mathbf{A}_{j,k}$ for $j=J-1,...,0$ and $k=0,...,2^j-1$:

    If $\Delta_{j,k} \geq \Delta_{j+1,2k} + \Delta_{j+1,2k+1}$ Then $\mathbf{A}_{j,k} = \mathbf{B}_{j,k}$,
    Else $\mathbf{A}_{j,k} = \mathbf{A}_{j+1,2k} \oplus \mathbf{A}_{j+1,2k+1}$ and set $\Delta_{j,k} = \Delta_{j+1,k} + \Delta_{j+1,2k+1}$.

Step 4: Rank in descending order the complete orthogonal basis functions found in *Step 3* according to their discrimination power.

Step 5: Use k (much less than n) most discriminant basis functions for constructing classifiers.

If we start from the last level of decomposition (which is usually the case), i.e., $J = n_0$, in step 3, there will be no summation and $\Delta_{j,k}$ will simply be the elements of level n_0. During step 3, a complete orthogonal basis with a fast computation of $O(n)$ is built. The orthogonality of bases is imposed in the algorithm to ensure that wavelet coefficients used as features during classification are uncorrelated. After this stage, one can simply select k highest discriminant bases in step 5 and use the corresponding coefficients as features in a classifier. To reduce the dimensionality of the problem even further, it is also possible to use first a statistical method such as Fisher's criterion and then input them into a classifier [23].

In brief, LDB algorithm starts by comparing the discriminatory power of the nodes at the highest scale, as "children" nodes, with their "parent" node residing one scale lower. For example, in a two-level decomposition of wavelet packet tree (Fig. 3), the algorithm first compares the discriminant power evaluated for the coefficients of the training data in different classes at node $\Omega_{1,1}$ with those nodes of $\Omega_{2,2}$ and $\Omega_{2,3}$. If the entropy of $\Omega_{1,1}$ is larger, the algorithm keeps the bases belonging to this node and omits the other two; otherwise it keeps the two and disregards the node $\Omega_{1,1}$ bases. This process is applied to all nodes in a sequential manner up to the scale $j = 0$. At this stage a set of complete orthogonal basis having the highest discriminatory power is obtained, which can be sorted out further at the second stage and used for the classification of designated classes.

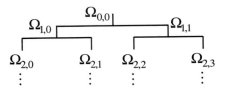

Fig. 3. Decomposition tree in a wavelet packet.

8 Shortcomings of LDB

Despite the potent capabilities of LDB, it encounters drawbacks. For a two-class case, for example, in step 2 of the LDB algorithm (i.e., in the last level of wavelet packet decomposition), LDB basically finds the relative entropy of two positive scalars, instead of two sequences. It is noted that each node in the last level of decomposition includes only one base. Thus, each node contains one coefficient only; as a result, relative entropy is derived between two scalars. To satisfy condition (1), it is necessary that these two scalars be unity; however, such a circumstance corresponds to a trivial case of entropy evaluation in which no useful information can be extracted.

Moreover, according to the definition (4), when $s^{(1)}/s^{(2)} < 1$ (which is highly probable), relative entropy is negative. Then in step 3 of LDB, during comparison of the sum of the discriminant measures of the two children nodes with their parent node, we may add up a negative number with a positive number and compare the sum with the discriminant measure of the parent node. In this situation, we will not have an effective measure of "distance" between two classes, since distance is to be a positive number.

Furthermore, in both best-basis algorithm and LDB, the energy of every node is normalized by the norm of the original signal; therefore, only in the root of the tree (the signal itself) condition (1), i.e., $\sum_i s_i = 1$, holds true. Consequently, in different levels of wavelet packet decomposition, the relative entropy measure is not always positive, since condition (1) is not satisfied. These conditions cause inaccuracies in the true interpretation of entropy and its use for discriminatory measure. On the other hand, the interpretation of entropy in both best-basis algorithm and LDB does not comply with the standard definition of entropy, in which (relative) entropy is applied on sequences of numbers that do not constitute a probability density function (pdf), in the sense implied by the condition (1).

As described in *lemma* in Section 6, the symmetric relative entropy measure is always non-negative, regardless of whether or not condition (1) holds (the sums of sequences are one). However, this is not the case for relative entropy, as mentioned above.

To overcome these shortcomings of LDB, the next section outlines a new methodology.

9 Cross-Data Entropy Method

To resolve the above-mentioned shortcomings, an approach is proposed here, in which for constructing the entropy measure we consider the individual coefficients derived for each training data in contrast to the use of the sum of coefficient energies of all training data in each class, and at each node, as used in LDB. As a result, the role of every single data is taken into account in the sense that the relative entropies of each element in the wavelet packet matrix — derived for all training data — are used to find the appropriate bases. With this approach we deviate from the concept of "averaging of data" as is the case in the LDB method. We gain two advantages by using the proposed scheme.

1) The averaging of all training data as used in the LDB method essentially uses the *first order* statistics of the data only and does not take into consideration second *order* statistics in which the dispersion of data is masked. This is considered a limitation of the LDB method. The proposed scheme eliminates this limitation by considering all training data in which coefficients are obtained and used for each and every training data.

2) In the proposed algorithm, at all nodes, including the last level of wavelet packet tree, the evaluation of entropy is carried out on a sequence of scalars (coefficients) derived from the training data rather than on a single scalar (the average value). The scheme can then be interpreted as a cross-data entropy evaluation or

cross-data energy map approach. Since we still use relative entropy, the discriminatory bases will be derived as before. Under this scheme, the relative entropy of the distributions of the coefficients in different classes is taken into account; that is, discriminant information of every data (mutual discrimination among all data) is considered. For this reason, we will refer to this method as a *cross-data entropy* or a *mutual-based approach*. The cross-data entropy approach eliminates the shortcoming of the standard relative entropy measure used in the LDB method.

The proposed method can also be used in conjunction with the main idea of other searching algorithms. In the following section, the formalization of the extended version of the LDB method referred to as *mutual local discriminant bases* (MLDB) is given. We define the following notations before describing the method.

Let "*map*" be the wavelet packet coefficients of each training data $\mathbf{x}_i$, for $i = 1,\ldots,$ N, which can be illustrated by a set of N matrices of size $n \times (\log_2 n + 1)$, where n is the signal length, N is the number of training data, and J is the scale index of the decomposition level. Let $C^{i=1:N}(j,k,m) \equiv (\mathbf{b}^T_{j,k,m}.\mathbf{x}_{i=1:N})^2$ be the energy map of each training data derived by squaring each element of the *map* matrices, where $C^{1:N}$ is used to denote N energy map matrices, each of the size of *map*. (Matrices $C^{1:N}$ can also be viewed as a 3D-array, *e-map*, of size $n \times (\log_2 n + 1) \times N$.)

Recall that N_l is the number of training data in class l, where $N = \sum_{l=1}^{L} N_l$ is the total number of training data in all classes. If $C_l^{1:N_l}$ are energy maps of each training data in class l then $[C_l^{1:N_l}(j,k,m)]$ can be defined as a vector consisting of N_l number of element (j,k,m) of $C_l^{1:N_l}$:

$$[C_l^{1:N_l}(j,k,m)] = [C_l^1(j,k,m),...,C_l^{N_l}(j,k,m)]^T \qquad \text{for } l = 1,...,L$$

Similarly, we can think of $C_l^{1:N_l}$ as a 3D-array *e-map*$_l$. The process used in the MLDB is described next.

9.1 Mutual Local Discriminant Bases

Consider a time-frequency dictionary such as wavelet packet transform. For a training dataset consisting of L classes of signals $\{\{\mathbf{x}_i^{(l)}\}_{i=1}^{N_l}\}_{l=1}^{L}$, MLDB can be implemented by induction on scale j, as follows:

Step 1: Expand each training signal into a dictionary of orthogonal bases (*map* matrices) to obtain coefficients.

Step 2: Find the energy map of the coefficients, $C^{1:N}$, composed of squared values of each element of *map* matrices.

Step 3: Normalize matrices $C^{1:N}$.

Step 4: Find the discriminant power (by applying Eqs. 2 or 4) amongst L vectors $[C_l^{1:N_l}(j,k,m)]_{l=1:L}$ for $j = 0,1,...,J$, $k = 0,1,...,2^j - 1$, and $m = 0,1,..,2^{n_0-j} - 1$,

Entropy Measure and Energy Map in Machine Fault Diagnosis 255

where n_0 is the maximum level of wavelet packet signal decomposition, with $n_0 = \log_2 n \geq J$. Call the resultant matrix *ent_map*.

***Step 5*:** As initial values for the algorithm, suppose that $\mathbf{A}_{J,k} = \mathbf{B}_{J,k}$ and set $\Delta_{j,k} = \sum_{m=0}^{2^{n_0-J}-1} ent\_map(J,k,m)$ for $k = 0,...,2^J - 1$ (if we start from the last level of decomposition, i.e., $J = n_0$, there is no summation and $\Delta_{J,k}$ is simply the elements of level n_0).

***Step 6*:** Set $\Delta_{j,k} = \sum_{m=0}^{2^{n_0-J}-1} ent\_map(j,k,m)$ and search for the best subspace $\mathbf{A}_{j,k}$ for $j = J-1,...,0$ and $k = 0,...,2^j - 1$:

If $\Delta_{j,k} \geq \Delta_{j+1,2k} + \Delta_{j+1,2k+1}$,
Then $\mathbf{A}_{j,k} = \mathbf{B}_{j,k}$,
Else $\mathbf{A}_{j,k} = \mathbf{A}_{j+1,2k} \oplus \mathbf{A}_{j+1,2k+1}$ and set $\Delta_{j,k} = \Delta_{j+1,2k} + \Delta_{j+1,2k+1}$.

***Step 7*:** Rank in descending order — according to their discriminant power — the complete orthogonal basis functions found in *Step 6*.

***Step 8*:** Use k most discriminant basis functions for constructing the classifiers. Note that k is usually much less than n.

The computational efficiency of MLDB is similar to the LDB method. As expected, the new method has greater data storage requirements, since the energy map of each training data must be saved for the evaluation of the entire relative entropy map.

9.2 Advantages of MLDB

The usage of the normalized and squared coefficients (not the "average") for the evaluation of entropy and for the application of discriminatory measure for identifying suitable basis offers the following advantages.

1. Normalized squared coefficients are positive and satisfy the condition (1), which are the requirements for a sequence to be a pdf function. In this case normalized squared coefficients play the rule of probability mass function (pmf) since the coefficients assume discrete values.

2. While the normalized squared coefficients of two classes do not represent the true probability mass function of the respective coefficients and it may be interpreted as not being significant for a statistical discrimination of the two classes, we note that the differences between individual coefficients of two classes are indicative of the class separation. Accordingly, they are reflective of true discriminatory measure between two classes. When they are close to each other and in

256 R. Tafreshi et al.

extreme case are identical, the two classes are labeled as identical. Therefore, the MLDB algorithm provides us a valid discriminatory measure between different classes, with statistically significant discriminant information, which does not impose extra computational time for the pdf evaluation. It is worth noting that pdf evaluation cost is usually in the order of $O(n (1 +\log n))$ [28].

The following presents some results and an analysis of MLDB.

10 Data Analysis Results

To assess the effectiveness of the proposed algorithm and to compare it with that of standard LDB, we applied the algorithm on a set of data gathered from a single cylinder spark ignition research engine. More information about data collection details can be found in [6]. The objective of the experiment was to collect acceleration data at the cylinder head position with three different ignition timings of -23 (normal), -33 (advance), and -10 (retard) degrees under stoichiometric conditions assimilating healthy and faulty conditions. The numbers denote ignition timings measured as angles before top dead center. A data- acquisition system capable of acquiring the data of 16 consecutive cycles of engine operation was used. For each of the engine conditions (classes), three consecutive 16-cycle data were collected to provide sufficient data (48 sets) for implementing the algorithm and evaluating the accuracy of the classification results. We used the first 32 data cycles in each class as *training* and the remaining 16 as *testing* data sets.

The data size for one full cycle of engine operation corresponding to two revolutions of the crank shaft (720 degrees of crank angle) varied for different cycles and was 1975±15 sample points for data collected at a sampling rate of 25 KHz. The variation in number of samples, caused by changes in engine RPM, was considered to be insignificant (less than 1%). In this experiment, our concern was to study the combustion event under different engine operating conditions as affected by different ignition timings. As such, we considered data belonging to the combustion zone only for the analysis. This was done by defining a window of data, the size of which was carefully selected for investigating different ignition timings. The window was to be wide enough to cover all of the characteristics of the combustion event under different advance or retard spark timing conditions, but not too wide to overlap with other engine events.

Using a narrow window size was also considered to be significant to maintain low computational time. Here we use a segment of data belonging to -15 to +31 degrees of crank angle, which covers the combustion event. For the acceleration data, this corresponds to data points ranging from 945 to 1072 for a 25 KHz sampling rate. Accordingly, the data size was chosen as 128 sample points. The maximum heat-release of the engine, as a measure of combustion quality, occurred at (-5.5) – (32), (-14.5) – (26.5), and (9) – (47) degrees of crank angle for -23, -33, and -10 degrees of ignition timing, respectively. The numbers in parenthesis correspond to the 5% and 95% of heat-release pairwise, which indicates the amount of chemical energy of the fuel released by the combustion process at the given crank angle degree [31]. With regard to the heat release where, for practical considerations, we need an identical interval for all of the data categories, the above window size of -15 to +31 was found to be an appropriate selection.

An analysis of the acceleration data indicates that the horizontal vibration data (direction perpendicular to the piston movement) do not carry useful information about the combustion event [6], in contrast to data in the vertical direction, which contain necessary information on the combustion event. Fig. 4 shows sample acceleration data acquired by the sensor in the vertical direction with -23, -33, and -10 degrees of crank angle. We refer to these data as class 1, class 2, and class 3 engine conditions, respectively. Different segments of the acceleration signal with high magnitudes correspond to different machine events: from left to right, exhaust valve closing (EVC), intake valve closing (IVC), combustion, exhaust valve opening (EVO), and intake valve opening events (IVO).

A histogram of the acceleration data indicates that distribution of the training data (Fig. 5) follows approximately the Gaussian distribution. Classes 1 and 3 exhibit highly clustered data patterns when the mean value of data in each class is drawn against their standard deviation.

A spectral analysis of the data for three classes indicates that the frequency bandwidth with high spectral energy is almost identical for all the classes, with some variations in spectral amplitudes (Fig. 6). This analysis indicates that the spectral features cannot be directly used for the discriminatory classification of different engine conditions, particularly for classes 1 and 3. An inspection of the acceleration data showed that the engine was running relatively smoothly with minor background noise, during which noise reduction was unnecessary.

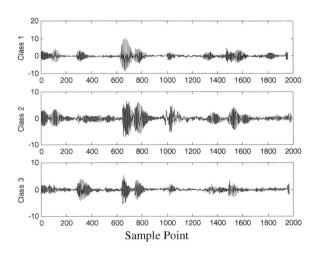

Fig. 4. One cycle of three classes of vertical vibrations with spark timings of -23, -33, and -10 degrees of crank angle, in stoichiometric conditions, 1500 RPM and 25KHz sampling rate. Vertical axis unit is "g."

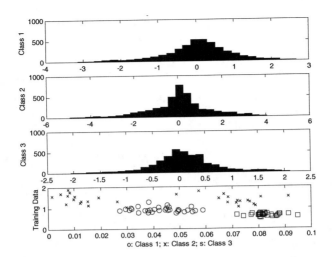

Fig. 5. Histograms and mean-std plots of training and testing data for three classes.

10.1 MLDB Classification

The wavelet packet indices associated with bases derived by applying the LDB algorithm on the set of data are (for more detail please refer to [6]):

Scale	5	4	6	5	4	6	7	5
Oscillation	0	2	12	9	2	16	34	7
Translation	3	6	0	0	7	0	0	3

and the ones for MLDB are:

Scale	4	7	4	4	4	5	7	7
Oscillation	3	33	3	2	3	9	35	34
Translation	3	0	4	6	5	0	0	0

where the first to third rows are scale, oscillation, and translation indices of wavelet packet, respectively. Fig. 7 plots the corresponding wavelet bases of MLDB. By comparing the above indices with indices obtained from LDB, we observe that the first basis of LDB indexed by (5,0,3) has not been selected by MLDB. This basis belongs to the interval (0, $\pi/32$) frequency band of wavelet packets, which corresponds to (0, 0.4) KHz frequency band of the signal as 12.5 KHz / 32 = 0.4 KHz.

The typical signal spectrum plotted in Fig. 6 shows that this frequency band is not located in a dominant frequency band of the signal with high spectral energy; thus, it does not carry considerable energy. Since the combustion event can be viewed as a set of impulses, which is accompanied by a high-energy release, a superior searching algorithm should readily pick those bases of wavelet packets that correspond to nodes with high signal energy. As a result, the selection of this basis by LDB has not been a good choice.

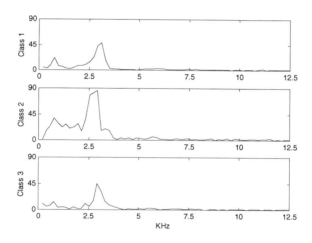

Fig. 6. FFT of vertical vibrations in combustion zone with 25 KHz sampling rate and three classes of spark timing: -23, -33, and -10 degrees of crank angle.

On the other hand, the fourth and eighth bases of MLDB (Fig. 7) have also been selected by LDB. A similar frequency analysis shows that these bases, along with the first three bases chosen by MLDB, carry a significant amount of energy and have been suitable selections. In fact, the frequency band of wavelet packet nodes belonging to the bases 2 and 4 are located in the middle of the dominant frequency band of the signal (around $\pi/4$ or 3.1 KHz), and the one associated with bases 1 and 3 is placed at the beginning of the dominant frequency band (around $3\pi/16$ or 2.3 KHz).

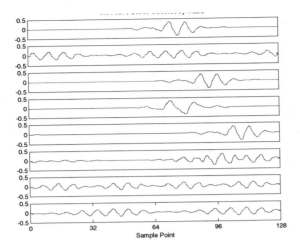

Fig. 7. The first 8 bases selected by MLDB using Coiflet1.

Fig. 8 illustrates the discriminant measure of all 128 complete orthogonal bases as selected by MLDB and ranked in decreasing order. The figure shows a sharp drop of the measure. Only bases with the largest discriminatory measures are considered for classification purposes, as the discriminant measures of the other bases are insignificant.

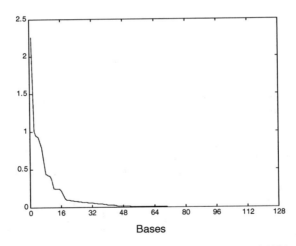

Fig. 8. Discriminant measure of the complete orthogonal 128 bases.

To examine the effectiveness of this modification, numerous trial runs were carried out on the experimental data introduced earlier. In order to ascertain and generalize the effectiveness of the proposed method, a wide range of data analysis using different wavelets was also planned and performed. These included the use of 32 different analyzing wavelets from the family of orthogonal, biorthogonal, symmetric as well as selected wavelets from Battle-Lamorie spline functions as follows:

1-Haar, 2-Beylkin, 3-Coiflet1, 4-Coiflet2, 5-Coiflet3, 6-Coiflet4, 7-Coiflet5, 8-Daubechies2 (Db2), 9-Db3, 10-Db4, 11-Db5, 12-Db6, 13-Db7, 14-Db8, 15-Db9, 16-Db10, 17-Db20, 18-Db40, 19-Db45, 20-Bior22, 21-Bior31, 22-Bior68, 23-Symmlet4 (Sym4), 24-Sym5, 25-Sym6, 26-Sym7, 27-Sym8, 28-Sym9, 29-Sym10, 30-Vaidyanathan, 31-Battle3, 32-Battle5.

Fig. 9 compares LDB and MLDB classification errors for the above 32 analyzing wavelets. It shows that for most of the analyzing wavelets, MLDB performs better or as good as LDB, which shows the overall superiority of the proposed approach.

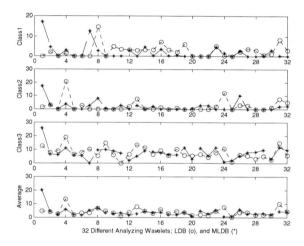

Fig. 9. LDB classification percentage error vs. MLDB for the 32 different analyzing wavelets listed above.

11 Different Neural Network Algorithms

Backpropagation algorithms applied on a multilayer perceptron network [32] have been extensively used in classification applications. Backpropagation is an extension of least mean square (LMS), which in turn is an approximation of the steepest descent algorithm [33] that is a simple, but slow, minimization method.

There are two different approaches for improving the convergence speed of the backpropagation learning rule [34]. The first approach employs heuristic methods, such as variable learning rate, to determine the length of steps in the steepest decent algorithm. The other approach focuses on numerical optimization techniques, since training feedforward neural networks to minimize square error is basically an optimization problem. For the latter approach, the focus of study by several research groups, several variations of the main backpropagation algorithm have been developed. For example, the conjugate gradient algorithm and Newton's method provide faster convergence.

The study of 12 different neural network backpropagation algorithms shows that the Levenberg-Marquardt algorithm is both fast and accurate. This inference was the overall observation using multiple runs with different analyzing wavelets. Table 1 shows the classification errors of the proposed cross-data entropy algorithm on the experimental data for each of these 12 methods, with Coiflet1 as the analyzing wavelet.

Table 1. Classification error using different backpropagation algorithms.

Algorithm	Error (%)
Basic gradient descent	12.77
Gradient descent with momentum	11.18
Adaptive learning rate	5.22
Resilient backpropagation	3.81
Fletcher-Reeves conjugate gradient	4.49
Polak-Ribiére conjugate gradient	4.77
Powell-Beale conjugate gradient	4.05
Scaled conjugate gradient	4.24
BFGS quasi-Newton	4.08
One step secant	4.12
Bayesian regularization	4.84
Levenberg-Marquardt	2.95

12 Conclusions

The transient nature of the machine vibration signals requires the use of basis functions that can capture the localized features of the signals. It was shown that wavelets with finite support width, both in time and frequency domain, are highly suitable for the analysis of these signals in diagnostic applications. Along this line, using wavelet packet and redundant signal decomposition, the local discriminant basis algorithm (LDB) enabled us to select a subset of basis with the highest discriminatory power to classify different engine operating conditions.

This paper presents a novel method, referred to as the mutual or cross-data entropy approach. In conjunction with this approach, LDB, a well-known discriminant algorithm used for classification, was modified. We showed that the classification results of MLDB are superior under a wide range of analyzing wavelets.

The MLDB algorithm attempts to select a set of orthogonal bases from a wavelet packet dictionary that best discriminates different states of the system. Wavelet coefficients constructed by projecting data onto the selected bases are used as feature variables and as inputs to a backpropagation neural network (NN) classifier. We used the MLDB algorithm for data analysis with three classes of ignition timings.

To perform fault diagnosis in internal combustion engines, we used cylinder head vibration data to characterize the underlying mechanical and combustion processes.

As the use of pressure signals for fault detection in internal combustion engines is prohibitive, due to their costly retrofit, the suitability of using acceleration data for engine diagnosis was investigated. Furthermore due to the large data size, a direct application of NN classifier to data in different classes is very time-consuming. Noting that sensor data usually contain redundant information, the direct use of sensor data may also lead to the dilution of information about engine faults and produce unacceptable classification results. Alternatively, the transformation of the data into wavelet domain and the use of wavelet coefficients as feature variables will reduce the data dimensionality considerably.

Choosing the number of features is also an important task; this number should be neither too large to dilute the information nor too small to miss important discriminant information. Roughly speaking, we considered a value around $\log(n)$ for the number of feature variables, where n is the length of the data or the space dimension.

References

1. Samimy, B., Rizzoni, G.: Mechanical signature analysis using time frequency signal processing: Application to Internal Combustion Engine Knock detection. Proc. of IEEE 84(9) (September 1996)
2. Zheng, G.T., McFadden, P.D.: A time-frequency distribution for analysis of signal with transient components and its application to vibration analysis. Trans. ASME 121 (July 1999)
3. Samimy, B., Rizzoni, G., Sayeed, A.M., Jones, D.L.: Design of training data–based quadratic detectors with application to mechanical systems. In: Proc. of ICASSP 1996, Atlanta, GA (1996)
4. Millo, F., Ferrro, C.V.: Knock in S.I engines: a comparison between different techniques for detection and control. SAE technical paper series, # 982477 (1998)
5. Huang, Q., Liu, Y., Liu, H., Cao, L.: A new vibration diagnosis method based on the neural network and wavelet analysis. SAE technical paper series, 2003-01-0363 (2003)
6. Tafreshi, R., Sassani, F., Ahmadi, H., Dumont, G.: Local discriminant bases in machine fault diagnosis using vibration signals. Journal of Integrated Computer-Aided Engineering 12(2) (2005)
7. Worret, R., Benhardt, S., Schwarz, F., Spicher, U.: Application of different cylinder pressure based knock detection methods in spark ignition engines. SAE papers, 2002-01-1668 (2002)
8. Carstens-Behrens, S., Bohme, J.F.: Fast knock detection using pattern signals. IEEE International Conference on Acoustics, Speech, and Signal Processing 5, 3145–3148 (2001)
9. Tafreshi, R., Ahmadi, H., Sassani, F., Dumont, G.: Informative wavelet algorithm in diesel engine diagnosis. In: The 17th IEEE International Symposium on Intelligent Control, ISIC 2002, pp. 361–366 (2002)

10. Rubini, R., Meneghetti, U.: Application of the envelope and wavelet transform analyses for the diagnosis of incipient faults in ball bearings. Mechanical Systems and Signal Processing 15(2-3), 287–302 (2001)
11. Tse, P.W., Peng, Y.H., Yam, R.: Wavelet analysis and envelope detection for rolling element bearing fault diagnosis- their effectiveness and flexibilities. Journal of Vibration and Acoustics, ASME 123 (2001)
12. Ahmadi, H., Dumont, G., Sassani, F., Tafreshi, R.: Performance of informative wavelets for classification and diagnosis of machine faults. International Journal on Wavelets, Multiresolution and Information Processing (IJWMIP) 1(3) (2003)
13. Wang, W.J., McFadden, P.D.: Application of wavelets to gearbox vibration signals for fault detection. Journal of Sound and Vibration, 927–939 (1996)
14. Chen, D., Wang, W.J.: Classification of wavelet map patterns using multi-layer neural networks for gear fault detection. Mechanical Systems and Signal Processing 16(4), 695–704 (2002)
15. Zheng, H., Li, Z., Chen, X.: Gear fault diagnosis based on continuous wavelet transform. Mechanical Systems and Signal Processing 16(2-3), 447–457 (2002)
16. Liu, B., Ling, S.F.: On the selection of informative wavelets for machinery diagnosis. Mechanical Systems and Signal Processing 13(1) (1999)
17. Burrus, C.S., Copinath, R.A., Guo, H.: Introduction to wavelets and wavelet transforms. Prentice Hall, Englewood Cliffs (1998)
18. Chui, C.K.: An introduction to wavelets, vol. 1. Academic Press, London (1992)
19. Strang, G., Nguyen, T.: Wavelets and filter banks. Wellesley-Cambridge Press (1996)
20. Daubechies, T.I.: lectures on wavelets. SIAM CBMS-NSF conference series (1992)
21. Mallat, S.: A wavelet tour of signal processing. Academic Press, London (1999)
22. Coifman, R.R., Meyer, Y., Wickerhauser, M.V.: Wavelet analysis and signal processing. In: Jones, Barlett, Ruskai (eds.) Wavelets and their Applications, Boston, pp. 153–178 (1992)
23. Saito, N., Coifman, R.R.: Local discriminant bases and their applications. J. Mathematical Imaging and Vision 5(4), 337–358 (1995)
24. Gray, R.M.: Entropy and information theory. Springer, Heidelberg (1990)
25. Watanabe, S., Kaminuma, T.: Recent developments of the minimum entropy algorithm. In: Proceedings of the International Conference on Pattern Recognition, pp. 536–540. IEEE, New York (1988)
26. Coifman, R.R., Wickerhauser, M.V.: Entropy-based algorithm for best basis selection. IEEE Transactions on Information Theory 38, 713–718 (1992)
27. Englehart, K., Hudgins, B., Parker, P.A., Stevenson, M.: Classification of the myoelectric signal using time-frequency based representations. Medical Engineering and Physics, special issue: Intelligent data analysis in electromyography and electroneurography 21, 431–438 (1999)
28. Saito, N., Coifman, R.R., Geshwind, F.B., Warner, F.: Discriminant feature 24 using empirical probability density estimation and a local basis library. Pattern Recognition 35(12), 2841–2852 (2002)
29. Englehart, K., Hudgins, B., Parker, P.A.: A wavelet based continuous classification scheme for multifunction myoelectric control. IEEE Transactions on Biomedical Engineering 48(3), 302–311 (2001)
30. Englehart, K., Hudgins, B., Parker, P.A., Stevenson, M.: Improving myoelectric signal classification using wavelet packets and principal component analysis. Proceedings of IEEE (1999)

31. Heywood, J.B.: Internal combustion engine fundamentals. McGraw-Hill Inc., New York (1988)
32. Ripley, B.D.: Pattern recognition and neural networks. Cambridge University Press, Cambridge (1996)
33. Fausett, L.: Fundamentals of neural networks, architectures, algorithms, and applications. Prentice Hall, Englewood Cliffs (1994)
34. Hagan, M.T., Demuth, H.B., Beale, M.: Neural network design. PWS Publishing Company (1996)

Enterprise Information Management for the Production of Micro- and Nanolayered Devices

Rainer Brück

University of Siegen, Hölderlinstr. 3, D-57068 Siegen, Germany
`rainer.brueck@uni-siegen.de`

Abstract. Many modern high-tech products gain their functionality from structured nanoscale layers. This is the case for e.g.nanoelctronic circuits, MEMS and NEMS sensors and actuators, photovoltaic cells, but also macroscopic structures with functional nanolayer coating. The production of these nanolayers is accomplished by chemical/physical processes performed in clean rooms under extremely controlled conditions. IT support for nanolayer production is currently available for the fabrication machinery on the one hand (MES - Manufacturing Execution Systems) and for the global controlling of enterprise operations (ERP - Enterprise Resource Planning) on the other hand. There is a gap between these two areas covering the whole field of project and product related planning, design, optimization and verification of a specific fabrication flow. Currently first PDES (Process development execution Systems) are available to close this gap and hence achieve a holistic enterprise information management (EIM). This article will give a brief introduction to EIM for nanolayered devices and will then present the benefits of PDES systems by the example of the first comprehensive solution in this field – the XperiDesk suite. This software system has originally been designed at Siegen University and is now available on the market place from ProcessRelations GmbH in Dortmund, Germany.

Keywords: Micro technology, nano technology, product engineering, enterprise information management, process design execution system.

1 Micro- and Nanolayered Devices – Properties and Challenges

Many of the modern high-tech areas are characterized by the fact that the functionality of technical artifacts is obtained by generating and structuring very thin layers of various materials. The thickness of such functional layers is typically between several nanometers and several microns.

1.1 Technologies Making Use of Nanolayered Devices

Many of the technologies that are generally perceived as "high-tech" are based on the paradigm of generating functionality from structuring thin layers. The following examples list some of the most prominent of those technologies.

Microelectronics: Microelectronic circuits are produced by deposition, structured removal and modification of thin material layers on solid silicon wafers. Electronics devices are formed by specific combination of such structured layers. The thickness of such functional layers is between few nanometers and few microns.

MEMS/NEMS: Micro and nano electro mechanical systems add mechanical functionality to microelectronic circuitry. The production processes are similar to those used for microelectronics. Mechanical functionality that requires movement of certain structures is obtained by sacrificial layers that are tentatively deposited and later in the production process removed again by specially adapted etching processes. In this manner micro and nano sensors and actuators are produced.

Photovoltaics: Photovoltaic elements are basically microelectronic circuits with very large lateral extension. The photovoltaic effect is gained by specific combinations of thin functional layers. Multi-layered photo cell structures today show a high degree of efficiency.

3D System Integration: Mounting thinned silicon wafers with microelectronics circuitry on top of each other and establishing electrical connectivity between them by so-called TSVs (Through silicon vias) is a promising technology for integration in the spatial volume. It allows the combination of circuits from various fabrication technologies, even the combination of microelectronics and MEMS structures in one single stack. The wafer structures that are mounted on top of each other have typically been thinned down to 20 to 50 microns.

Functional coating of macroscopic devices: Coating ordinary macroscopic devices like glass panes or turbine blades with nanoscaled functional layers is a promising technology to enhance the properties of functional surfaces, e.g. making use of the lotus effect to make surfaces hydrophobic.

1.2 Properties and Challenges of Nanolayered Device Fabrication

There is a wide variety of specific techniques used in order to deposit, to structure and to remove thin layers to obtain technical artifacts with the technologies mentioned above. The interested reader might e.g. refer to [1] and [2] for a thorough introduction to these techniques.

As manifold as they are, all of these technologies bear some commonalities that are most significant from a product development point of view:

- *Structure generation by layer deposition and lithography induced structuring*. Structure generation hence is a "negative" process, first covering the whole artifact to be handled with a complete layer, which afterwards is covered by a resist structure in specific areas and then selectively removed again to form functional structures by appropriate etching processes.
- *Device structuring by indirect physical/chemical processing*. No direct manipulation using "mechanical tools". All processes used in the fabrication of nanoscaled functional layers are "indirect" physical processes. No direct creation or modification of these layers making

Enterprise Information Management   269

> use of nano scale equivalents to mechanical tools can be used in this context.
>
> - *Batch production of many devices at the same time.* This is one of the most important success factors for this kind of technologies, especially for microelectronics and probably also for the increasing success of MEMS/NEMS structuring e.g. for the fabrication of sensors and actuators for automotive applications.

The control and the performance of production facilities for nanolayered devices is a highly complex task, associated with a large set of challenges, that even today, after fifty years of experience in the commercial use of these technologies are still severe and still require large effort to keep the yield of production at a commercially reasonable level. The following examples can only give a selection of these challenges:

- High sensitivity to process parameter variations
- High sensitivity to environmental conditions
- Highly complex machinery with a high degree of automation provides a high dimensional parameter space that is computationally complex to manage
- Contamination avoidance required
- Only statistical process control
- Extreme amount of data

Traditional approaches to design e.g. in microelectronics try to provide means to keep the designer from caring for this sort of problem. As the dimensions of the functional structures shrink and as the complexity of the systems increases more and more over the time, this seems no longer possible.

Designers of nanolayered devices require precise process data to take care of all relevant effects that might influence the product during production and use, but usually lack expertise in process issues. Traditional software tools to support "Computer Aided Design" or "Computer aided Engineering" use to fail to address this issue. A more generalized approach of product engineering as presented in section 2 promises to provide a way to overcome this problem. With the XperiDesk suite furthermore a first approach to close the gap in software tools between the pure design centered systems and the production support systems is available now and will be presented in section 3.

2 From Product Engineering to Enterprise Information Management

While the classical view of CAD or CAE is very narrowly dedicated towards providing specific tools to give assistance to routine tasks during product design and construction we take a broader view covering the whole process from the product idea to the final fabricated product instances. We use the term "product engineering" for this process oriented view of product development and fabrication (see 2.1).

A product engineering methodology for MEMS/NEMS products is currently developed within the EU funded project CORONA[1] presented in section 2.1. One major result from the research work in CORONA is that a holistic approach to managing the large amount and the wide variety of data that arise during the engineering process is the key to successfully giving IT support to the product engineering process. This aspect is covered by section 2.3.

2.1 Product Engineering – Definitions

To give a common-sense definition of what is meant by product engineering the definition of this term given by Wikipedia is quite useful [3]. An excerpt of this definition covers our view very clearly:

- "Product engineering is an engineering discipline that deals with both design and manufacturing aspects of a product.
- Product engineering refers to the process of designing and developing a device, assembly, or system such that it be produced as an item for sale through some production manufacturing process.
- Product engineering usually entails activity dealing with issues of cost, produceability, quality, performance, reliability, serviceability and user features. [...] It includes design, development and transitioning to manufacturing of the product. [...]
- After the initial design and development is done, transitioning the product to manufacture it in volumes is considered part of Product Engineering"

The two most important aspects for an appropriate methodology to support the product engineering process a very clearly addressed by this definition:

- Product engineering covers both design and manufacturing aspects and has a clear emphasis on the transition from the design phase to the manufacturing phase.
- Product engineering is strongly influenced by non-functional constraints like cost, quality, performance, reliability, etc.

The second aspect, taking into account non functional constraints, is the major issue with current approaches to support product engineering, where these constraints are usually not taken into account appropriately. So we adhere to the following more tight definition of product engineering:

- **Product Engineering** is the process of product development with special emphasis on taking into account non-functional requirements.

Fig. 1 shows this specific view of product engineering. A first comprehensive approach to cover this aspect of product engineering is covered in the CORONA project presented in the next section. Here cost and time-to-market as well as produceability are the central classes of non functional requirements handled.

---

[1] The CORONA Project is funded by the EU under grant No. CP-FP-213969-2.

Fig. 1. Product Engineering with non functional requirements

2.2 Distributed Product Engineering – The CORONA Vision

This section will give a short introduction to the research work done in the CORONA project, (for more details on CORONA refer to e.g. [4], [5]). This will provide a good motivation to understand the increasing importance of enterprise information management (EIM) for modern high-tech products.

The area on MEMS/NEMS is different from the "classical" microelectronics field or other areas of high-tech which are mostly dominated by large companies that cover the whole process from the idea of a product to its final volume fabrication in one house. In MEMS/NEMS a regime of distributed operation of several companies, each specialized on a specific part of the overall product engineering chain is employed.

CORONA Product Engineering is based on a generic product engineering methodology and a variety of software tools to assist the product engineer in employing the methodology. Fig. 2 gives an overview of CORONA product engineering.

The generic product engineering methodology is based on an approach that has originally been invented by Ortloff in [6] for fabless MEMS companies. The methodology realizes a combination of a StageGate (for details refer to [7] and [8])product development methodology with a PRINCE2(for details refer to [9] and [10])project management regime. The Ortloff methodology has successfully been applied to various industrial MEMS development projects [11].

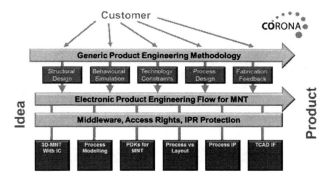

Fig. 2. The CORONA vision

In CORONA this methodology has been extended by a different view of-StageGate gates that allows a fine grained distribution of the product engineering flow among an arbitrary set of distributed cooperating partners.

The execution of the product engineering methodology is supported by a comprehensive IT infrastructure, the Electronic Product Engineering Flow that guides the different users in the various cooperating partner companies or departments through their respective subsequences of the overall product engineering process.

A distributed product engineering process is in general executed making use of a public network infrastructure. A major issue of concern in this context is access rights and IPR protection. Within the CORONA software a specific middleware (Distributed Project Binder) takes care of this kind of issues.

For all stages of the overall product engineering procedure the CORONA software provides interfaces to link into appropriate sets of design and production support tools.

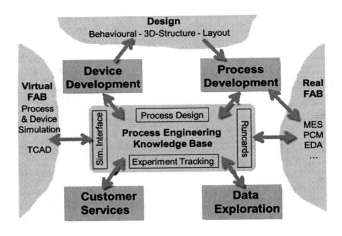

Fig. 3. The CORONA software architecture

Fig. 3 gives a different view of the CORONA software architecture from the product engineers' perspective. The CORONA methodology allows with one set of tools either to perform a virtual fabrication by extensive use of device and process simulation tools for MEMS products or to control real fab operation by linking into the appropriate MES software.

The user is given access to device development tools that assist in the product design stage. As in MEMS a strict one-product-one-process regime must be followed, for every product an adapted (manufacturing) process development must be provided.

Enterprise Information Management 273

The CORONA software includes four classes of tools centered around a complex process engineering knowledge base:

- Device development tools assist the user in the development of his target product. This includes all sorts of simulation tools as well as the known design automation software systems. The CORONA software suite provides interfaces into many common tools of this class.
- Process development tools assist in designing an appropriate manufacturing process specifically suited to the product to be realized. PDES-tools like the one presented later in this article as well as tools interfacing between device development and process development are in this class of tools. Tools for automatically synthesizing manufacturing process flows from device layout data are one example of such tools currently developed at Siegen university [12], [13].
- Data exploration tools allow a back annotation from real fab data collected by MES systems into the product and process design stage. In this manner the current state of a manufacturing plant can be taken into account during product and process development.
- Customer services allow the customer to take charge of the overall product development process by providing assistance in data and IPR management for a distributed product development procedure where an arbitrary number of locally distributed parties may be involved.

In the heart of this tool suite there is a complex process engineering knowledge base that contains all data required to perform the product engineering process and that provides furthermore interfaces to all relevant tools.

2.3 From Product Engineering to Enterprise Information Management

As we have seen in the previous section in the heart of the product engineering process there is a huge amount of data (the process engineering knowledge base) of various kinds that must be managed in an intelligent manner.

This data can be divided in three categories:

- *Hard data from manufacturing*: Manufacturing execution systems permanently collect data from the running production facilities. This includes settings of all machines, metrology results from the material in production, environmental monitoring data, etc. In modern semiconductor fabs gigabytes of data of this type per week are collected.
- *Project management and controlling data*: This is front end data from the company management, concerned with project and company economics and monitoring of the project operations.
- *Informal knowledge and experience*: This is data from former projects. Recipes, rules of thumb and other informal data that have been collected from earlier projects and that give hints on how future projects might be performed successfully. There is frequently little structure and mostly no

formalized method of storing and transferring this sort of knowledge. From handwritten notes to orally communicated hints everything seems possible here.

Currently there is no complete and comprehensive framework to keep and manage all three kinds of knowledge. Whereas project management and controlling data is handled in ERP systems, and manufacturing data is managed in MES systems, there is no link between these two.

Product Engineering Knowledge Management links the front end (management oriented) information related processes to the back end (engineering oriented) information related processes. In the area of MEMS/NEMS development a first approach to close this gap has been proposed by ProcessRelations GmbH, Dortmund, based on an experimental system developed at Siegen University [14].

The resulting Process development execution system (PDES) closes the gap between front end and back end information processing (see. Fig. 4) and opens a way to easily transform experience and informal knowledge into a form that helps to handle it in a formal manner.

Fig. 4. PDES for EIM

In this manner PDES along with ERP and MES systems realizes a comprehensive Enterprise Information Management (EIM). For the area of high-tech manufacturing process development with an emphasis on MENS/NEMS process the XperiDesk system is a first commercially available PDES tool suite.

3 The XperiDesk Tool Suite

The XPeriDesk tools suite from ProcessRelations GmbH in Dortmund, Germany is the first commercially available PDES system for high-tech processes. Fig. 5 shows the development circle for manufacturing processes that is supported by XperiDesk. The following sections give a brief introduction into XperiDesk use in the context of EIM. For more details on the overall capabilities and usage scenarios refer to e.g. [15], [16].

3.1 The XperiDeskDevelopment Circle

Starting from the idea of a new product first product design is performed. From the result of this stage assisted by knowledge and experience from previous development projects a new recipe for manufacturing is developed. In the heart of XperiDeskthere is a 3-stage verification process for this new recipe that will be presented in section 3.2.

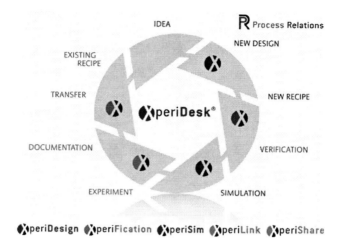

Fig. 5. The XperiDesk circle

The results of the verification process that after several iterations will lead to a new consolidated manufacturing recipe will be thoroughly documented. XperiDesk provides a comprehensive data management that allows storing and linking together nearly arbitrary types of documents. In this manner even informal knowledge and experience data can be handled.

The new and validated recipe will then finally be transferred into real production and will be available as a new existing recipe for the current and for other future product development projects.

3.2 Stage Process Verification

The development of a new process sequence to be used for a specific product is performed using the XperiDesign module. Based on an extensive knowledge base of existing process steps along with their specific properties and parameters a proposed complete process sequence is developed. This is currently performed manually making use of the XperiDesign process editor. In the future this step can be performed (semi-) automatically using the process synthesis tools currently under development in the CORONA project.

The result of this design step is a tentative process sequence that will be transferred to the verification process.

Process verification is then performed in three stages as shown in Fig. 6.

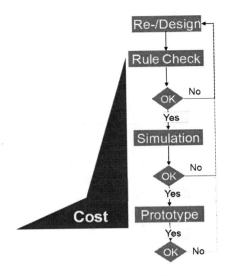

Fig. 6. The XperiDesk 3-stage-verification process

The first verification stage is performed by XperiFication. It is a verification of principle manufacturability based on a consistency check of the process sequence. This consistency check is basically a rule check. Consistency rules are associated with all process steps in the process knowledge base. XperiFication checks, whether for the whole process sequence all of these rules are obeyed. Inconsistencies detected during this process are marked as shown in Fig. 7. This is a very fast verification step; however, it will detect many problems with process designs that previously could only be detected by costly experimental production.

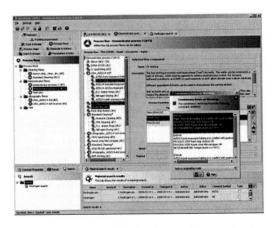

Fig. 7. Process verification by consistency check

Enterprise Information Management

The second verification stage is performed by XperiSim. The verification method here is simulation. XperiSim is not a simulator in its own right but a simulation interface that links into arbitrary device and process step simulators. Unlike most of these tools that can only simulate single aspects of devices or process steps, XperiSim is capable of automatically setting up simulation runs for a complete process step sequence.

The final verification stage is based on design of experiments. The XperiLink module sets up experiments for real fab runs of the recipe that is represented by the process step sequence given. It links into the MES and allows comprehensive analysis and comparisons of the experimental fab runs.

The advantage of this 3-stage-procedure of process verifications is in the fact that the more expensive verification stages 2 (simulation) and 3 (fab experiment) are only performed very rarely, as most of the process design inconsistencies and errors can be detected in the initial consistency checking stage. As back annotation from the real fab runs performed can be used to continuously improve the rule base used for consistency checking the power of this stage tends to continuously improve.

4 Conclusions

In many of today's high-tech areas functionality of technical devices is provided by structured nanoscaled layers. Nanofabrication in these fields is complex, only feasible by indirect chemical and physical procedures, and extremely prone to various kinds of parameters and environmental influences. To keep track of all these influences requires a huge amount of very diverse information to be handled.

In all common nano fabrication areas there is currently no comprehensive enterprise information management available. There are tools and methods however for front end management oriented information and for back end production-oriented information, respectively. A link is required to couple front end to back end enterprise knowledge management. We could show here that process development execution systems (PDES) can serve as such a link.

In this article we presented XperiDesk. It is the first comprehensive PDES suite adapted to MEMS/NEMS technologies. It turned out to be capable to provide the link between and towards nearly all specific EIM related tools for micro and nano fabrication. In this manner XperiDeskcould help realize a novel distributed product engineering methodology for MEMS/NEMS that is currently being implemented in the EU-funded CORONA project.

XperiDesk is available from Process Relations GmbH, Dortmund, Germany. My special thank is dedicated towards Dr. Dirk Ortloff from ProcessRelations who provided me with valuable material on XperiDesk and its applications.

References

1. Madou, M.: Fundamentals of Microfabrication, 2nd edn. CRC Press, Boca Raton (2002)
2. Garrou, P., Bower, C., Ramm, P. (eds.): Handbook of 3D Integration, vol. 1 & 2. Wiley-VCH, Chichester (2008)

3. Wikipedia issue 14, 23:58 (June 2010),
   `http://en.wikipedia.org/wiki/Product_engineering`
4. Ortloff, D., Popp, J., Schmidt, T., Brück, R.: Process Development support environment: a tool suite to engineer manufactoring sequences. In: Int. J. Computer Mater. Sci. and Surf. Eng. 2009 (IJCMSSE 2009), vol. 2, pp. 312–334 (2009)
5. Hahn, K., Schmidt, T., Mielke, M., Ortloff, D., Popp, J., Brück, R.: Comprehensive design and process flow configuration for micro and nano tech devices. In: Proc. of SPIE: Nanosensors, Biosensors and Info-Tech Sensors and Systems, vol. 7646, pp. 337–340 (2010)
6. Ortloff, D.: Product engineering for silicon-based MEMS IP, doctoral thesis, Siegen University (2006),
   `http://dokumentix.ub.uni-siegen.de/opus/volltexte/2006/228/index.html`
7. Cooper, R.G.: Product leadership – Creating and Launching Superior New Products, 1st edn. Basic Books (1998)
8. Product Development Institute. Helping Innovators Innovate (June 2005),
   `http://www.prod-dev.com/products.shtml`
9. Great Britain Office of Government Commerce, Managing Successful Projects With Prince 2, Stationery Office Books; 5th Revised edition (2005)
10. Spoceltd. PRINCE2 Introduction (2005),
    `http://www.spoce.com/PRINCE2%20P2Introduction.htm`
11. Ortloff, D., Veenstra, B., Coojmans, F.: A Systematic Approach Towards Reproducibility and Tracking of MEMS Process Development. In: Proceedings of the 10th International Conference on the Commercialization of Micro and Nano Systems, Baden-Baden (2005)
12. Schmidt, T., Hahn, K., Brück, R.: A Knowledge Based Approach for MEMS Fabrication Process Design Automation. In: 33rd IEEE/CPMT International Electronics Manufacturing Technology Symposium 2008, Penang (2008)
13. Schmidt, T., Mielke, M., Hahn, K., Brück, R., Ortloff, D., Popp, J.: A visual approach on MEMS process modeling using device cross-sections. In: Proceedings of Microtech 2010, Anaheim (2010)
14. Wagener, A., Schmidt, T., Popp, J., Hahn, K., Brück, R., Ortloff, D. (eds.): Proceedings of SPIE: Micromachining and Microfabrication Process Technology XI, Photonics West 2006, Process Design and Tracking Support for MEMS, vol. 6109 (2006)
15. Ortloff, D., Popp, J.: The challenges of statistical experiment design in process development. In: Proceedings of NSTI TechConnect World (2010)
16. Ortloff, D., Popp, J.: Development Support for Manufacturing Process Design. GSA Forum 15(1) (March 2008)

An Interdisciplinary Study to Observe Users' Online Search Tasks and Behavior for the Design of an IS&R Framework

I-Chin Wu

Department of Information Management, Fu Jen Catholic University, Taipei, 242 Taiwan
Icwu.fju@gmail.com

Abstract. The Web is the world's largest data repository, and search engines are regarded as key tools for finding and extracting useful information from the tremendous amount of available data. In recent years, user-oriented Information Seeking (IS) research methods rooted in the social sciences have been integrated with computer science-based Information Retrieval (IR) approaches to capitalize on the strengths of both fields. The concept is called the Information Seeking and Retrieval (IS&R) framework. Pilot research models in the IS research area show that workers' information seeking activities exhibit common patterns. In this study, we will systematically identify and categorize important theoretical frameworks of IS&R. Based on the observations of IS&R studies, we propose implicit and explicit evolving topic-needs determination methods. The methods consider personal factors, and content factors simultaneously in order to fulfil the user's evolving information needs precisely. The research contributes to design the retrieval functions based on the IS&R models for supplying documents for knowledge-intensive tasks.

Keywords: Evolving topic-needs, Information search process, Information seeking and retrieval, Knowledge retrieval.

1 Introduction

Seeking and retrieving information is an essential aspect of knowledge workers' behavior during problem-solving and decision-making tasks. It is generally agreed that seeking and retrieving information is not an easy task; and in the long run, it can be a complex process for workers during a task's execution (Bates, 1989; Hider, 2006; Kuhlthau, 1993). Knowledge workers are often unable to express their information needs precisely in short query terms (Pons-Porrata et al., 2007). Normally, users seek information from a variety of sources to satisfy their needs, which are often vague initially and evolve during the search process (Bates, 1989; Kuhlthau, 1993). Bates' well known "berrypicking" model (1989) is based on the notion that that users may start with a broader topic, adjust their information needs based on the pieces of information they encounter and then form a new query.

A user may repeat an information retrieval activity and revise a query several times until his/her information needs are fulfilled. Hider (2006) observed that users may change the search goal, either consciously or unconsciously, while they are interacting with systems. Accordingly, he proposed an IR model based on search goal revision that provides a perspective of user–system interaction.

IR can be regarded as a process of IS, so the latter may include one or more information retrieval activities (Ingwersen and Järvelin, 2005; Wilson, 1999). In the past, researchers emphasized various aspects of computer science and designed algorithms for information retrieval systems (IRS). Interestingly, there has been much less interest in using Information Retrieval (IR), Information Seeking (IS) and Computer-human Interaction (CHI) (Belkin et al., 2004; Ingwersen, and Järvelin, 2005; Kelly et al., 2009) to design IRSs that satisfy users' information needs. Recently, however, user-oriented Information Seeking (IS) research methods rooted in the social sciences have been integrated with computer science-based Information Retrieval (IR) approaches to capitalize on the strengths of both fields. The concept is called the Information Seeking and Retrieval (IS&R) framework. In this paper, we present evolving topic-needs determination methods based on a synthesis of IS&R models proposed in previous studies (Byström and Järvelin, 1995; Kuhlthau, 1993; Vakkari, 1998, 2001; Wang and Soergel, 1998; Wang and White; 1999).

The proposed model observes the user's search behavior pattern to determine his/her cognitive status from a macro viewpoint (i.e., changes in problem stages) and a micro viewpoint (i.e., changes in the topics needed across problem stages) simultaneously. The objective is to capture changes in the user's information needs precisely so that evolving information needs can be satisfied in a timely manner. We propose a multi-level, topic-based taxonomy generated from our domain to determine variations in a worker's topic needs and help him/her locate needed information in terms of topics. Evolving topic-needs across problem stages can be identified implicitly by indicators of "generality" and "specificity" based on variations in the proposed taxonomy. The indicators have different functionalities in different problem stages, which influence the final interactive topic-needs identification process. Based on the user's current profile and problem stage, the system shows the relevant topics in the topic-based taxonomy and interactively maps his/her information needs to the specific level of topics. In this way, the system can provide relevant documents in the long-term based on the proposed evolving topic-needs determination method and the results of the interactive topic-needs identification process.

2 Overview of the Information Seeking and Retrieval Model

In this project, we conducted a longitudinal case study to understand researchers' Information Seeking and Retrieval (IS&R) problem-solving behavior. IS&R studies focus on observing users' search behavior and provide useful suggestions for the design of intelligent IR systems for academic researchers. Our objective is to provide an effective method for supplying documents in professional task contexts. In the Information Search Process (ISP) framework in Table 1, we find that users' information seeking activities exhibit common patterns and we identify key factors to determine users' information needs. Figure 1 shows our research model,

An Interdisciplinary Study to Observe Users' Online Search Tasks 281

which is synthesized from the models listed in Table 1. Our goal is to deliver relevant and pertinent information (Kuhlthau, 1993) that facilitates the execution of professional tasks.

The types of tasks in Figure 1 include daily life asks, genuine work tasks or general search tasks. Our task definition is closer to that of a genuine work-task, which requires problem and task solving knowledge as well as search task solving knowledge to execute complex, long-term tasks (Ingwersen and Järvelin, 2005, p.287). Based on the observations of previous studies (see Table 1), our research model considers personal factors (e.g., the user's problem stage) and content factors (e.g., the evolving topics needs in the research domain), simultaneously. Our rationale is that both types of factors influence the user's behavior in selecting and using documents during a task's performance.

Table 1. Information Search Process (ISP) Models

Research	Search Process	Observations	Information Types	Data Collection	Research Purpose
Kuhlthau's model (1993)	Task Initialization Topic Selection Prefocus Exploration Focus Formulation Information Collection Search Closure	Feelings Thoughts Actions	Relevant Information Pertinent Information	Observations Questionnaires	Understand students search process
Wang and Sogergel (1998)	Selecting Reading Citing	Decision criteria and rules	Content information situational information	Observations Questionnaires	Design intelligent document selection assistant
Vakkari's theory (1998)	Pre-focus Pre-focus Pre-focus Focus formulation Post-focus Post-focus	Search terms Operator types Search tactics	General background information Faceted background information Specific information	Observations Questionnaires	Understand human search process
This research	Pre-focus Pre-focus Pre-focus Focus formulation Post-focus Post-focus	Changes in stages, evolving topic-needs	General topic Specific topic	Automatic system tracking, interactive with humans (Interface support)	Supply relevant and pertinent documents

Following Vakkari (1998, 2001), we divide the search process into three stages: pre-focus, focus formulation, and post-focus stages. Clearly, Vakkari's task-based information retrieval theory can be implemented by an information retrieval system more easily than Kuhlthau's ISP model. Furthermore, the system will show the portion of the topics in the topic-based taxonomy based on a user's particular problem stage to help him/her evaluate topic-needs by interactively mapping information needs to the specific level of topics in the taxonomy.

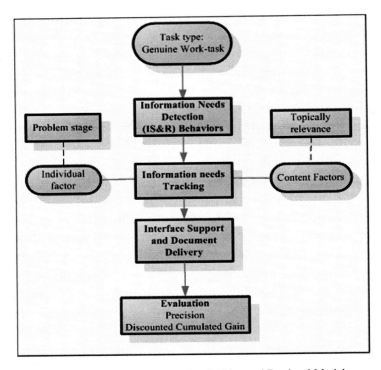

Fig. 1. The Proposed Information Seeking and Retrieval Model

3 Supporting Search Activities Based on a User's Evolving Topic-Needs

3.1 Overview of the Proposed Approach

We propose an *evolving topic-needs determination technique* for identifying a user's up-to date information needs. The steps of the proposed approach are as follows.

Exploration of content and usage information: To gather data about workers' search behavior, i.e., usage information, we observed how they used the proposed system and recorded their feedback (e.g., search behavior and relevance feedback on knowledge items) To track a user's access behavior pattern over time, we model his/her information needs as a set of topics and associated relevance degrees via the topic taxonomy. The topic taxonomy is extracted from a set of documents and expressed as a hierarchy of topics and subtopics. The change degree in topics is determined by the indicators of "generality" and "specificity" associated with the topics in the different levels of the taxonomy.

Determining evolving topic-needs: In this work, we use a topic-based information search method to overcome the limitations of keyword search methods. It is easier to express information needs by identifying topics rather than by keyword-based queries (Pons-Porrata, 2007). We propose an on-line problem stage identifier that determines a worker's problem stage by analyzing his/her access pattern based on topics in the research domain. Since our objective is to identify variations in a worker's topic-needs, we label the level of his/her topic-needs changes with indicators of "generality" and "specificity" to select general and specific topics of interest. In the taxonomy, the leaf nodes, i.e., task-level topic nodes, represent specific task topics; and the non-leaf nodes, i.e., field-level topic nodes, represent general topics, as shown in Figure 2. The system also adjusts the relative importance of general and specific topics based on the user's problem stage.

Identifying users' topics of interest interactively: The system provides an explicit feedback mechanism to help users identify topics of interest interactively. Specifically, we identify a user's precise topic-needs by interactively mapping his/her information needs to the specific level of topics in the taxonomy. Initially, the system only shows the portion of the topic taxonomy relevant to the value of the indicators for the associated topics and the user's current problem stage. The steps of the basic algorithm are detailed in Figure 5, Section 3.5.

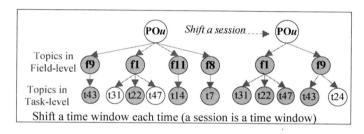

Fig. 2. The Evolution of a User's Topic Ontology over Two Sessions

3.2 Search Behavior Tracking

To identify top-relevant topics(*TRTs*), we selected the top-k task-relevant topics based on the similarity between the user's temporal search profile and topics in the taxonomy. Let *TRTWs* denote the set of top-k relevant topics with the associated weights derived by similarity calculation. *TRTWs* are expressed as a set of (*task_id, relevance degree*) pairs. The *TRTWs* with their associated degrees of relevance are recorded in the user's profile to model his/her information needs over time.

Example 1: A set of *TRTWs* with their associated weights is derived by similarity calculation and expressed as (*task_id, relevance degree*) pairs. The top-4 relevant topics of each session in transaction 2 are listed below. A session and a task transaction are used to analyze a user's feedback on knowledge items, i.e., documents or notes. A session is defined as a sequence of user feedback actions related to knowledge items during a single visit to the system. A task transaction records the number of times the user accesses the knowledge repository over several sessions. The time interval between two consecutive upload events (i.e., documents the worker wants to share in the system) is defined as a task-transaction in the investigated domain.

$$TRTWs^t (Trans_2^{S1}) = \{(t_{14}, 0.340), (t_{43}, 0.301), (t_{22}, 0.292), (t_7, 0.291)\}$$

$$TRTWs^t (Trans_2^{S2}) = \{(t_{43}, 0.727), (t_{22}, 0.589), (t_{47}, 0.488), (t_{31}, 0.466)\}$$

$TRTWs^t (Trans_i^{S_j})$ denotes the top task-relevant topics in session j of transaction I; and the superscript t in $TRTWs^t (Trans_i^{S_j})$ represents the task-level topics in the topic-based taxonomy. We express the relation between the top task-relevant topics and the field-level topics in the topic-based taxonomy as (*field_id, relevance degree*) pairs. For example, task 22, task 31 and task 47 belong to field *1*, as shown in Figure 2. Therefore, the top task-relevant topics can be aggregated to form the following field-levels:

$$TRTWs^t (Trans_2^{S1}) = \{(f_{11}^*, 0.340), (f_9^*, 0.301), (f_1^*, 0.292), (f_8^*, 0.291)\}$$

$$TRTWs^t (Trans_2^{S2}) = \{(f_9^*, 0.727), (f_1^*, 0.589), (f_1^*, 0.488), (f_1^*, 0.466)\}$$

From the above example and Figure 2, it is clear that the user's information needs focus increasingly on topic *t43*, i.e., Applying Topic Maps and Data Mining to Deploy Composite E-service Platform and topic *t22*, i.e., Towards a Framework for Discovering Project-Based Knowledge Maps. As most of the task-relevant topics belong to the field, *f1*, it appears that the user's information needs in the field focus increasingly on the topics in *f1*, i.e., Information Retrieval and Organization Impact, as shown in the figure. Meanwhile, the user has two new emergency topic-needs for *t31* and *t47*, which belong to *f1*. The above example is a simple case that explains the usage pattern modeling process. Based on the proposed idea, the system can track and identify the user's task-needs on specific topics from different abstraction levels of the multi-level topic-based taxonomy.

3.3 Personal Ontology Formulation

As mentioned earlier, the evolution of a user's information needs can be determined by examining the variety of topics selected from the topic-based taxonomy. Once the top task-relevant topics with their associated weights (*TRTWs*) have been identified, a user's personal ontology for a specific transaction can be formulated intuitively. Each user's information needs are represented by a personal ontology, Ψ_u. *Definition 1* defines the personal topic profile of a user u. We represent the user's information needs in terms of topics in the research domain, instead of using keyword sets. The rationale is that topics provide a more expressive and less abstract means of representing a user's information needs.

Definition 1: The topic ontology of a user u, denoted by $\Psi_u = \{<topic_j, w_p(topic_j)>\}$, contains a set of topics (field- or task-level nodes in the taxonomy) with their associated degrees of relevance to the target task in a specific time period; $w_p(topic_j)$ represents the relevance degree of $topic_j$ to the target task at time p, from the perspective of u. The associated degree of relevance indicates the similarity between a topic and the target task in a specific time period. Let *FS* denote the set of topics at the field level, and let *TS* denote the set of topics at the task level. An ontology threshold value δ can be defined by the user to generate his/her personalized ontology for the target task by filtering out irrelevant fields or task topics whose values are lower than the threshold. Accordingly, $\Psi_u = \{< topic_j,, w_p(topic_j,)> | w_p(topic_j,) \geq \delta$ and $topic_j, \in FS \cup TS \}$. The result forms the user's personalized topic ontology for the target search task. Figure 3 shows a user's personal topic ontology in a transaction based on Example 1. The coloured nodes in the tree represent the topics that are relevant to the user's current information needs, i.e., transaction i.

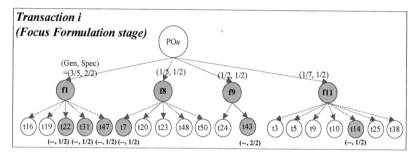

Fig. 3. A User's Topic Ontology for a Specific Transaction

3.4 Determining Evolving Topic-Needs at Different Abstraction Levels of Topics

We use a top-down examination procedure with two indicators to filter and extract the user's specific topic-needs and update his/her personal ontology. First, we check the nodes at the field level to assess the generality of the topic-needs; then we check the nodes at the task level to determine the specificity of the topic-needs. The output of the procedure categorizes a user's information needs for specific topics with "generality" and "specificity" indicators. Figure 3 shows an example of a user's personal topic ontology with two indicators for a specific transaction.

Generality of relevant topics: The higher the generality value, the greater will be the user's interest in the topics belonging to the same field. We calculate the generality of information-needs for topics across sessions within the same transaction. A field-level topic may include one or more task-level topics; therefore, the generality is the ratio of the top task-relevant topics at the task level to all nodes in a specific field-level topic, as shown in Equation (1).

$$Gen(f_l)_{field-level} = \frac{N_{TRT}^{f_l}(Trans_i)}{N^{f_l}} \quad , \tag{1}$$

where N^{f_l} denotes the number of task-level topics belonging to field l in the topic-based taxonomy; and $N_{\tau\tau\tau}^{l}(Trans_i)$ is the number of distinct task-level topics belonging to field l and the *TRTW*s of transaction i.

Specificity of relevant topics: The higher the specificity value of a field or set of task-level nodes, the more the worker will focus on a specific topic node. Equations (2) and (3) show the specificity of topic f_l at the field level and the specificity of topic t_k at the task level respectively.

$$Spec(f_l)_{field-level} = \frac{\sum\limits_{session\ j} B_{i,j}^{f_l}}{S_i} \tag{2}$$

$$Spec(t_k)_{task-level} = \frac{\sum\limits_{session\ j} B_{i,j}^{t_k}}{S_i}, \tag{3}$$

where S_i is the number of sessions in a transaction i. $B_{i,j}^{f_l} = 1$ if f_l is a top relevant topic of session j in transaction i; otherwise it equals 0. Similarly, $B_{i,j}^{t_k} = 1$ if t_k is a top relevant topic of session j in transaction i; otherwise it equals 0. Note that the system uses *TRTWs(Trans)*, described in Section 4.2, to determine whether a topic $(t_k$ or $f_l)$ is a top relevant topic of transaction i. The summation of $B_{i,j}^{t_k} / B_{i,j}^{f_l}$ indicates the number of sessions in which the topic $(t_k$ or $f_l)$ is a top relevant topic. The specificity of a topic $(t_k$ or $f_l)$ represents the ratio of top-relevance occurrences of the topic in the sessions of a transaction.

3.5 Identifying Personal Topic-Needs Interactively

To map a user's information needs to the specific level of topics in the topic-based taxonomy interactively, the system presents general or specific topics to users based on the results reported in the previous section. Figure 4 shows the procedure for identifying a user's topic-needs interactively by using stage information.

The function of the identification procedure is twofold: (1) to support search activities by determining variations in topic-needs and problem stages, as shown Steps (1) and (2) in Figure 4; and (2) to update the user's personal ontology, $\mathbf{\Psi}_u$, based on his/her feedback on the topics, as shown in Step (3). The variable, *I_Topic* is an array that stores the user's topics of interest derived from the user feedback on the topics during the interactive topic identification process. If the user is in the task pre-focus stage, the system checks whether he/she has general or specific topic-needs. The system will then display the appropriate part of the topic taxonomy to fulfil the user's information needs. It will also update the user's $\mathbf{\Psi}_u$, to help the user conduct future searches. The mechanism enables the user to expand his/her topic-needs if the information needs are vague (i.e., in the task pre-focus stage), as shown in the Step (1). As the task progresses, the mechanism guides the user to identify specific topic-needs (i.e., Step (2) in Figure 4).

3.6 Profile Adaption, and Document Supply Strategies

We designed strategies to supply task-relevant documents based on the user's evolving topic-needs throughout the work-task information seeking process. The user's problem stage, the variation in topics of interest and the interactive topic-mapping process influence the profile adaptation process. From a technical perspective, the system delivers task-relevant documents based on the topic identification results, after which the new task profile of the target task, denoted as $\vec{S}_{p+1}$, is generated by Equation (5). The equation considers the user's problem stage as well as the *"generality"* and *"specificity"* of the topics of interest.

$$\vec{S}_{p+1} = \alpha \vec{S}_p + \lambda \vec{R} + (1-\lambda)\vec{Temp}_{u,p}$$
$$\vec{R} = w_{gene} \sum\nolimits_{\forall t_j \in Gene.\ topic} Gene(topic_j) \times w_{p+1}(topic_j) \overrightarrow{topics}_j +$$
$$w_{spec} \sum\nolimits_{\forall t_j \in Spec.\ topic} Spec(topics_j) \times w_{p+1}(topic_j) \overrightarrow{topics}_j \qquad (4)$$

Input:
(a) **Gen(f_l), Spec(f_l), Spec(t_k)** : The user's personal ontology with "generality" and "specificity" indicators
(b) **Ψ_u**: The user's personal ontology in *Trans$_i$*
(c) **Trans$_i$.stage** : The problem-stage of the *i*th transaction
(d) **θ, δ**: The threshold of generality or specificity
Output:
(a) Show and update the **Ψ_u** to provide user feedback on relevant topics
(b) Update the user's task profile $\vec{S}_p$ to retrieve relevant documents
Begin
(1) If *Trans$_i$.stage* in the pre-focus stage then
(1.1) **If** *Spec(t_k)*>= δ **then**
    // The user has specific task-needs, but is in the pre-focus stage
    (1.1.1) List the t_k 's related topics in the task-level to help the user identify topics of interest, *I_Topic*
    (1.1.2) Set all topics in the *I_Topic* = $\delta\delta$
(2) **Else if** *Trans$_i$.stage* is not in the pre-focus stage **then**
(2.1) **If** (*Gen(f_l)* >θ and *Spec(f_l)* < δ) then
    //The user does not have specific field level topic-needs.
    (2.1.1) List the topics in the task-level of *field l* to help the user identify topics of interest, *I_Topic*
    (2.1.2) Set all topics in *I_Topic* = $\delta\delta$
(2.2) **Else if** (*Gen(f_l)* > θ and *Spec(f_l)* >= δ) **then**
    // The user has specific field level topic-needs, but not specific task level topic-needs.
    (2.2.1) If there are no topics in *field l* with *Spec(t_k)*>= $\delta\delta$ then
        (2.2.1.1) List the topics in the task-level of *field l* to help the user identify topics of interest, *I_Topic*
        (2.2.1.2) Set all topics in *I_Topic* = δ δ
Else No Action
(3) Update Ψ_u and $\vec{S}_p$
End.

Fig. 4. Procedure for Updating a User's Topic Ontology by Identifying Topic-needs Interactively

where $\vec{S}_p$ denotes the task profile of the target task at time p. The task-relevant feature vectors are derived from the task corpora of relevant tasks sets, while the temporal file, $\vec{Temp}_{u,p}$, isgenerated based on feedback analysis. In addition, $w_{p+1}(topic_j)$ denotes the associated relevance degree of task t_k or field f_l to the target task; $Gene(topic_j)$ and $Spec(topic_j)$ are derived from the evolving topic analysis, as mentioned earlier; and α, and λ are tuning constants. The parameter α is the value of the correlation between transactions $corr_u(Patt_{Trans_{i-1}}, Patt_{Trans_i}^{S_1})$,as shown in Equation (1). The parameter λ is used to adjust the relative importance of relevant topics and the temporal search profile.

4 Experiment Setup and Results

4.1 Experiment Procedure

We conducted experiments to assess the effect of a user's topic-needs and that of interface support across different problem stages. The five methods are the *Linear-0*, *Topic*, *Topic (G, S)*, *Stage-Topic (G,S)*, and *Stage-Topic-Interaction* methods. The *Linear-0 (baseline) method* is our baseline method. It is based on the traditional relevance feedback (RF) technique in the vector space model (Salton and Buckley, 1990). The technique is widely used in information filtering studies to learn a user's current information needs from feedback about recommended information (i.e., documents), and updates the user model for future information filtering. The *Topic* method is similar to the traditional *incremental learning technique* in that it also considers a user's implicit feedback on relevant topics, i.e. the parameter λ is set to 0.5. Our proposed *Topic (G, S)* method is based on the *Topic* method, but it also considers the degree of topic-needs variation in the incremental learning process to adjust the user's profile. That is, the profile adaptation equations consider the generality value, $Gen(f_l)$, and the specificity values, $Spec(f_l)$, and $Spec(t_k)$. Finally the *Stage-Topic-Interaction method* identifies a user's topic-needs interactively by considering the user's problem stage and variations in topic-needs. The system then adjusts the user's personal ontology and the relevance degrees of topics based on the user's feedback results.

4.2 Evaluation Metrics

Generally, users do not care if an IR system can retrieve all relevant documents; however, it is important that users can get highly relevant results in a few documents. Hence, the recall rate may not be important in an interactive or web-based information retrieval system. In this work, we use the precision rate and the discounted cumulated gain (DCG) to compare the performance of topic-based and stage-topic based methods.

Precision: The precision rate for an evaluation task e_r is the ratio of the total number of relevant documents retrieved to the number of top-N support documents in the presented system.

An Interdisciplinary Study to Observe Users' Online Search Tasks

Discounted Cumulated Gain: The Cumulated Gain (CG) and Discounted Cumulated Gain (DCG) measures were proposed by Järvelin and Kekäläinen (2002). The DCG measure evaluates an IR method's ability to retrieve highly relevant documents as a ranked list. If we sort the relevance scores (similarity values) of documents in decreasing order, the lower the ranked score of a relevant document, the more important it is for the user. The CG function, shown in Eq. (5), progressively sums the relevance scores of the documents from ranked position 1 to n; thus, each document has an associated accumulated gain value. Furthermore, the DCG function reduces a document's score as its rank increases, but not too steeply in comparison with CG function, by dividing the log of its rank, as shown in Eq. (6). In this work, we set b=2.

$$CG[i] = \begin{cases} G[1] & if \ i = 1 \\ CG[i-1] + G[i] & , otherwise \end{cases} \tag{5}$$

$$DCG[i] = \begin{cases} CG[i] & if \ i < b, \\ CG[i-1] + \dfrac{G[i]}{\log_b^i} & , i \ge b \end{cases} \tag{6}$$

4.3 Experiment Results

Table 2 shows the precision and DCG scores for each case under the four methods. The results were derived by averaging the scores of the three stages for various numbers of documents.

Table 2. Comparison of the Four Methods

Stage		Linear-0		Topics		Topic (G,S)		Stage-Topic (G,S)		Stage-Topic-Interaction	
nth Stage	*Top-N*	*Precision*	*DCG*	*Precision*	*DCG*	*Precision*	*DCG*	*Precision*	*DCG*	*Precision*	*DCG*
1st stage (Pre-focus)	Top-5	0.267	1.333	0.333	1.333	0.133	2.144	0.667	2.764	**0.733**	**3.097**
	Top-10	0.367	3.434	0.467	4.434	0.433	5.767	**0.667**	**6.201**	0.633	6.100
2nd stage (Focus formulation)	Top-5	0.267	1.000	0.133	0.667	0.037	0.144	**0.200**	0.620	**0.200**	**0.810**
	Top-10	0.233	1.767	0.333	2.867	0.300	2.534	**0.367**	**3.434**	0.333	3.100
3rd stage (Post-focus)	Top-5	0.533	2.620	0.267	1.144	0.667	2.953	0.800	2.954	**0.933**	**4.097**
	Top-10	0.600	6.100	0.333	2.201	0.567	5.767	0.567	5.201	**0.633**	**6.100**
Total Average		0.378	2.709	0.311	2.108	0.356	3.218	0.545	3.529	**0.578**	**3.884**

Observation 1: On average, the proposed *Stage-Topic-Interaction* method yields better *precision* and *DCG* scores than the other three methods. Specifically, the

proposed *Stage-Topic-Interaction* method achieves the best *precision* and *DCG* scores in the early problem stage. The *Stage-Topic (G, S)* method yields the best performance in the focus formulation stage, while the *Stage-Topic-Interaction* method achieves the best performance in the pre-focus and post-focus stages. In addition, the two stage-topic-based methods outperform the topic-based method, i.e., *Topic (G, S)*. The results indicate that considering personal factors (e.g., changes in the user's cognitive status), and content factors (e.g., changes in topic-needs) simultaneously can improve the topic-based technique and satisfy the user's evolving information needs more precisely.

Observation 2: The *Stage-Topic-Interaction* method achieves the best DCG performance on the top-5 documents in all problem stages. This result demonstrates that the interactive topic identification process helps the user by providing information search suggestions based on his/her problem stage. It also shows the users can find highly relevant documents in a small number of retrieved documents, i.e., they can find pertinent documents, without checking a lot of documents.

5 Conclusion and Future Work

Because of the rapid advances in applying search engine techniques to support enterprise search activities, users can employ search engines to find useful information and search for interesting topics when their information needs are vague or evolving. Most current user modeling approaches focus on the analysis of changes in a user's information needs in his/her daily work life, instead of considering the user's information needs for a specific long-term task. To address this issue, we propose a topic-needs variation determination technique based on an analysis of information seeking and retrieval (IS&R) theories. The evaluation results demonstrate that the proposed technique is effective in finding relevant documents because it considers a user's information needs for topics across problem stages. Among the compared models, the proposed *Stage-Topic-Interaction* method achieves the best performance on average, especially in the early and late problem stages. The results also show that the method can help users find relevant topics via an interactive topic identification process.

There is increasing demand for more effective enterprise content (document) management(ECM) systems that go beyond the basic document management and retrieval functions of current knowledge management systems (Päivärinta and Munkvold, 2005; Smith and McKeen, 2003). To address this demand and facilitate effective ECM, there is a growing trend in user-oriented information retrieval (IR) research to apply information seeking and retrieval techniques in social and organizational contexts (Ingwersen, and Järvelin, 2005; Tyrväinen et al., 2006). This fact motivated us to develop a novel IS&R model that considers the user's search pattern and the project or task contents. In the future, we will investigate the issue of collaborative information seeking support system to leverage collative intelligence for effective enterprise knowledge retrieval (Golovchinsky et al., 2009; Hyldegård, 2009).

References

1. Bates, M.J.: The Design of Browsing and Berrypicking Techniques for the Online Search Interface. Online Review 13(5), 407–424 (1989)
2. Belkin, N., Dumais, S., Scholtz, J., Wilkinson, R.: Evaluating Interactive Information Retrieval Systems: Opportunities and Challenges. In: proceedings of CHI 2004, pp. 1594–1595. ACM, New York (2004)
3. Byström, K., Järvelin, K.: Task Complexity Affects Information Seeking and Use. Information Processing and Management 31(2), 191–213 (1995)
4. Golovchinsky, G., Qvarfordt, P., Pickens, J.: Collaborative Information Seeking. IEEE Computer 42(3), 47–51 (2009)
5. Hider, P.: Search Goal Revision in Models of Information Retrieval. Journal of Information Science 32(4), 352–361 (2006)
6. Hyldegard, J.: Beyond the Search Process: Explore Group Members' Information Behavior in Context. Information Processing and Management 45, 142–158 (2009)
7. Ingwersen, P., Järvelin, K.: The Turn: Integration of Information Seeking and Retrieval Context. Springer, Heidelberg (2005)
8. Järvelin, K., Kekäläinen, J.: Cumulated Gain-based Evaluation of IR Techniques. ACM Transactions on Information Systems (TOIS) 20(4), 422–446 (2002)
9. Kelly, D., Dumais, S.T., Pedersen, J.O.: Evaluation Challenges and Directions for Information-Seeking Support Systems. IEEE Computer 42(3), 60–66 (2009)
10. Kuhlthau, C.: Seeking Meaning: A Process Approach to Library and Information Services. Ablex Publishing Co., Norwood (1993)
11. Päivärinta, T., Munkvold, B.-E.: Enterprise Content Management: an Integrated Perspective on Information Management. In: Proceedings of the 38th Hawaii International Conference on System Sciences, IEEE, Los Alamitos (2005)
12. Pons-Porrata, R., Berlanga-Llavori, R., Ruiz-Shulcloper, J.: Topic Discovery based on Text Mining Technique. Information Processing and Management 43(3), 752–768 (2007)
13. Salton, G., Buckley, C.: Improving Retrieval Performance by Relevance Feedback. Journal of the American Society for Information Science 41(4), 288–297 (1990)
14. Smith, H.A., McKeen, J.D.: Enterprise Content Management. Communications of the Association for Information Systems 11, 438–450 (2003)
15. Tyrväinen, P., Päivärinta, T., Salminen, A., Iivari, J.: Characterizing the Evolving Research on Enterprise Content Management. European Journal of Information Systems 15, 627–634 (2006)
16. Vakkari, P.: Growth of Theories on Information Seeking: an Analysis of Growth of a Theoretical Research Program on the Relation Between Task Complexity and Information Seeking. Information Processing and Management 34(2-3), 361–382 (1998)
17. Vakkari, P.: A Theory of the Task-based Information Retrieval Process: A Summary and Generalization of a Longitudinal Study. Journal of Documentation 57(1), 44–60 (2001)
18. Wang, P., Sogergel, D.: A Cognitive Model of Document Use during a Research Project. Study I. Document Selection. Journal of the American Society for Information Science and Technology 49(2), 115–133 (1998)
19. Wang, P., White, M.D.: A Cognitive Model of Document Use during a Research Project. Study II. Document at Reading and Citing Stage. Journal of the American Society for Information Science and Technology 50(20), 98–114 (1999)

A Generic Knowledge Integration Approach for Complex Process Control

Stefan Berlik

University of Siegen, D-57068 Siegen, Germany
`berlik@informatik.uni-siegen.de`

Abstract. Current manufacturing processes are characterized by their high complexity and require an increased control effort. Operating them effectively and efficiently is crucial and knowledge integration methods can make a valuable contribution to this. Presented here is a generic model predictive system that enables the integration of different sources of knowledge. In addition, the system is adaptive and allows for a self-adaptation to changing operating conditions and a self-optimization. The implementation of an inferential control mechanism finally ensures continuous process control in the absence of primary measurements.

Keywords: Model Predictive Control, Inferential Control.

1 Introduction

Numerous machines and processes for production of fibers, yarns and textile structures require a great know-how of the operating personnel. This applies to the specific choice of the appropriate machine parameters at product changes or the introduction of new products or materials. Often, expensive trial-and-error experiments are necessary to establish meaningful machine settings. Suitable machine settings are not only important for economic reasons, but also for environmental protection.

Against the background of the increasing relocation of production facilities abroad, this expertise is often locally not available any longer. Therefore, systems that help the operator in finding the optimal setting of the machinery and equipment are becoming increasingly important. For many years now systems based on artificial neural networks are developed and used industrially for this.

The big advantage of such systems lies in the ease of use. Their big disadvantage is that contexts that are analytically studied and thus quantifiable or able to be formulated in rules cannot be regarded. Thus often a significant part of the available operator's knowledge is not included in the recommended setting. In addition, the acquired knowledge is present only implicitly as a black box.

The use of systems based on fuzzy logic promise to remedy this [1]. It allows formulating fuzzy rules based on known coherencies. Through a combination of such rules the fuzzy system can make a statement on the optimal setting of the machine or system. In contrast to a prediction of an artificial neural network this prediction is immediately replicable by the operator since the knowledge is present here explicitly. Its acceptance is therefore considerably higher.

Problems often to be encountered in practice are disparate types of measurements and different rates at which they occur. Frequently parameters defining the objective function are available only sporadic due to necessary analyses from laboratory spot tests. On the other hand online measurements stemming from machine sensors are continuously available; however their expressiveness might be unknown at first. Special care has to be taken in these situations.

For a machine in the mentioned environment an advanced control system shall be developed. To be able to use both, operator's expertise and knowledge stemming from measured data a fuzzy model shall shape the core of the system. Hence it seems reasonable to use a model predictive control scheme. To enable continuous optimization of the fuzzy model an adaptive system is developed [2]. The contribution of this paper is the development of a consistent draft of such a system and the presentation of first results of its partly implementation.

This paper is organized as follows. First, some related work is sketched in the next section. The architecture and some details of the system are discussed in Section 3. Section 4 outlines the current state and Section 5 finally presents our conclusions.

2 Related Work

A modern method for the control of complex processes is *model predictive control* (MPC), also referred to as *receding horizon control* (RHC) [3,4]. Model predictive control uses a time-discrete dynamic model of the process to be controlled to calculate its future states and output values as a function of the input signals. Using this prediction, in particular suitable input signals for desired output values can be found. While the model behaviour will be predicated several steps ahead till a certain time frame, the input signal is usually only searched for the next time step and then the optimization is repeated. For the calculations of the next time step then the actual measured state is used, resulting in a feedback and thus a closed loop. Model predictive control technology offers a number of advantages that have made it one of the most widely used advanced control methods in the process industry: It can calculate control variables where classical nonlinear control techniques fail, is easily extended to multivariable problems, can take restrictions on the actuators into account, permits the operation near the constraints, allows a flexible specification of the objective function, delivers optimum control devices and is last not least model based.

Fuzzy set theory provides structured ways of handling uncertainties, ambiguities, and contradictions which made systems based on fuzzy set theory the approach of choice in many situations. Since its introduction in 1965, fuzzy set theory has found applications in a wide variety of disciplines. Modeling and control of dynamic systems belong to the fields in which fuzzy set techniques have received considerable attention, not only from the scientific community but also from industry. Their effectiveness together with their ease of use compared to systems based on classical two-valued logic paved the way to countless practical applications of fuzzy systems. Also in the textile industry fuzzy set techniques have successfully been used, e.g. in the area of textile technology, textile chemistry, textile software, and textile economy [5-8].

2.1 Fuzzy Model Predictive Control

Generally, a distinction is made between linear and nonlinear model predictive control. Linear model predictive control uses linear models for the prediction of system dynamics, considers linear constraints of the states and inputs, and a quadratic objective function. Nonlinear model predictive control is based on nonlinear models and can include non-linear constraints and / or general objective functions. When historically first linear model predictive control strategies have been examined and used, nowadays nonlinear control is of increasingly interest. A major reason for this is that many processes are inherently nonlinear and are often run over a wide operation range, linear regulators therefore ultimately achieve only a low control quality or stability problems may also arise. Particularly suitable for the creation of the nonlinear models and thus the core of the model predictive control is the fuzzy set theory.

Compared with already investigated controls based on neural networks, the use of fuzzy models for control of the machine offers significant advantages: Already existing expertise can be directly fed into the system, in contrast to neural networks the knowledge is explicitly, thus self-explanatory and the acceptance of the procedure higher.

The modeling and control of nonlinear systems using fuzzy concepts is described in [9]. Current techniques on the one hand and the expertise on the other, but also mixed forms of both approaches. Basis of the data-driven approach is a clustering algorithm whereby fuzzy models are derived, preferably of the Takagi-Sugeno type. The obtained models form the basis of the desired nonlinear control, whether using the inversion of the model or model predictive control.

Characteristic of the project are the different rates at which the output variables are measured and the input variables can be varied. Generally, processes as such are termed *dual rate systems*. In the present case, the primary variables, i.e. the yarn parameters are available solely after laboratory tests and this only sporadically. For continuous adjustment of the process they are only suitable to a limited extent. What can be used in addition then is an approach known under the name *inferential control*. Online measured secondary variables of the process are in the absence of primary values used to estimate the actual disturbances affecting the process [10].

3   System Development

Figure 1 shows the interaction of the model predictive control system and the production process together with the appertaining data flow. It can be seen that the system operates the production process via the machine settings and receives feedback in form of machine readings and yarn properties. Both have to be considered adequately to improve overall system performance.

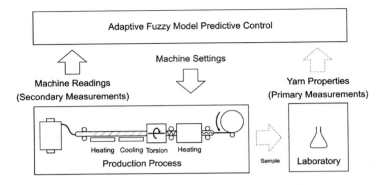

Fig. 1. Interaction of the control system and the production process

A more detailed view of the mentioned controller is given in Figure 2. On the left-hand side one can see the set point controller using the fuzzy model to find optimal machine settings together with the higher level static operating point optimization. On the right-hand side the estimator of the inferential control component is shown. Build upon the fuzzy model and the secondary measurements it estimates the disturbances on the primary variable. By feeding back the primary variable into the fuzzy model automatic adaptation of the model becomes possible. The following subsections describe the treated aspects in greater detail.

3.1 Development of the Initial Fuzzy Model

Basis of the initial fuzzy model is the acquired knowledge of a domain expert. By means of interviews, his knowledge is formulated as a linguistic fuzzy model in form of fuzzy rules and appropriate fuzzy sets. In combination these model the known relationships of the application domain. According to the expected complexity of the different control ranges of the fuzzy controller and the available data gaps in the data set are identified. Missing data sets are generated by practical experiments.

The collected data provide the alternative to automatically deduce a fuzzy model using clustering techniques - preferably in the form of Takagi-Sugeno type. After transformation of also the linguistic model into a Takagi-Sugeno model both can be checked by experts for consistency. Should significant differences appear the causes and effects are to question. Often this approach reveals previously unknown relationships and the process expert gains a deeper understanding of the process. Newly acquired knowledge is incorporated into the fuzzy model.

A Generic Knowledge Integration Approach for Complex Process Control

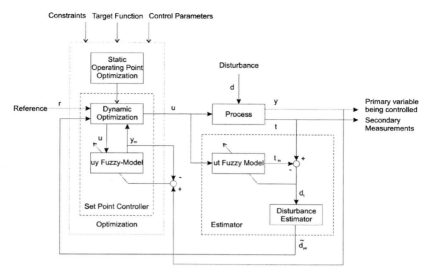

Fig. 2. Adaptive fuzzy model predictive control system with inferential control element

For fine tuning the initial system the authors developed an evolutionary optimization module based on the Covariance Matrix Adaptation.

Evolution Strategy [11,12]. The module was successfully applied in earlier projects, e.g. in the area of dyeing polyester fibers [13]. To verify that the resultant rule-based model maps the relationship between process parameters and yarn properties sufficiently precise, its forecast quality is reviewed for an independent data set [14].

3.2 Implementation of the Nonlinear Fuzzy Model Predictive Controller

Expertise in the field of yarn processing lies predominantly in the direction 'When machine setting, then yarn property', so according to the causal relationship. Consequently, the initial fuzzy system uses as inference direction a forward chaining of facts from the domain of machine settings to draw conclusion in the domain of yarn properties. To answer the more relevant question of appropriate machine settings for desired yarn properties the other inference direction is necessary. For this purpose with the model predictive control an appropriate solution is available that uses the controller structure known as *internal model control*, see Figure 3.

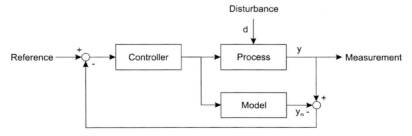

Fig. 3. Internal model control scheme

3.3 Integration of an Inferential Control

As already mentioned, the primary variables, i.e. the yarn parameters are available solely after laboratory tests and this only sporadically. For continuous adjustment of the process they are only suitable to a limited extent. Promising seems to be the additional use of an inferential system whereby an estimate of the influence of non-measurable disturbances on the controlled variables is returned to the control loop, see Figure 2. The aim of the inferential control scheme is therefore, to estimate the disturbance d even if there are no primary values y but only secondary values t available. This is a major difference to procedures such as Kalman filters that use the secondary values in order to predict the primary values and to regulate it - and thus in particular violate the feedforward nature of the inferential control.

To be solved is essentially the identification problem of the form: Find an estimator for a given set of past primary values y and secondary values t, that estimates the influence of noise d_{ye} on the primary values from the influence on the secondary values d_t. Used for this purpose is first a linear regression model.

3.4 Extension to an Adaptive Fuzzy Model Predictive Control

The fuzzy model predictive control is based at first solely on the expert knowledge and a static set of observed historical data with which it supplies yarn parameter forecasts with respect to given machine settings. The sporadic incoming laboratory results on the other hand describe the effective, current correlation of the input and output variables, so this knowledge is usefully used to the continuous adaptation of the fuzzy system. Integration of an adaptation mechanism allows for two things: First, in the steady state of the process model discrepancies can be identified and eliminated. Ideally, it is possible to reconstruct an accurate, error-free process model over time. Furthermore, the model can adjust itself to (temporarily) changing process conditions and also determine optimal control variables in these situations. For this purpose the controller structure is extended so that the difference between model and process is returned as a new control variable into the fuzzy model.

3.5 Integration of a Nonlinear Estimator and a Static Working Point Optimization

The previously integrated inferential scheme can be extended to a nonlinear estimator - similar to the automated procedure that has been used to derive the initial TS fuzzy model. Necessary data for this will have been collected at this point already during former work packages of the project. Expected is an increase in the quality of the estimates on the error signals acting on the primary values. The new estimator is evaluated and compared with the linear estimator.

In addition to the previously discussed dynamic optimization of the operating point, i.e. the search of appropriate control variables for given yarn quality, the chosen architecture allows with only slight modifications also for a static operating point optimization. While the former is more likely to be assigned to the process control, the latter belongs to the plant management level. Although it is in our case not desired to improve the output parameters by multi-objective optimization, but rather to keep them constant, the process can be optimized with respect to higher goals, such as to increase throughput, minimize energy use, et cetera. For this purpose, simply the objective function has to be modified adequately and possibly extended by new constraints.

4 Current State

The presented system is still under construction and right now only partly implemented using the technical mathematical software Matlab. However, first results have been achieved. For the problem given, a proof-of-concept has been done with a prototypic rule base containing seven rules. The results the rule base gave for a set of test data are quite promising. Also the tuning module for the rule base works stable. Next steps are to incorporate the measured data by means of a clustering algorithm as described in Section 3.1. The model predictive control algorithm can then be applied to the machine. At this preliminary stage, it corresponds to the status quo of the control of the machine. As the feedback of the measured yarn parameters in the machine settings is sporadic, the control with lengthy periods rather equals to an open-loop control.

Consequently, the next milestone is the integration of an inferential control scheme in order to regulate the process in times without primary measures. Expected is a significant improvement in the management of the production process. The next aim is the structural extension of the control scheme to an adaptive fuzzy model predictive control to ensure continuous improvement of the underlying fuzzy model as well as an automatic adjustment to changing process conditions.

Finally, the nonlinear estimator has to be introduced into the inferential scheme and the static operating point optimization has to be added.

5 Conclusion

Presented has been an architecture that addresses the design of an adaptive fuzzy model predictive control system, comprising a data driven optimization component supporting disparate types of measures. Since many production processes are characterized by producing dual rate data, including an inferential control mechanism to subsume laboratory spot tests - at least partly - by online available measurements is of great help. The cycle time of control can be reduced and costs saved.

References

1. Hopgood, A.A.: Intelligent Systems for Engineers and Scientists. CRC Press, Boca Raton (2001)
2. Cordón, O., Herrera, F., Hoffmann, F., Magdalena, L.: Genetic Fuzzy Systems. World Scientific Publishing Company, Singapore (2001)

3. Maciejowski, J.: Predictive Control with Constraints. Prentice Hall, Englewood Cliffs (2001)
4. Brosilow, C., Joseph, B.: Techniques of Model-BasedControl. Prentice Hall, Englewood Cliffs (2002)
5. Thomassey, S., Happiette, M., Dewaele, N., Castelain, J.M.: A short and mean term forecasting system adapted to textile items sales. The Journal of the Textile Institute 93, 95–104 (2002)
6. Lennox-Ker, P.: Using fuzzy logic to read image signatures. Textile Month (1997)
7. Kuo, C.F.J.: Using fuzzy theory to predict the properties of a melt spinning system. Textile Research Journal 74(3), 231–235 (2004)
8. Kim, S., Kumar, A., Dorrity, J., Vachtsevanos, G.: Fuzzy modeling, control and optimization of textile processes. In: Proceedings of the 1994 1st International Joint Conference of NAFIPS/IFIS/NASA, San Antonio, TX, USA, pp. 32–38 (1994)
9. Babuška, R.: Fuzzy Modeling for Control. Kluwer Academic Publishers, Norwell (1998)
10. Joseph, B.: A tutorial on inferential control and its applications. In: Proceedings of the American Control Conference, San Diego (June 1999)
11. Hansen, N., Ostermeier, A.: Adapting arbitrary normal mutation distributions in evolutionstrategies: The covariance matrix adaptation. In: IEEE International Conference on Evolutionary Computation, pp. 312–317 (1996)
12. Hansen, N., Ostermeier, A.: Completely derandomized self-adaptation in evolution strategies. Evolutionary Computation 9(2), 159–195 (2001)
13. Nasiri, M., Berlik, S.: Modeling of polyester dyeing using an evolutionary fuzzy system. In: Carvalho, J.P., Dubois, D., Kaymak, U., da Costa Sousa, J.M. (eds.) Proc. of the 2009 Conf. of the International Fuzzy Systems Association (IFSA) and the European Society for Fuzzy Logic and Technology (EUSFLAT), Lisbon, Portugal, July 20-24, pp. 1246–1251 (2009)
14. Siler, W., Buckley, J.J.: Fuzzy Expert Systems and Fuzzy Reasoning. Wiley-Interscience, Hoboken (2005)

A Novel Hybrid Adaptive Nonlinear Controller Using Gaussian Process Prior and Fuzzy Control

H.R. Jamalabadi, F. Boroomand, C. Lucas, A. Fereidunian,
M.A. Zamani, and H. Lesani

University of Tehran, Iran
h.jamalabadi@ece.ut.ac.ir, f.boroomand@ece.ut.ac.ir,
lucas@ut.ac.ir, arf@ece.ut.ac.ir,
ma.zamani@ece.ut.ac.ir, lesani@ut.ac.ir

Abstract. Control of an unknown nonlinear time-varying plant has always been a great concern for control specialists, thus an appealing subject in this discipline. Many efforts have been dedicated to explore the various aspects of this problem. This research has led into introducing many new fields and methods. These methods can be categorized into two general classes as: data-driven and model-driven. Model driven methods in spite of having rigorous analytical basis are not employed as frequently as data-driven ones. On the other hand data-driven methods are suitable to be employed in nonlinear and dual controller design; however they are slightly unsuccessful in handling missing data, moreover these methods are in need of considerable amount of computation. This paper introduces a novel hybrid nonlinear controller which aggregates Gaussian Process Prior as a data-driven and a Fuzzy Controller as a model-driven method .This special structure brings model-driven and data-driven advantages altogether, thus naturally lead into a robust and adaptive controller. Since no prior knowledge of the plant is used to issue the control law, the proposed hybrid controller can also be regarded as a dual controller.

Keywords: Gaussian Process Prior, Fuzzy Control, Adaptive dual control.

1 Introduction

Control of an unknown nonlinear time-varying plant has always been a great concern for control specialists, thus an appealing subject in this discipline [1-4]. Considerable research has been dedicated to propose methods to deal with these sorts of plants, leading to establishment of new fields in control domain, such as predictive control, adaptive control, robust control, and dual control [5-8]. These methods are classified into two general categories: data-driven, and model-driven.

An extensive amount of research has been performed to model nonlinear plants and many authors have presented effective methods to model a plant with a desired degree of accuracy [9]. Nevertheless, these methods usually describe the models as a finite number of parameters; usually combined linearly. In this case, all nonlinearities will be represented in parameter uncertainties; even, when flexible tools like artificial neural networks are used, the model uncertainty is represented in parameter uncertainty [10]. Fuzzy control, on the other hand, has proved itself as a promising tool in many applications. As fuzzy controllers could be viewed as a sort of expert systems [11], they have the potential to correctly model many systems, without parameterizeing the model.

Having all of these efforts, another choice is using non-parametric methods. Non-parametric methods, which include many new established tools, demonstrate very interesting features such as robustness, high stability properties and an elegant ability to deal with unknown model parameters, as well as handling missing data. Although it must be stated that once having a complete model of a plant, model-driven methods usually respond better, yet this is a rare case.

This paper introduces a novel hybrid nonlinear controller which aggregates Gaussian Process Prior as a data-driven and a Fuzzy Controller as a model-driven method. This special structure brings model-driven and data-driven advantages altogether, thus naturally lead into a robust and adaptive controller. Since no prior knowledge of the plant is used to issue the control law, the proposed hybrid controller can also be regarded as a dual controller. Simulation results show high regulatory and satisfactory trajectory tracking performance in the presence of Gaussian noises with different amplitudes and variances. The above ideas are closely related to the work done on dual adaptive control and supervisory control [12-14].

The reminder of this paper is organized as follows: a brief background is presented on dual control, supervisory control and Gaussian Process for the reader who is new to these concepts; then, the proposed methodology, results, comparison and discussions on the results are presented.

2 Background

This section is intended to briefly introduce the main concepts of dual control, supervisory control and Gaussian Process, in order to make this paper self-explanatory.

2.1 Dual Control

The concept of Dual Control developed in early 1960's, by Alexander Aronovich Fel'dbaum [13], intended to control the systems whose characteristics are initially unknown. The main idea is to use a twofold strategy which addresses two fundamental issues: first, system identification, and second, calculating proper control law to achieve regulatory or trajectory tracking goals.

Many methods are developed to get to these goals, yet still, a vast amount of research is needed to get to a satisfactory solution in terms of computational efficiency, stability issues and scope of validation [15]. This field is also in close relation with adaptive control, which tries to regulate control the performance of systems with little piece of knowledge at the initial steps. The adaptive control methodologies include Direct Adaptive Control (DAC) and Indirect Adaptive Control (IAC) algorithms [35-38].

This paper employs the main idea of adaptive Dual Control to construct a controller. Figure 1 shows the conceptual framework of the dual controller.

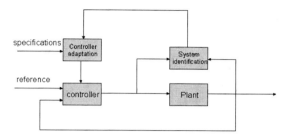

Fig. 1. Dual control framework

2.2 Supervisory Control

In most real industrial applications with high degree of uncertainty and nonlinearity, it is almost impossible to meet the satisfactory control specification using only one single controller. Typical examples such as chemical processes and electric power plants clearly show incapability of one single control loop to meet performance criteria, highest product quality, and economic operations. In these sorts of applications, stability is of high importance. To ensure an acceptable control performance, another 'level' of control is usually added. This is commonly called as "Supervisory Control". As Yazdi [16] defines "a supervisory system is a system that evaluates whether local controllers satisfy pre-specified performance criteria, diagnoses causes for deviation from the performance criteria, plans actions, and executes the planned actions."

Supervisory control concept makes it possible to design several controllers, each responsible for a particular control specification. This special structure provides the possibility of designing several different controllers, regardless of the stability issues which ultimately will lead to more desirable control performance [17]. Figure 2 shows the main concepts of supervisory control [12].

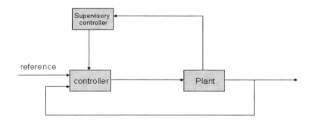

Fig. 2. Supervisory control framework

This paper uses a Gaussian prior structured controller as the main controller (and also dual controller) and a fuzzy controller as the supervisory controller. Besides, in this special design, the fuzzy controller would also bring the required initial training data for the main controller.

2.3 Gaussian Process

Gaussian Process (GP) is a flexible non-parametric method first introduced by O'Hagan and later reviewed by Williams [18-20]. GP's have been successfully adopted from their statistics origins by the neural network community [2]. GP is moderately simple to implement and use, which has been reported successful in many applications, as it could easily be used for prediction [21-23]. A comprehensive comparison between GP and artificial neural network has shown that GP could better handle missing data once encountering less amount of training information [24-25]. Many uses of GP in control context have been reported [26-29].

Although this is an essential perquisite for proposed materials, only just a very brief review is presented, due to the space limitation. The readers could find more details on GP in [10], [30].

In the following, the matrix containing the system states and control law is shown by "ϕ" .The output is represented by "y", the control law by "u" and the state vector by "X", which here is assumed to be

$$X(t) = [y(t), y(t-1),..., y(t-n)] \tag{1}$$

The aim is to predict N2 desired points when having N1 training points. In this method, instead of formulating the system with an explicit equation or a parameterized model such as:

$$y(t) = F(\Phi(t)) \tag{2}$$

We define a prior over the space of functions, which we guess F belongs to. GP provide the simplest form of prior over functions. It is common to assume a zero-mean multivariate normal distribution as follows: (in fact even if the mean is not zero we can subtract the offset and do the prediction and then add the offset to the predicted data.).

$$\begin{bmatrix} y_1 \\ y_2 \end{bmatrix} \sim N(0,\Sigma), \Sigma = \begin{bmatrix} \Sigma_1 & \Sigma_{12} \\ \Sigma_{21} & \Sigma_2 \end{bmatrix} \tag{3}$$

Where Σ shows the covariance matrix for ϕ and $\sum 12$ and $\sum 21$ show the cross covariance matrices for $\phi 1$ and $\phi 2$.

Like Gaussian distribution, a Gaussian Process can be fully described with its mean and a covariance function. The value of the covariance function shows the expected value of covariance between $\phi 1$ and $\phi 2$; so without explicitly determining a function relating y to ϕ, it's possible to calculate the value of y given ϕ [10].

Joint probability is divided into two parts: marginal Gaussian process and conditional Gaussian process. The conditional part also known as marginal Gaussian process, similar to that of other regression methods gives the desired output given a special ϕ as follows [10]:

$$P(y_1 | \Phi_1) = (2\pi)^{\frac{-N_1}{2}} |\Sigma_1|^{\frac{-1}{2}} e^{(-\frac{1}{2}y_1^T \Sigma_1^{-1} y_1)} \qquad (4)$$

And the following equation shows conditional Gaussian process [10]:

$$P(y_2 | \Phi_1, y_1, \Phi_2) = \frac{P(y_2, y_1)}{P(y_1)} = \frac{e^{-\frac{1}{2}(y_2 - \mu_{21})^T \Sigma_{21}^{-1}(y_2 - \mu_{21})}}{(2\pi)^{\frac{N_2}{2}} |\Sigma_{21}|^{\frac{1}{2}}} \qquad (5)$$

The conditional part very similar to other kind of regression gives the posterior density function. So the desired output will be calculated as follows [10]:

$$y(\Phi_2) = \mu_{21} = \Sigma_{21}^T \Sigma_1^{-1} y_1 \qquad (6)$$

2.4 Nonlinear Gaussian Regression Example

To show the ability of Gaussian process to predict and handle missing data a one dimensional nonlinear example is used to train:

$$Y = \sin(\frac{\pi \cos 2(x-1)^2}{2}) \qquad (7)$$

In order to show how well Gaussian process can deal with a small amount of data just 20 training points have been chosen uniformly in [-2 2]. Figure 3 shows the result.

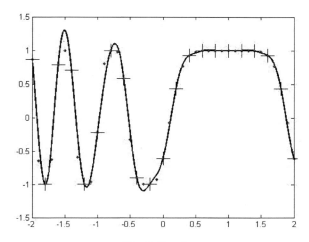

Fig. 3. Result

3 Methodology

This paper presents a hybrid nonlinear control based on a Gaussian process controller (GPC) and a general well-established fuzzy PID controller as the supervisory control level. The proposed controller called Gaussian Process based Hybrid Controller with Fuzzy Supervisor, abbreviated as GPHCFS.

3.1 Outline of GPHCFS

GP has been proved a promising tool to implement controllers [26-27]. In detail; GP can exceptionally handle missing data and this nominates it as a powerful tool when employing data-driven methods to implement dual controllers. In fact, most of the times there is not a model of the controlled plant and in those rare cases which a model exists, there are many uncertainties. Besides in some plants there is no precise estimation of the controlled plant state space order which makes the control task complicated when employing the classical control methods.

On the other hand, Johnstone and Phillips approved that under certain conditions, it is possible to estimate different conditions using Gaussian models. [39-40]

Many GP based methods have been proposed to meet different control performance criteria. However, all these methods suffer from low computational efficiency [10]. In detail; they all need to inverse high dimensional covariance matrices when analytically obtaining the control law which has become an end for most of these methods. This problem can better be understood when noticing that the task of matrix inversion should be done in the time intervals between two sampling times.

This hybrid controller on the other hand, while taking advantages of Gaussian Process for prediction has surmounted this problem. GPHCFS uses a simple numerical method instead of high computational-needed analytical methods to achieve the control law and employs the features of fuzzy controllers to guarantee the stability and adaptiveness [11]. GPC uses the input-output pairs obtained from the fuzzy controller to be trained. In fact GPC sees the whole system including the plant and all environmental noises as a unique system and then can "infer" the right control law.

Figure 4 demonstrates the conceptual framework of GPHCFS.

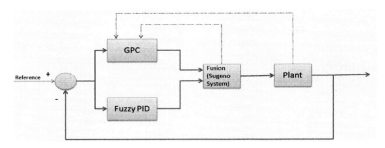

Fig. 4. GPHCFS conceptual framework

Fuzzy controller employed in this control architecture is a well known fuzzy PID controller as implemented by Zhao, Tomizuka, & Isaka [41].

This controller undertakes several important tasks:

Unlike other GP-based controllers the required data to train the GPC is not derived from some analytical formulas which in many cases try to minimize a kind of cost function but from the real input-output pairs obtained from the fuzzy PID controller.

In that situation which system is under a noise with such high amplitude which GPC can't issue the control signal correctly and the output deviation from the reference signal increases to a pre-defined value the fuzzy PID will temporarily undertakes the control task. This will lead to better training the GPC and also guarantee the stability of the whole system.

In those cases which the reference signal has changed and GPC needs to be re-trained, the fuzzy controller will obtain the required training set.

The fusion (aggregation) system placed immediately after the GPC and fuzzy controller is the main part which determines the correct and reliable value of the control law issued to the plant.

The fusion part is a simple Sugeno fuzzy system [42]. Any time the deviation between the control law and reference signal increases to a pre-defined value; the fusion system will switch the control task to the fuzzy controller. These cases often arise when a high-amplitude noise has affected the system or the reference signal has changed (e.g. the reference signal has changed from a sinusoid signal to a step).

The Sugeno system will lead to a smooth and ripple-free switch between controllers and minimizes the overshoots and errors made when noises affect system.

4 Implementation and Results

A simple unstable system is employed to verify the feasibility of the proposed method to implement GPHCFS. The off-equilibrium points are in each stage used to train the GPC. This is due to two general statements. First it's a well-known statement which once a data-driven model is selected, choosing an appropriate training set can effectively improve system's intelligence [32-33], and The second is Shannon Coding Theory [34].

In this paper in all simulations the squared exponential (SE) covariance function (also called the radial basis function (RBF) or Gaussian covariance function) is used which mathematically might be described by [30].

"For this particular covariance function, we see that the covariance is almost unity between variables whose corresponding inputs are very close, and decreases as their distance in the input space increases"[30]. This is exactly what is required to get to our ultimate aim. In fact we need the GPC controller to issue the same signals as the supervisory control had issued when the system state is the same and can deduce the required signal when the system states has changed. In the following the fuzzy PID controller is used when start-up to control the system and also provide enough pairs to train GPC. Here we have limited the total training points to 150 input-output pair due to computational efficiency. However the training wouldn't be done in a uniform manner. The first 100 pairs have been employed at the first moments of training and the other 50 points have been outspread to guarantee the proper functionality of GPC.

We have demonstrated the feasibility of our proposed method by two tests.

4.1 Regulatory Performance

At this test we have exerted the system with a unity step. And the environment is corrupted with a zero mean Gaussian noise with variance of $\sigma^2 = 0.001$. During the simulations we limited the control law to an upper limit to prevent an aggressive control law to be issued. Note that the output of the hybrid controller is much smoother than that of the single fuzzy controller. The small error can also be removed at the cost of a more aggressive control law.

Figure 5 shows the regulatory performance of GPHCFS under a zero-mean normal (Gaussian) noise with $\sigma^2=0.001$ variance.

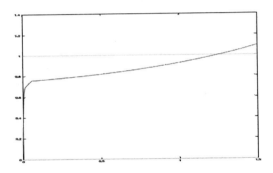

Fig. 5. GPHCFS regulatory performance

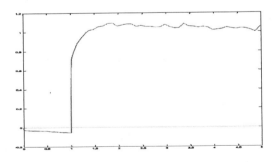

Fig. 6. Fuzzy PID control response is shown to

4.2 Tracking Performance

At this test we have exerted the system with a unity sinusoid signal. Here we limited the number of training points to 70 and as can be seen from figure that the hybrid system could handle the trajectory tracking with an acceptable degree of accuracy. The less number of training points used here could simply be explained due to the nature of two signals here. In fact an ideal step is comprised of infinitely sinusoidal signals, so the task of training an expert system needs reasonably a higher number of training pairs. Besides as we are obtaining our training pairs mostly from off-equilibrium points of the output and input signal the sinusoid signal is not generally a good candidate to be used, however the control performance is generally satisfactory.

Figure 7 shows the tracking performance of the system in tracking a sinusoid signal.

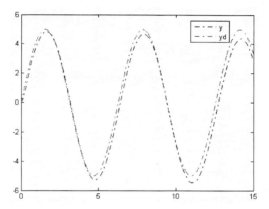

Fig. 7. GPHCFS trajectory tracking performance

5 Discussion

A novel adaptive nonlinear hybrid controller is presented. The hybrid controller made up of two model-based and data-based controller shows high stability and adaptiveness. Simulations on linear and nonlinear plants demonstrate high capability of GPHCFS in dealing with environmental noises and meeting control specification.

Both sub-controllers show consistent and robust behaviour. Besides using a fuzzy controller to aid GPC has led to a naturally robust controller. In fact a fuzzy controller can generally be view as a special expert system.[11] On the other hand GP as a flexible and powerful machine learning tool [30] has been employed to predict the suitable control law at each moment. At this special point of view GPHCFS can be viewed as a powerful expert system realized to control an unknown nonlinear plant. Note that in spite of using the Gaussian distribution when calculating the control law, the same approach is equally applicable when encountering non-Gaussian distribution of data. This claim is supported by two general statements. First is the Central Limit Theorem (CLT) which justifies that once having a large-sample set, one may approximate it with a normal distribution [43]. Second, Johnstone and Phillips approved that under certain conditions, it is possible to estimate different conditions using Gaussian models [39-40]. Besides, since GPHCFS uses no prior knowledge of the controlled system it could be nominated also as a powerful dual controller.

The special structure of GPHCFS has also surmounted the fundamental problem of all GP-based controllers. In detail the need to inverse a high-dimensional covariance matrix which arises when trying to train more data sets to GPC in order to solve some analytical formulas, has been removed by employing a fuzzy controller whenever need for more and new training pairs. The satisfactory functionality of GPHCFS in handling the control tasks with very high precision also suggests some good applications of this controller. In detail in an industrial environment which many controllers are being used to do the same task, keeping a fluid level constant for instance, GPHCFS is capable of handling all control tasks together with just one supervisor controller.

The following further works are under our study or are already studied and are under review for publication, as the continuation and improvement of this present paper:

- Employing well-established control structures like Internal Model Control when implementing GPC.
- Improvement of the supervisory control properties and adaptation features.

6 Conclusion

A novel adaptive nonlinear hybrid controller was introduced which shows interesting adaptive ness and dual control properties. Result show high stability and a high performance in trajectory tracking. Many issues which are less obvious but not less serious problems [31] such as choosing the near-to-best training set and supervising the stability of the proposed controller has been addressed and the computational complexity of all GP-based controllers is elegantly removed.

References

1. Tao., G., Kokotovic, P.V.: Adaptive Control of Plants with Unknown Dead-Zones. IEEE Transactions on Automatic Control 39(1) (January 1994)
2. Cho, J., Lan, J., Principe, J.C., Motter, M.A.: modeling and Control of Unknown chaotic systems via multiple models. In: IEEE Workshop on Machine Learning for Signal Processing (2004)
3. Whidborne, J.F., Murad, G., Gu, D.-W., Postlethwaite, I.: Robust control of an unknown plant—the IFAC 93 benchmark. International Journal of Control, 1366-5820 61(3), 589–640 (1995)

4. Graebe, S.F.: Robust and adaptive control of an unknown plant: A benchmark of new format. Automatica 30(4), 567–575 (1994)
5. Sastry, S., Bodson, M.: Adaptive Control: Stability, Convergence, and Robustness. Prentice-Hall, Englewood Cliffs (1989-1994)
6. Wang, L.: model predictive control system design and implementation using MATlAB. Springer, Heidelberg
7. Ackermann, J.: Robust Control Communications and Control Engineering, 2nd edn., vol. XIII, p. 483, 321. Springer, illus, Heidelberg (2002)
8. Astrom, K.J., Wittenmark, B.: Adaptive Control. Addison-Wesley Longman Publishing Co., Inc., Boston (1994)
9. Johansen, T.A., Foss, B.A.: Identification of non-linear system structure and parameters using regime decomposition. Automatica 31, 321–326 (1995); Also in Preprints 10th IFAC Symposium on Identification and System Parameter Estimation, Copenhagen (SYSID 1994), Vol.1, pp.131-137 (1994)
10. Smith, M., Sbarbaro, D.: Nonlinear adaptive control using non-parametric Gaussian Process prior models. In: 15th Triennial World Congress of the International Federation of Automatic Control, Barcelona, Spain, 21-26 July (2002)
11. Klira, G., Yuan, B.: Fuzzy Sets and Fuzzy Logic, Theory and Applications. Prentice Hall P R T, Englewood Cliffs (1995)
12. Wittenmark, B.: Adaptive Dual Control Methods: An Overview. In: 5th IFAC symposium on Adaptive Systems in Control and Signal Processing (1995)
13. Feldbaum, A.: Dual Control Theory I-IV. Automation Remote Control 21, 22, 874–880, 1033–1039, 1–12, 109–121 (1960)
14. Zhangab, H.-T., Lib, H.-X., Chenc, G.: Dual-mode predictive control algorithm for constrained Hammerstein systems. International Journal of Control 11(12), 1–17 (2008)
15. Král, L., Punčochář, I., Duník, J.: Functional Adaptive Controller for MIMO Systems withDynamic Structure of Neural Network. In: 10th International PhD Workshop on Systems and Control Young Generation Viewpoint Hlubok a nad Vltavou, September 22-26 (2009)
16. Yazdi, H.: Control and supervision of event-driven systems., Ph.D. Thesis, Department of Chemical Engineering, Technical University of Denmark (1997)
17. Wang, L.: A Course in Fuzzy Systems and Control, 424 pages. Prentice Hall, Englewood Cliffs (1997)
18. : On curve fitting and optimal design for regression. Journal of the Royal Statistical Society B40, 1–42
19. O'Hagan, J.M., Bernardo, J.O., Berger, A.P.: Some Bayesian numerical analysis. In: Dawid, Smith, A.F.M. (eds.) Bayesian Statistics, pp. 345–363. Oxford University Press, Oxford
20. Williams, C.K.I.: Prediction with Gaussian processes: From linear regression to linear prediction and beyond. In: Jordan, M.I. (ed.) Learning and Inferenceing Graphical Models, pp. 599–621. Kluwer, Dordrecht (1998)
21. Neal, R.M.: Bayesian Learning for Neural Networks. Springer, New York (1996)
22. Williams, C.K.I., Rasmussen, C.E.: Gaussian process for regression. In: Touretzky, D.S., Mozer, M.C., Hasselmo, M.E. (eds.) Advances in Information Processing Systems, 8th edn., pp. 111–116. MIT Press, Cambridge (1996)
23. Gibbs, M.N.: "Bayesian Gaussian Processes for Regression and Classification",PhD thesis, Department of Physics, University of Cambridge (1997)

24. Thomas Miller III, W., Sutton, R.S., Werbos, P.J.: Neural Networks for Control, p. 542. Short, MIT Press, Cambridge (1991)
25. Fyfe, C., Wang, T., Chuang, S.: Comparing Gaussian Processes and Artificial Neural Networks for Forecasting. JCIS (2006)
26. Kocijan, J., Smith, R., Edward Rasmussen, C., Likar, B.: Predictive control with Gaussian process models. In: The IEEE Region 8 EUROCON 2003: Computer as a Tool: Proceedings, pp. 22–24.
27. Kocijan, J., Murray-Smith, R., Rasmussen, C.E., Girard, A.: Gaussian process model based predictive control. In: American Control Conference, June 30 - July 2, Boston, Massachusetts (2004)
28. Rasmussen, C.E., Williams, C.K.I.: Gaussian Processes for Machine Learning. The MIT Press, Cambridge (2006)
29. Daimler-Benz: Neuro Control Towards An Industrial Control Methodology. Wiley, Chichester (1997)
30. Akhavan, E., Haghifam, M.R., Fereidunian, A.: Data-Driven Reliability Modeling, Based on Data Mining in Distribution Network Fault Statistics. In: Proc. Of the IEEE PowerTech, Bucharest, Romania (July/June 2009)
31. Haghifam, M.R., Akhavan, E., Fereidunian, A.: Failure Rate Modeling: A Data Mining Approach, Using MV Network Field Data. In: Proceedings of the IEEE Canada EPEC 2009, Montreal, Quebec, Canada, October 22-23 (2009)
32. Mitezenmatcher, M., Upfal, E.: Probability and Computing. Cambridge University Press, Cambridge (2000)
33. Wang, C.-H., Liu, H.-L., Lin, T.-C.: Direct Adaptive Fuzzy-Neural Control With State Observer and Supervisory Controller for Unknown Nonlinear Dynamical Systems. IEEE Transactions on Fuzzy Systems 10(1) (February 2002)
34. Sastry, S., Isidori, A.: Adaptive control of linearization systems. IEEE Trans. Automat. Contr. 34, 1123–1131 (1989)
35. Marino, R., Tomei, P.: Globally adaptive output-feedback control on nonlinear systems, part I: Linear parameterization. IEEE Trans. Automat. Contr. 38, 17–32 (1993)
36. : Globally adaptive output-feedback control on nonlinear systems, part II: Nonlinear parameterization. IEEE Trans. Automat. Contr. 38, 33–48 (1993)
37. Johnstone, Function Estimation and Gaussian Sequence Models, Department of Statistics, Stanford University (2002)
38. Philips, Yu, Exact Gaussian Estimation of Continuous Time Models of The Term Structure of Interest Rates. Yale University, Department of Economics (2000)
39. Zhao, Tomizuka, & Isaka 1993. Z.Y. Zhao, M. Tomizuka and S. Isaka , Fuzzy gain scheduling of PID controllers. IEEE Transactions on Systems, Man and Cybernetics 23 5 (1993), pp. 1211–1219.
40. Takagi, T., Sugeno, M.: Fuzzy identification of systems and its applications to modeling and control. IEEE Trans. Syst., Man, Cybernetics SMC-15(1), 116–132 (1985)
41. Rice, J.: Mathematical Statistics and Data Analysis, 2nd edn. Duxbury Press (1995); ISBN 0-534-20934-3

Image Based Analysis of Microstructures Using Texture Detection and Fuzzy Classifiers

Lars Hildebrand and Thomas Ploch

Technical University of Dortmund, Computer Science I,
Otto-Hahn-Str. 16, 44221 Dortmund, Germany
lars.hildebrand@tu-dortmund.de, thomas.ploch@tu-dortmund.de

Abstract. Modern steel products are manufactured using many different production steps. One step is the hardening by heat treatment. To assess the quality of the processed steel products, the microstructure can be analyzed. This article describes all necessary steps, beginning from image capturing from microscopes to the classification using fuzzy logic based classifiers.

Keywords: Image processing, texture analysis, Haralick-features, microscopic analysis, fuzzy c-means classifier, steel quality assessment.

1 Introduction

The quality of modern steel products is determined by the composition of the alloy, i. e. the compound of the different additional elements, as well as the heat treatment after the alloying process. A large variety of physical properties can be achieved, depending on the demands of the product. One way to assess the quality of the steel is the analysis of the microstructure. Directly after the production, a sample is cut and the surface is grinded and etched. A microscope is used to reveal the properties of the microstructure. Based on large sets of sample images, the properties of the microstructure are assessed.

The process of using sample images shows two main problems: The assessment is based on human estimation and the differences between two quality classes can sometimes not be seen by the naked eye. The human assessment is subjective; two steel samples can be assessed in two different qualities by two different humans. Even if the same human supervisor assesses the same sample on different days, the assessment of the quality may vary.

The proposed approach is not based on human assessment. Texture based quality assessment is used to measure micro structural properties. This analysis results in 26 different real-valued measures. The classification of the samples, based on these 26 measures can be done using fuzzy c-means classifier. Two commonly used steel variations (X38CrMoV5-1 and X40CrMoV5-1) are used to demonstrate the assessment of the steel qualities according to STAHL-EISEN-Prüfblätter (SEP) 1614, a frequently used set of sample images for 30 different quality classes.

2 Image Processing and Feature Analysis

The sample has to be analyzed suing a microscope and a digital camera, taking magnified images of the sample surface. The sample has to be prepared using mechanical grinding and chemical treatment [1, 2].

2.1 Sample Preparation

The sample is cut from the block of material, where the direction of the cut has to be parallel to the main direction of forming. The size of the sample surface has to be at least 10mm by 20 mm. The sample has to be grinded and polished, the final step of sample preparation is etching the sample with a 3% nitric acid to reveal its microstructure.

2.2 Image Processing

The microscope allows a 500 times magnification of the microstructure and uses reflected light from a light source above the sample. The light is not homogeneous, so a shading correction has to be used to allow the use of texture analysis in a further step. An example of an image capture with non homogeneous shading is shown in figure 1:

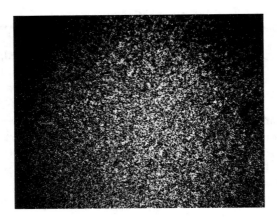

Fig. 1. Sample image

Image Based Analysis of Microstructures Using Texture Detection 315

The degree of shading can be measured using a pure white sample without a structure. The following figure shows the light intensity, related to the position of the sample area.

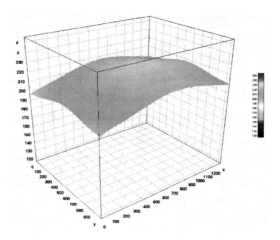

Fig. 2. Light distribution of the optical system

The light distribution can be used to calculate the correction matrix for shading correction [3]. The original sample, shown in figure 1 can be expressed as follows:

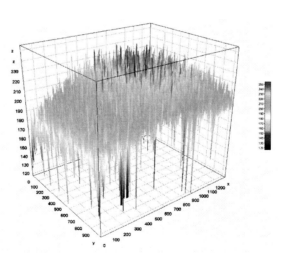

Fig. 3a. Sample image, depicted as light distribution

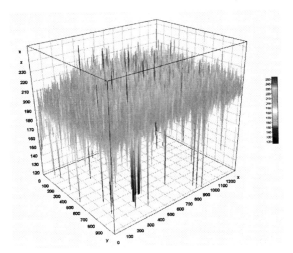

Fig. 3b. Sample image, depicted as light distribution after shading correction

The correction of the shading effects is very important, due to the fact that texture analysis is based on the distribution of grey values. Not corrected sample image will result in wrong grey values and influence all further processing steps in a negative way. An example of a shading corrected sample image is shown in figure 4:

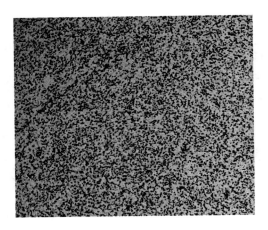

Fig. 4. Sample image after shading correction

2.3 Haralick Features

Haralick defined an almost complete set of different features that are able to describe the distribution of grey values in a digital image [4]. The calculation of these features is not directly based on the grey values of the digital image. The cooccurrence matrix is used instead. The cooccurrence matrix is based on the grey value of a pixel and also on the grey values of the surrounding pixel in varying neighborhood sizes.

The cooccurrence is calculated as follows [5, 6]:

$$C_{\Delta x, \Delta y}(i, j) = \sum_{p=1}^{n} \sum_{q=1}^{m} \begin{cases} 1, & if \ I(p,q) = i \\ & and \ I(p + \Delta x, q + \Delta y) = j \\ 0, & otherwise \end{cases} \quad (1)$$

The calculation is based on the grey value of the pixel $I(p,q)$ and the grey values of the pixels the neighborhood $I(p + \Delta x, q + \Delta y)$. Based on the matrix, the Haralick feature can be calculated as stated in [4].

As a result of calculating the Haralick features, a set of 14 measures is determined, that describe the microstructure of the sample.

3 Classification

The whole set of features consists of too many dimensions to be handled by humans. A classification algorithm has to be found, that is able to differentiate between the different classes of the microstructure. We decided to use a classification algorithm, that is based on fuzzy logic, due to the fact, that small variations of the texture features should not result in the assignment of a different class. These small variations may develop because of small changes in the light or the orientation of the sample and do not reflect different classes.

The used classification algorithm is the fuzzy c-means algorithm. A detailed description can be found in [7, 8, 9, 10]. The difference between binary classifier (or classifiers, that are based on binary logic) and classifiers based on fuzzy logic can be seen in the figures 5 and 6. The classifier used for the classification depicted in figure 5 is based on binary logic and shows the discrimination line, that separates both classes. The distance of the samples to the class center is not used and the class assignment of point 22 is questionable.

The second example classification shows the classification result of the fuzzy c-means algorithm. In addition to the class assignment, the membership value of each point to the class is specified. It can be seen, that the point 22 belongs to both classes now, with a slightly larger membership value related to class red.

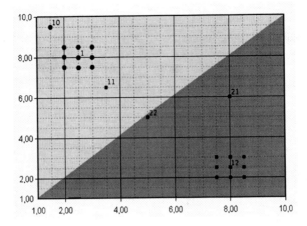

Fig. 5a. Classifier, based on binary logic

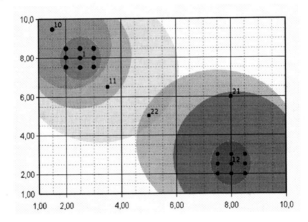

Fig. 5b. Classifier, based on fuzzy logic

4 Experiments

The processing pipeline of the microstructure analysis consists of the following steps for the learning part of the analysis:

1. Sample preparation
2. Image capturing
3. Shading correction
4. Calculation of cooccurrence matrix
5. Calculation of Haralick features
6. Supervised classification

During the learning part, samples with known class have to be used (supervised learning). After learning, the classifier can be used on samples with unknown quality.

Experiments have shown, that a classification rate of 82% correct classifications can be achieved.

5 Summary

Microstructure analysis based on image processing, Haralick feature analysis and fuzzy logic based classification is able to automate a quality control process, that is typically performed by humans. The automated process reduces the risk of subjective and varying quality assignments.

References

1. Griffiths, B.: Manufactoring surface technology - Surface integrity and functional performance. Penton Press, London (2001)
2. Dornfeld, D., Lee, D.-E.: Precision Manufacturing. Springer Science and Media, New York (2008)
3. Schmidt, K., et al.: Gefügeanalyse metallischer Werkstoffe. Carl Hanser Verlag, München (1985)
4. Haralick, R.M., Shanmugam, K., Dinstein, I.: Textural features of image classification. IEEE Transactions on System, Man and Cybernetics SMC-3, 610–621 (1973)
5. Haralick, R.M., Shanmugam, K.S.: Combined spectral und spatial processing of ERTS imagery data. Remote sensing of environment 3, 3–13 (1974)
6. Tahir, M.A., Bouridane, A., Kurugollu, F.: An FPGA Based Coprocessor for GLCM and Haralick Texture Features and their Application in Prostate Cancer Classification. In: Analog Integrated Circuits and Signal Processing, vol. 43(2), pp. 205–215. Springer, Heidelberg (2005)
7. Bezdek, J.: Pattern Recognition with Fuzzy Objective Function. Plenum Press, New York (1981)
8. Dunn, J.: A fuzzy relative of the isodata process and its use in detecting compact, well-seperated clusters. Journal of Cybernetics 3(3), 32–57 (1973)
9. Gustafson, D., Kessel, W.: Fuzzy clustering with a fuzzy covariance matrix. In: Proc. IEEE CDC 1979, San Diego, USA, pp. 761–766 (1979)
10. Kaymak, U., Babuska, R.: Compatible cluster merging for fuzzy modelling. In: Proceedings of the Fourth IEEE International Conference on Fuzzy Systems, Yokohama, Japan, vol. 2, pp. 897–904 (March 1995)

Fulfilling the Quality Demands of the Automotive Supply Chain through MES and Data Mining

Ralf Montino and Peter Czerner

Elmos IT, Heinrich-Hertz-Str. 1, 44227 Dortmund, Germany
`ralf.montino@elmos.eu, peter.czerner@elmos.eu`

Abstract. Usage of electronics and sensors in modern vehicles is constantly increasing. The quality demands of the automotive supply chain require complex manufacturing execution systems (MES) and data analysis tools (data mining). On one hand, the constant demand for increase in efficiency leads to bigger wafer sizes and more automation in the fabs. On the other hand this causes problems with products of small quantities, where the meaning of "small" is continuously rising with wafer size and automation. New products such as MEMS devices and microsystems increase these problems, because the processes must here also be adapted to the products.

The established tools are not sufficient to comply with the continuously increasing demands. The use of modern production control systems and methods are thus presented using the ELMOS group as an example.

Keywords: Semiconductor, ASIC, MES, zero defect, data mining, PAT.

1 Products and Technologies

Electronics and sensory properties play an increasing role in modern vehicles. From airbag control to pressure sensors in tires and power window control, there are numerous applications that either save lives or that simply make it more comfortable.

The ELMOS group has specialized in the development and production of ASICs and MEMS for the automotive industry and has aligned the development and production to the high quality standards of our customers. This applies to the developmental procedure as well as to the applied technologies and processes and the broad scope of production control.

2 Quality Demands

2.1 Functionality under Extreme Conditions

For use in vehicles, functionality in the temperature range of -40°C to 125°C must be ensured. Often enough, the upper limit is not sufficient, i.e. for application in oil circulation functionality above 150° C must be guaranteed.

2.2 Zero Defect

The maximum field failure probability demanded by the automotive suppliers means practically a 'zero defect' default for product quantity.

2.3 Full Traceability

Consistent traceability of each die is increasingly postulated as standard. This means that each individual product must be sufficiently identified and that the manufacturing history of the product can be called up based on this identification. Moreover a suitable data archive is needed. The data is automatically stored during production and can be easily accessed during the full life cycle of the product.

3 CMOS: Manufacturing for High Quality

3.1 Complexity of Manufacturing

CMOS manufacturing distinguishes itself by especially high complexity. Manufacturing ASICs entails hundreds of production steps. In the wafer process alone there are approximately 300 to 600 steps. For almost each process there is a possibility of destroying the material or – even worse – of causing field failure risk.

3.2 Test

The high quality demands of the automotive industry are reflected in the production. Multiple tests are conducted for all products in which comprehensive characteristics are measured and registered. These tests are conducted under various external conditions in order to guarantee the functionality of the products in vehicles, mentionably numerous temperature, pressure and acceleration tests.

These tests result in huge amounts of data that must be provided for all possible types of evaluation. This in turn results in great demands on the data storage and archiving systems.

4 MEMS: Challenges

Micro systems (microelectromechanical systems) mean additional challenges for semiconductor production. Whereas process and product development for the manufacture of ASICs is largely independent, this is not the case for MEMS. Here, the process depends on the product, in manufacturing as well as in testing and assembly.

Fig. 1. Various packages of pressure sensors requires different test equipment (Source: SMI, ELMOS Group)

This means product specialization of equipment and procedures, which in turn must be illustrated in the MES.

In addition, the database of the process and test measuring values is often insufficient for significant statistical evaluation, due to the low quantities.

5 MES, the Remote Controlled Production

5.1 Global Data Storage

For many years, ELMOS has maintained a central database system with an area-spanning relational database model. This is the technical core of the MES and an interface for all manufacturing execution systems. Additionally it also serves different other systems needing data.

Fig. 2. ELMOS company database model

5.2 Automation

Where a false parameter entry on the production machine can lead to the destruction of numerous lots, automation is a must.

When starting the process manually, the operator must identify the material to be processed, enter the process recipe and parameters, based on the paper lot traveler. Also the post-processing data must be entered manually into the lot traveler to be transferred to controlling procedures. All these steps contain unnecessary high error probability: A to high failure ratio encurs.

On the other hand, with automatic processing the operator only needs to scan the material barcode from the wafer box. The MES system connected to the equipment takes care of all data transfers, choice of recipe and process release. The failure ratio has proven to decrease drastically after equipment automation. The machine data, the control parameters as well as the process measurement values are logically stored for later evaluation.

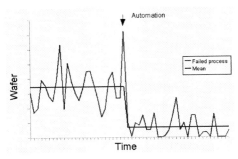

Fig. 3. Reduction of the failure ratio of a production machine due to automation

A fabrication or even enterprise-spanning tool connect system also offers additional advantages. Central control with the possibility of fluent data exchange and correlation diagnosis can only be realized with such a system. There is also a uniform development suite for most machine types, which simplifies the equipment integration. A software library reduces developing time and reusable user interfaces reduces support and initial training effort needed by operators.

5.3 Recipe Management

A central recipe management ensures a secure reproduction of the product and process designs on the production procedures and is an important aspect of quality assurance. Master data management and paperless release of recipes is executed in the central ELMOS MES using standard software and user interfaces. This accelerates the workflows and is safer due to the stored data dependencies. Moreover, storing design, product and process information in the same relational database system enables multiple crossover checks and guarantees consistent and error free recipes. Last but not least, paperless recipes are also environment friendly.

5.4 Die Traceability

Consistent identification of the production history of a product is only possible with a global MES. The chain of evidence is only complete when all data, from the recipe to the production parameters, tests and assembly of the product can be distinctly allocated.

In the ELMOS MES, each die centrally receives a unique identification number, which is made available to the staff and machines for product identification. This unique identification number is saved into the ASIC Non Volatile Memory and can be accessed even after the die has been delivered as a part of an application to the final customer.

The central database and homogeneity of the data enable simple and standardized data access for the die identification data and storage thereof.

6 Data Mining

In a semiconductor production that specializes in the automotive industry, systematic data analysis for failure diagnosis and process control is an important part of quality assurance. Based on the central database, ELMOS operates various area-specific and area-spanning data evaluation systems.

6.1 SPC/SYA

Statistic process controls (SPC) and yield analysis (SYA) enable early recognition of abnormalities in measurement values and yield.

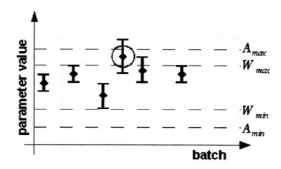

Fig. 4. Simply control chart: The value of the parameter must not be out of specification limits (W: Warn limit, A: Stop limit)

Selected equipment, test and material parameters will be gathered automatically, monitored and verified against defined warn and stop limits. Yield data can also be similarly recognized.

The sample base of the parameter must be well defined: E.g. to analyze material or process properties, a wafer can be used as a good sample unit. To analyze equipment parameters on the other hand, lot or sub lot data processed on the same machine will be taken. A mixed sample, e.g. containing more than one batch, causes overlaps of several distributions and makes the statistical analysis difficult.

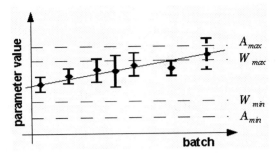

Fig. 5. Trend analysis. A problem case can be detected, before a parameter value exceeds the specification limit: The most correct value interpolates growth of the measured parameter values.

A large enough sample size is required and an estimated measurement error is useful to keep the influence of statistical and systematic errors on the analysis as low as possible. On the other side, noise effects render all the analysis results useless and trigger false MES alarms.

The analysis is based on the statistical trend analysis and statistical control charts [1][2].

Geometric distributions are also a valuable source of information and help with early recognition of problem cases. E.g. the capability indexes C_p (1) and C_{pk} (2) in terms of comparing the process spread to the specification spread will be used (see Fig. 6)

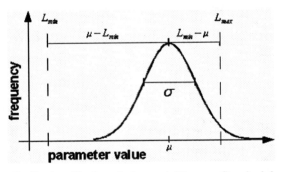

Fig. 6. Frequency distribution. Used symbols are μ: Mean, σ: Standard deviation, L: Specification limits.

$$C_p = \frac{L_{\max} - L_{\min}}{6\sigma} \quad (1)$$

$$C_{pk} = \frac{\min(\mu - L_{\min}, L_{\max} - \mu)}{3\sigma} \quad (2)$$

C_p and C_{pk} express the capability of a process or machine and are the fundamental numbers for quality analysis.

A Gaussian (normal) distribution, as shown in Fig, 6 is required for the statistical analysis described but is unfortunately rare. In reality, parameter distribution can be more or less approximately Gaussian, non-Gaussian or a superposition of more then one Gaussian and non-Gaussian distribution depending on the mixed sample. In such cases the first workaround is to compute robust mean (3) and standard deviation (4) based on quartiles Q_n.

$$\mu = Q_2(median) \qquad (3)$$

$$\sigma = \frac{Q_3 - Q_1}{1.35} \qquad (4)$$

This method works fine for most of real data samples.

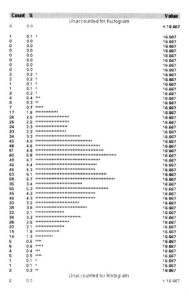

Fig. 7. Real parameter distribution

For user acceptance and fast benefit it is advisable to use a few simple statistical methods first. They have to be robustly implemented and well-integrated into the MES. Such a warn and stop system controls the material flow, triggers an additional analysis by the engineer and stops further processing of the conspicuous material.

6.2 PAT

Part Average Test is a method defined in the semiconductor production for quality assurance. The statistically interrelated measurements are thereby inspected and cleansed of the runaway value. Dice are thus sorted out that functionally comply with the specification, but can be classified as failure risky based on the statistical conspicuity [3].

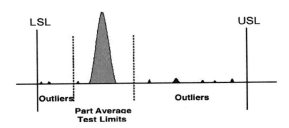

Fig. 8. Diagram of Part Average Test. USL and LSL are lower and upper specification limits Source: [3]

The basic PAT approach uses robust mean and standard deviation (3,4) to compute new parameter limits (PAT limits).

$$L_{PAT} = \mu \pm 6\sigma \qquad (5)$$

The same statistic sample used for the calculation of PAT limits can also be used to evaluate these limits (dynamic PAT). For the dynamic PAT, the dice must be identifiable, because the PAT outliers must be sorted out after the whole sample has been tested. This method can be used for wafer tests. Otherwise the static PAT limits can be used. Typically, a sample of 6 lots will be computed and applied to the material following to identify PAT outliers. With static PAT limits one must be careful of violating effects such the parameter drift or equipment caused mean variation.

The usual parameter distribution and any possible violating effects must be taken into consideration when selecting the parameters for PAT. Because the various parameters of PAT outliers correlate strongly, a good choice of parameters for PAT makes a small amount of them sufficient for effective outlier identification.

Furthermore, the PAT data can be used for other useful quality considerations. E.g., the PAT rate should be monitored, because a growth of the outlier count can be caused by a problem in the production process.

The PAT system is also integrated in the ELMOS MES. It is transparent in production procedures and provides data for production reports and later data analysis.

6.3 Complex Failure Cause Analysis and Yield Forecast

Where there are no evident connections between measurement and machine parameters and correlation analysis does not lead to any results, classic analysis tools are hardly of any help. ELMOS has therefore taken a new direction and has created an analysis method, based on a neuronal network and feature selection [4].

As opposed to the classic tools of the neuronal network, the system not only makes forecasts based on the trained procedures, but is also used for failure cause analysis in problem cases. The feature selection serves the purpose of iterative dimension reduction of the multi-dimensional parameter space and the neuronal network validates the selection. Furthermore, it provides indication of the correlations between the parameters.

7 Conclusion

Complex ASIC/MEMS production also requires a sophisticated MES which, in connection with the data analysis tools, guarantees product quality. For years, the ELMOS group has been developing and using central MES and analysis tools that have proven themselves in practice numerous times. Further development is continuously being made on these systems in order to keep up with the growth and further development of production.

References

1. Amsden, R.T., Butler, H.W., Amsden, D.M.: SPC Simplified, Productivity Inc. (1998)
2. Quesenberry, C.P.: SPC Methods for Quality Improvement. John Wiley & Sons, Inc., Chichester (1997)
3. AEC-Q001-REV-C: Guidelines for part average testing, AEC (July 18, 2003)
4. Montino, R.: Crystal Ball, extracting usable information from data objects with fuzzy relation (Die Gewinnung von verwertbarer Information aus Datenobjekten mit unscharfem Zusammenhang), Dissertation, Universität Siegen (2009)

Evaluation of Professional Skill and *Kansei* Based on Physiological Index: Toward Practical Education Using Professional *Kansei*

Koji Murai and Yuji Hayashi

Maritime Sciences, Kobe University, Kobe, Japan
`murai@maritime.kobe-u.ac.jp`, `hayashi@maritime.kobe-u.ac.jp`

Abstract. This paper describes the evaluation of skill and kansei of professionals for ship handling toward practical education using thier skill and kansei. We evaluate the professionals' performance using a ship bridge simulator; moreover, we ascertain their mental workload on board by a training ship. We show the effect of physiological indices from the professionals' performance and the characteristics of their mental workload with heart rate variability (R-R interval), nasal temperature and salivary amylase activity. We confirm whether the response to their performance for the ship handling is clear or not. Mental workload is useful to evaluate performance of ship bridge teammates: a captain, a duty officer, a helmsman, and a pilot.

Keywords: Ship handling, Simulator, Real ship, Mental workload, Physiological index.

1 Introduction

Physiological index, heart rate variability (R-R interval), facial (nasal) temperature and salivary amylase activity, are sufficient to evaluate the mental workload of a ship navigator [1]-[4]. The heart rate variability has advantage over some other indices in that it can be quick response from an event. The nasal temperature has in that it can be measured by non-contact with a subject. The salivary amylase activity is able to get the results in a short time and on the spot without requiring special sensors for the subject. In a recent study, researchers showed salivary amylase measurement reveals sympathetic nervous system activity. The research evaluated a motor vehicle driver's stress and developed a compact measuring instrument [5].

In this paper, we attempt to evaluate the mental workload using nasal temperature, heart rate variability (R-R interval) and salivary amylase activity simultaneously. The experiment is carried out using a simulator and a real ship. Firstly, we are carried out the experiment using the simulator with which we are easily to control some conditions- subjects, time, temperature, humidity, scenarios, etc., and confirm the effect. After that, we evaluate the three physiological indices using a real ship. The ship handling situation is a ship entering and leaving a port. The navigator has to make a lot of ship handling decisions as he moves the ship into/from a port. We show that salivary amylase activity, nasal temperature and R-R interval are good indices for effective navigation.

2 *Kansei* and Skill

Kansei is the art of "sensibility" or "individual sense". Just as an artist or musician has this sensibility when creating [6], [7], professional navigators also have it when steering a large ship [8]. Of course, sensibility (or *kansei*) is not identical for each of those three behaviors. Nevertheless, it exists; and the artist knows when he possesses it just as the professional navigator knows when he possesses it. We consider their *kansei* for ship handling includes skill. Our *kansei* is treated as a cognitive process of a high order.

Kansei information is divided into four kinds - 'symbol', 'parameter', 'pattern', and 'image'. We can express the symbol in an adjectival word like 'warm'. We can express the parameter by a vector in the coordinate data made by adjectival words using factor analysis. Pattern is difficult to describe in the same way that it is difficult to describe the feel of fabric or the tonal quality of an instrument. Image is difficult to exteriorize just as a mental picture or inspiration [6], [7].

Research on *kansei* attempts to identify a person's sensibility while performing a certain activity. Psychological and physiological methods are used to determine the kinds of *kansei* information processing. The psychological method measures the relationship between an impulse (e.g., light, sound, heat, music, and picture) to the human and the response to it [9], [10]. Emotion is evaluated by Semantic Differential assessment [11]. The physiological method measures the heart rate, respiration, skin temperature, blood pressure, etc. of the subject of an investigation [12], [13]. The idea of *kansei* is not the same as human error [14], [15] as revealed in human performance models. Research to identify human factors in transport systems except ships has been successful, especially in the aerospace and space fields [16]. However, a ship is different from other systems because the navigator is never in a fixed operational seat and members of the Bridge Resource Management (BRM) team move freely on the bridge in order to keep a sharp lookout around the ship, 360 degrees [17], [18]. The maritime traffic system is most difficult to navigate for large number and kind of ship: oil tanker, LNG: Liquefied Natural Gas/LPG: Liquefied Petroleum Gas ship, car carrier, container, tall ship, ferry boat, battle ship, motor boat, fishing boat, and so on, and an infrastructure is not sufficient on the sea and is never under control for small vessels.

3 Ship Bridge Team

The ship bridge team consists of a captain, a duty officer, a helmsman, and sometime includes a ship's pilot. Team members depend on the geographical and weather conditions. The conditions are divided into four sections depending on fog, heavy traffic, entering a channel, harbor or restricted area, heavy weather and fire, flooding, or other emergency. Each watch condition is as follows [19].

Evaluation of Professional Skill and *Kansei* Based on Physiological Index 333

- *Watch Condition 1*:

All clear conditions for maneuverability, weather, traffic and systems. A deck officer and a helmsman can handle the bridge watch, and sometimes a deck officer does it.

- *Watch Condition 2*:

Somewhat restricted visibility, constrained geography and congested traffic. A deck officer and a helmsman can handle the bridge watch.

- *Watch Condition 3*:

Serious poor visibility, close quarters - in bay and approach channels, and heavy traffic. A captain, a deck officer and a helmsman can handle the bridge watch.

- *Watch Condition 4*:

On berthing and anchoring, a captain, a deck officer, a quartermaster and pilot at special ports on the bridge can handle the bridge watch. A chief officer and bosun are at the bow station, and a second officer and deckhand are at the stern station.

We select "Watch Condition 4" upon entering and leaving a port for the experimental situation during daytime. On leaving and entering port, a captain pays attention to the control of the ship's course as it approaches a berth. He must control the rudder, the engine motion and the thruster at the same time. Furthermore, he must give orders to the bow and stern stations as wind and current often affects the ship's control. Also, a ship's rudder effect is worse at low speed. We can say that he needs to multitask in the short time.

4 Experiment

The experiment was carried out on the Training Ship and Ship Bridge Simulator (simulator) of Kobe University, Graduate School of Maritime Sciences.

4.1 Simulator Experiment

An evaluation of the mental workload for the ship's navigator is carried out on the simulator. The indices are the nasal temperature and the heart rate variability. The scenario is a ship entering a port (Figure 1), where the ship is a container ship (6,000 TEU, able to take 6,000 containers on board), with 70,000 Gross Tonnage, following a traffic route; from entrance to traffic lane to berth (RC5) on *Rokko* Island. The captain (subject) needs to pass another stationary ship at a neighboring berth. There are breakwaters before the approach to the berth, so he needs to pass in a narrow sea space and control his own ship under the worst rudder answer for decreasing the ship's speed. The navigator can use two tug boats with 3,500 PS. The tug boats are used for controlling the position of the ship by pushing or pulling. The weather is fine; sea and wind conditions are calm. The scenario is repeated five times.

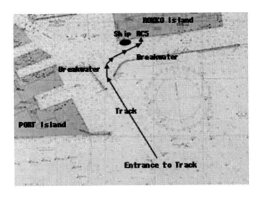

Fig. 1. Outline of the experiment for entering a port

The subject is a real captain who has eight and half years experience on board. A tester measures the facial temperature including the nasal temperature and the heart rate variability at the same time using thermography and a heart rate monitor. Moreover, we record the track passed and the ship's performance (course, speed, rudder angle, etc.) using the simulator system and the subject's behavior, conversation (subject's performance) using a video camera. We also analyze his performance with work-sampling every second [18], [20].

4.2 Onboard Experiment

The experimental ship is *Fukae-Maru* (Figure 2). Her length is 49.95 meters, width is 10.00 meters, draft is 3.20 meters, tonnage is 449.00 tons. The subject is the Captain of Training Ship (male, 52 years old). Based on the pre-experiments using R-R interval and nasal temperature, entering a port is a common navigational situation; we have selected the "Entering a port" and "Leaving a port" for two ports: *Fukae* and *Miyazaki* port (Figure 3) like similator experiment.

Fig. 2. Training Ship *Fuka-maru*, Kobe University, Graduate School of Maritime Sciences

Fig. 3. Experiment position in west side of Japan

A tester measured the Salivary Amylase Activity (SAA) [kIU/l], the facial temperature including the nasal temperature and the R-R interval at the same time using a SAA monitor, a thermography and a heart rate monitor. Moreover, we record the ship's performance (course, speed, rudder angle, etc.), the subject's behavior, and conversation (subject's performance) using a video camera. We also analyze his performance with work-sampling every second [18], [20].

4.3 Salivary Amylase Activity

A tester measured SAA value using the SAA monitor which consists of a measurement part and a marker (Figure 4). The saliva is gathered with the marker (Figure 4(b)); the subject puts it under his tongue for thirty seconds; the amylase is extracted by the measurement part (Figure 4(a)) for thirty seconds.

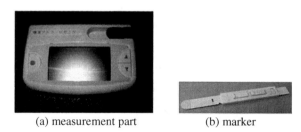

(a) measurement part   (b) marker

Fig. 4. SAA value measuring instrument

We measured the SAA value for one minute. The clean saliva comes from under the tongue because a salivary gland is located there. The captain can drink some water during the experiment to aid in getting enough saliva for the measuring instrument; however, he never needed any water during all the experiments.

4.4 R-R Interval

A tester measured the heart rate variability (R-R interval) with a tolerance of one millisecond. In the majority of the R-R intervals, the accuracy is one millisecond as well; 95 percent confidence interval is better than plus/minus 3 milliseconds. The heart rate monitor consists of a chest belt and a wrist watch (Figure 5).

Fig. 5. The heart rate monitor (chest belt and wrist watch)

The chest belt with sensor measures the R-R intervals. It sends the data to the wrist watch which has a memory. The memory of the wrist watch can keep 30,000 bits of data. We take the R-R interval data from the heart rate variability.

The R-R interval means the time interval from a peak point 'R' wave to the next peak point. The 'R' is one of the waves which consist of P, QRS and T of an electrocardiogram (Figure 6). The 'R' wave is easier to pick up than other waves because the amplitude is clear. We use the heart rate to evaluate the mental workload.

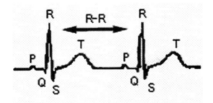

Fig. 6. Outline of the R-R interval

4.5 Facial Temperature

A tester measured the facial temperature at intervals of one second. To do the measurement, we have the subject stand in front of the ship's compass which is exactly one meter from where the thermography device is set on the onboard experiment [21]. Figure 7 shows outline of the experimental situation on the bridge. The position isn't influenced by the wind from the air conditioner in the bridge house.

On the simulator experiment, the thermography device mounts in front of the bridge house, and we can take a thermo image by moving a joystick while watching the real time thermo image.

Evaluation of Professional Skill and *Kansei* Based on Physiological Index 337

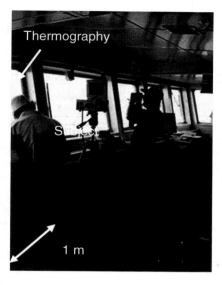

Fig. 7. Outline of the experimental situation on the bridge

5 Evaluation

5.1 Salivary Amylase Activity

We evaluate the mental workload using a reference value. The reference value is the mean value of three to five SAA values after the experiment. Research on motor vehicle drivers uses the mean value before the experiment [5]; however, a captain plans to guide the ship to the birth or to the open sea for entering/leaving a port before the experiment. In other words, he gets the mental workload for the ship handling before the experiment. We show the relationship between the reference value and the mental workload in equations (1) and (2).

$$\textit{If } \text{``Reference SAA value''} < \text{SAA value, } \textit{then } \text{``strain'' results} \quad (1)$$

$$\textit{If } \text{``Reference SAA value''} > \text{SAA value, } \textit{then } \text{``no strain'' results} \quad (2)$$

5.2 R-R Interval

We measure R-R interval, and R-R interval shows the heart rate on the spot. The heart rate is small for a long time R-R interval and large for a short time. For example, the R-R interval 1,000 [ms] means 60 heartbeats per minute. We show the relationship between the R-R interval and the mental workload in equations (3) and (4).

$$\textit{If } \text{``R-R interval decreases'', } \textit{then } \text{``strain'' results} \quad (3)$$

$$\textit{If } \text{``R-R interval value increases'', } \textit{then } \text{``no strain'' results} \quad (4)$$

On the other hand, we calculate "LF/HF value" every thirty seconds by Maximum Entropy Method (MEM) in order to evaluate the mental workload of the navigator based on the R-R interval. The thirty seconds produced by comparison between the LF/HF values and the professionals' subjective evaluation [22].

On the 'LF', 'HF' and 'LF/HF', Low Frequency (LF) and High Frequency (HF) are the frequency components of the R-R interval. The 'LF' value is from 0.04 to 0.15 Hz which reflects the sympathetic and the parasympathetic nervous systems. The 'HF' value is from 0.15 to 0.40 Hz which reflects the parasympathetic nervous system [23], [24]. The 'LF/HF' value is available to evaluate the sympathetic and parasympathetic nervous systems simultaneously.

We show the relationship between the LF/HF value and the mental workload in equations (5) and (6).

$$\textit{If } \text{``LH/HF value increases''}, \textit{ then } \textit{``strain'' results} \qquad (5)$$

$$\textit{If } \text{``LF/HF value decreases''}, \textit{ then } \text{``no strain'' results} \qquad (6)$$

5.3 Facial Temperature

We calculate the "Nasal-Forehead temperature (N-F temperature)" in order to evaluate the mental workload. The N-F temperature is the difference between the nasal temperature and the forehead temperature which are mean values of the data in one square centimeter (Equation (7)). In this paper, we use the difference between the forehead temperature and the nasal temperature as the N-F temperature, because our index should be available on a real ship where the subject walks in the bridge space. Our index is influenced by the temperature. We use the forehead temperature for the base value to evaluate the nasal temperature.

$$\text{N-F temp.} = \text{Mean (Nasal temp.)} - \text{Mean (Forehead temp.)} \qquad (7)$$

The part of the calculation areas are decided with the frame of the pair of spectacles as in Figure 8. We need to identify the measurement position in some way, and propose to utilize the frame of a pair of spectacles.

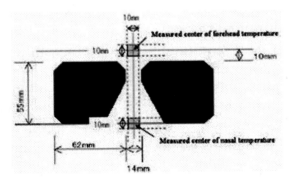

Fig. 8. The parts of the calculation areas to evaluate the N-F temperature utilizing the frame of a pair of spectacles

The size of the thermal image changes because the captain walks around on the bridge to do a careful lookout, and is never fixed in the cockpit like an airline pilot and car driver. It is difficult for us to fix accurately the distance between the captain (subject) and the thermography device, so we correct for this variable by using the frame of spectacles is worn by the subject.

We show the relationship between the N-F temperature and the mental workload in equations (8) to (10).

$$\textit{If } \text{``N-F temperature''} < 0, \textit{ then } \text{``strain'' results} \qquad (8)$$

$$\textit{If } \text{``N-F temperature''} \fallingdotseq 0, \textit{ then } \text{``normal'' results} \qquad (9)$$

$$\textit{If } \text{``N-F temperature''} > 0, \textit{ then } \text{``no strain'' results} \qquad (10)$$

The nasal temperature decreases during strain. Meanwhile, the forehead temperature doesn't change for various mental workload conditions [25], [26]. Therefore, the N-F temperature is better to evaluate the mental workload on the ship's bridge where it is difficult to control the temperature. We need a base value not influenced by the movement of the subject or space conditions.

6 Results

6.1 Simulator

We show the typical result of the N-F temperature and the LF/HF value when entering a port every thirty seconds where the LF/HF value is calculated every thirty seconds by MEM. This result is set for the same interval time (Figure 9).

In the Fig.9, the blue line and the pink line are the N-F temperature and the LF/HF value respectively, and 'A' to 'E' represents the events.

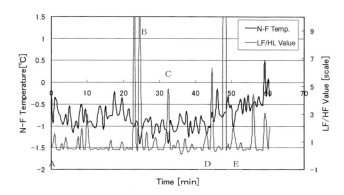

Fig. 9. The results of the N-F temperature and the LF/HF value while entering a port every thirty seconds

Events in Fig.9 are below;

A) Subject does sharp 360 degree lookout from his ship and checks the navigational instruments: radar, log speed, course, etc.

B) Subject's ship is at the entrance of harbor: it is between the breakwaters, and a narrow area; moreover, he needs to change the course by more than ninety degrees.

C) He turns the ship to the right with the rudder and the tug boat while checking his ship's position by radar.

D) The ship's engine motion is dead slow engine to decrease the ship speed; her heading turns right and she closes on the stationary ship a little bit.

E) He moves his ship near the berth using two tug boats while avoiding the other ship.

In Fig.9, the minimum N-F temperatures are taken at events B and C, when the captain begins to approach the port at the narrow area (event B) and control the approach and the speed while keeping a safe distance from the stationary ship (event C). In other words, the values of the N-F temperature show well when he makes decisions for safe navigation. It decreases at events D and E which is also a period of decision making for safe navigation. On the other hand, the LF/HF value is at maximum at event B and increasing value at events from C to E. Moreover, the values of the N-F temperature increase after event E, so the N-F temperature shows the broad trend of the mental workload.

6.2 Real Ship

We show the typical result of the SAA value, R-R interval, N-F temperature for *Miyazaki* leaving a port (Figure 10). SAA value is every two minutes, from 0 to 22 minutes.

In the Fig.10, the black line, the red line and triangle are the R-R interval, the N-F temperature and the SAA value respectively, and 'A' to 'D' represents the events. Events in Fig.10 are below;

A) Lets go all shore lines; leaves a birth at 3 minutes.

B) Fishing boat appears from the stern side at 10 minutes.

C) Pass the breakwater, port side at 16 minutes.

D) Pass the breakwater, starboard side at 20 minutes.

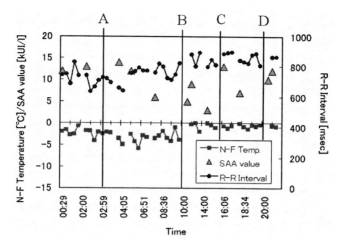

Fig. 10. The results of the SAA value, the N-F temperature and the R-R interval while leaving a port (*Miyazaki* port)

In Fig.10, the small values of N-F temperature and R-R interval are taken before 3 minutes including events 'A' when the captain begins to leave the port (event A). The SAA value also takes large values of more than 12 [kIU/l] (Around event 'A' in Fig.10). In event B, the captain's mental workload is stable- small SAA value, large N-F temperature and R-R interval, because the fishing boat does not create an avoidance situation. However, the captain's mental workload increases when the ship passes the narrow passage at the breakwater (event C). The event D is similar to C. The SAA value is taken at 13 [kIU/l] and the R-R interval decrease. The N-F temperature is stable (event D). Then, all indices show the tendency of decreasing the mental workload; the SAA value decrease; the R-R interval increase; the N-F temperature increase, from the port to open sea.

From the results, the SAA values, R-R interval and nasal temperature show well the mental workload when the navigator needs to make a lot of decisions for safe navigation. The trend is also similar.

7 Conclusions

We attempted to evaluate the mental workload of the navigator using the SAA value, nasal temperature and the R-R interval simultaneously on a real ship.

As a result, we can confirm the effect of the SAA value, the nasal temperature (N-F temperature) and the R-R interval as follows:

- The SAA value increases, and the N-F temperature and the R-R interval decrease when the navigator begins his mental workload for safe navigation: decision making for ship handling.

- The SAA value evaluates the mental workload of the ship navigator quickly on the spot in one to two minutes.
- The SAA value, the nasal temperature and the R-R interval are better indices as they are complementary to each other. These indices double/triple check each other, and show more accurate evaluation.

We confirmed that the SAA value, the nasal temperature and the R-R interval are better cross-indices for evaluating the mental workload of the ship navigator.

In future research; 1) we will evaluate the response time for the differences among individuals; 2) we will design student's training evaluation systems with which we can evaluate results in real time; and 3) we need more data to get more accurate results.

Acknowlegement

This research was supported by Japan Science and Technology Agency, ERATO, *Maenaka* Human-Sensing Fusion Project, University of Hyogo, Japan. We also thank the captain and the crew of the Training Ship *Fukae-maru* of Kobe University, Graduate School of Maritime Sciences, and Dr. Shin-ichi Wakida and Dr. Takashi Miyado of the National Institute of Advanced Industrial Science and Technology.

References

1. Kobayashi, H., Senda, S.: A Study on the Measurement of Human Mental Work-load in Ship Handling Using SNS Value. Journal of Japan Institute of Navigation (JIN) 98, 247–255 (1998)
2. Murai, K., Hayashi, Y., Nagata, N., Inokuchi, S.: The Mental Workload of a Ship's Navigator using Heart Rate Variability. Journal of Interactive Technology and Smart Education 1(2), 127–133 (2004)
3. Murai, K., Okazaki, T., Stone, L.C., Hayashi, Y.: A Characteristic of A Navigator's Mental Workload Based on Nasal Temperature. In: Proceedings of 2007 IEEE International Conference on Systems, Man and Cybernetics, pp. 3639–3643 (2007)
4. Murai, K., Wakida, S., Miyado, T., Fukushi, K., Hayashi, Y.: Basic Study of A Ship Navigator's Stress Using Salivary Amylase Activity. IEE of Japan Transaction on Electrical and Electronic Engineering 4(5), 680–682 (2009)
5. Deguchi, M., Wakasugi, J., Ikegami, T., Nanba, S., Yamaguchi, M.: Evaluation of Driver Stress Using Motor-vehicle Driving Simulator. IEE of Japan Transacton on sensors and micromachine 126, 438–444 (2006)
6. Inokuchi, S., Inoda, K., Tanabe, S., Nagata, N., Nakamura, T.: KANSEI Information Processing, pp. 1–2, 103–130. Ohmsha Ltd. (1994)
7. Gyoba, J., Hakoda, Y.: Psychology of Intellect and KANSEI, pp. 65–75. Fukumura Ltd. (2000)
8. Murai, K., Inokuchi, S.: New Concept of Navigational Information Including Human Navigator's KANSEI. Journal of Japan Society of Kansei Engineering 4(1), 71–76 (2004)

Evaluation of Professional Skill and *Kansei* Based on Physiological Index

9. Stevens, S.S.: Psychophysics. Transaction Books (1986)
10. Guilford, J.P.: Psychometric methods. McGraw-Hill, New York (1954)
11. Iwashita, T.: Measurement of image by SD assessment. Kawashima Books (1983)
12. Sayers, B.: Analysis of Heart Rate Variability. Ergonomics 16(1), 17–32 (1973)
13. Kato, Z., Okubo, T.: How to measurement of body function for beginner. Japan Publication Service (1993)
14. Rasmussen, J., Duncan, K., Leplat, J.: New Technology And Human Error. John Wiley & Sons, Chichester (1987)
15. Reason, J.: Human Error. Cambridge University Press, Cambridge (1990)
16. Campbell, R.D., Bagshaw, M.: Human Performance and Limitations in Aviation, 3rd edn. Blackwell Science, Malden (2002)
17. Murai, K., Hayashi, Y.: Evaluation of Ship's Navigator's Performance by Event Sampling Method. In: Abstract of 70th Japan Association of Applied Psychology Annual Conference, vol. 153 (2003)
18. Murai, K., Hayashi, Y.: A Few Comments on Lookout Method of Ship's Bridge Teammates by Work Sampling Method. In: Abstract of 71st Japan Association of Applied Psychology Annual Conference, vol. 83 (2004)
19. Bowditch, N.: The American Practical Navigator, 1995th edn., ch. 25, pp. 363–371. Deffense Mapping Agency Hydrographic/Topographic Center, Maryland (1995)
20. Murai, K., Hayashi, Y., Stone, L.C., Inokuchi, S.: Basic Evaluation of Performance of Bridge Resource Teams Involved in On-Board Smart Education: Lookout Pattern. Review of The Faculty of Maritime Sciences, Kobe University 3, 77–83 (2006)
21. Murai, K., Okazaki, T., Mitomo, N., Stone, L.C., Hayashi, Y.: Evaluation of Ship Navigator's Mental Workload Using Nasal Temperature and Heart Rate Variability. In: Proceedings of 2008 IEEE International Conference on Systems, Man and Cybernetics, pp. 1528–1533 (2008)
22. Moroi, T.: Analysis of Frequency Components for A Navigator's R-R Interval. Bachelor's Thesis, Kobe University of Mercantile Marime (2002)
23. Malik, M.: Heart Rate Variability. Circulation 93(5), 1043–1065 (1996)
24. Sayers, A.: Analysis of Heart Rate Variability. Ergonomics 16(1), 17–32 (1973)
25. Hayashi, Y., Takehara, T., Murai, K., Yano, Y.: Quantitative Evaluation of Ship Navigator's Mental Workload by Means of Facial Temperature. Journal of JIN 116, 213–218 (2007)
26. Sakamoto, R., Nozawa, A., Tanaka, H., Mizuno, T., Ide, H.: Evaluation of the Driver's Temporary Arousal Level by Facial Skin Thermogram -Effect of Surrounding Temperature and Wind on the Thermogram. IEE of Japan Transaction on Electronics, Information and Systems 126C(7), 804–809 (2006)

WLoLiMoT: A Wavelet and LoLiMoT Based Algorithm for Time Series Prediction

Elahe Arani, Caro Lucas, and Babak N. Araabi

School of Electrical and Computer Engineering, University of Tehran, Tehran, Iran
School of Cognitive Sciences, Institute for Research in Fundamental Sciences (IPM), Tehran, Iran
e.arani@ece.ut.ac.ir, lucas@ipm.ir, araabi@ut.ac.ir

Abstract. Time series prediction has been used widely in engineering, finance, economy, traffic and many other areas. It is an important tool in complex system identification. Many methods for identifying nonlinear systems by local linear model have been introduced; one of the most important methods used in these fields is LoLiMoT, which is an incremental learning algorithm. Wavelet analysis breaks the original signal into two parts: details and approximation. Wavelet analysis gives ability of using the prolonged intervals where more accurate low-frequency information is needed. Shorter intervals are used when we analize high-frequency information. In this paper an efficient method is presented to enhance accuracy of time series prediction that combines wavelet decomposition and Lo-LiMoT. The experimental results demonstrate the enhancement of accuracy of proposed method over the data set of sunspot number time series.

Keywords: Wavelet Decomposition, Local Linear Model Tree (LoLiMoT), Neurofuzzy Models, Time Series Prediction.

1 Introduction

Since only with a proper identification method can be modeled a phenomenon to control its efficiency, the identification of system plays an important role in control engineering. In other words, system identification allows us to build a mathematical model of a dynamical system based on measured data. In many proposed methods a set of input-output data is used offline to adjust the model parameters according to a performance index or a cost function. After that, the validity of yielded model is evaluated by some of validation tests. It is clear that the resulted model is time invariant.

On the other hand, whereas Local Linear NeuroFuzzy (LLNF) models as global estimators have the ability to solve stability/plasticity, so they are appropriate [1].

One of the most popular learning algorithms on the LLNF models is local Linear Model Tree (LoLiMoT) which is an incremental learning algorithm to adjusting the antecedent and consequent parameters of such models [2,3]. LLNF models with this incremental learning algorithm have been used for modeling and control of several complex systems and the this approach in description of complex phenomena has a very good performance. But LoLiMoT algorithm has some limitations in its structure. First of all, it is an incremental learning and one directional algorithm and there is no possibility to return former state during learning phase. Also, in this algorithm the area which should be divided into two areas is split from the middle.

In this article a novel learning method for offline system identification of time invariant nonlinear systems based on ordinary LoLiMoT algorithm is proposed. This method is based on LoLiMoT and multi-resolution analysis (MRA). To resolving the limitations of ordinary LoLiMoT algorithm in adjusting antecedent parameters of LLNF model, this algorithm uses wavelet transform. Therefore, using this algorithm, the efficiency of system identification and in result the performance of nonlinear time invariant control systems is improved. the advantage of proposed algorithm is that it has less number of local linear models in any part of principle time Series.

The organization of the paper is as follows: in sec. 2 we recall the main aspects of Local Linear NeuroFuzzy models and wavelet analysis and some limitations of Local Linear NeuroFuzzy models. In sec.3 we introduce the proposed prediction methodology to model time invariant nonlinear system. Sec.4 is dedicated to show the efficiency of proposed learning algorithm in modeling some nonlinear systems. We conclude with discussions in the last section.

2 Problem Definitions

In this section, first the offline system identification of nonlinear systems is briefly discussed. Then, Local Linear NeuroFuzzy (LLNF) with LoLiMoT algorithm is introduced as a powerful tool for the identification of nonlinear dynamics system. In the following we describe the multi-resolution analysis (MRA) technique to improve the prediction accuracy.

2.1 Motivation

In many applications, process has a time invariant nonlinear behavior. Several reasons such as system complexity might cause it to be nonlinear. If the system is strongly nonlinear, offline system identification methods which yield time invariant models, in specific time interval can be used for modeling such dynamics. However, such offline methods do not have any usage in fast changes of systems. Also in many applications, events of disturbances cannot be modeled or even be measured. Thus, being disturbances in such processes cause a complex nonlinear system is formed. In general if there is some of modeling data which are noisy (in the real issues there existed), is suitable to identify model with optimal methods. Practical time series are non static and traditional prediction model cannot give good prediction result. An efficient method to improve prediction accuracy is the multi-resolution analysis (MRA) technique. By wavelet transformation (WT), the component of time series is mapped to the different scales, to the some things which can be used in comparative prediction methods. The final prediction result ultimately is obtained by reconstruction of prediction result at different scales. The Local Linear NeuroFuzzy (LLNF) model with its incremental learning algorithm is a powerful tool for the identification of nonlinear systems which contains an offline training phase but it seems that is very time consuming. In this article, tried to propose an efficient training algorithm for LLNF models based on LoLiMoT learning algorithm which is suitable in application of chaotic systems.In general, this method improves the efficiency of LoLiMoT learning algorithm.

2.2 Local Linear NeuroFuzzy Models with LoLiMoT Learning Algorithms

The fundamental approach with Locally Linear NeuroFuzzy (LLNF) model is dividing the input space into small linear subspaces with fuzzy validity functions. Any produced linear part with its validity function can be described as a fuzzy neuron. Thus the total model is a NeuroFuzzy network with one hidden layer, and a linear neuron in the output layer which simply calculates the weighted sum of the outputs of locally linear neurons:

$$\hat{y} = \sum_{i=1}^{M} \hat{y}_i \, \emptyset_i(\underline{u}) \qquad (1)$$

Where

$$\hat{y}_i = w_{i_0} + w_{i_1} u_1 + w_{i_2} u_2 + \cdots + w_{i_p} u_p \qquad (2)$$

In this structure $\underline{u} = [u_1 u_2 \cdots u_p]^{\mathrm{T}}$ is the model input, M is the number of LLM neurons and w_{ij} denotes the LLM parameters of the ith neuron. The validity functions are chosen as normalized Gaussians; normalization is necessary for a proper interpretation of validity functions:

$$\emptyset_i(\underline{u}) = \frac{\mu_i(\underline{u})}{\sum_{j=1}^{M} \mu_j(\underline{u})} \qquad (3)$$

where

$$\begin{aligned}
\mu_i(\underline{u}) &= \exp(-\frac{1}{2}(\frac{(u_1 - c_{i_1})}{\sigma_{i_1}^2} + \ldots + \frac{(u_p - c_{i_p})}{\sigma_{i_p}^2})) \\
&= \exp(-\frac{1}{2}\frac{(u_1 - c_{i_1})}{\sigma_{i_1}^2}) \times \ldots \times \exp(-\frac{1}{2}\frac{(u_p - c_{i_p})}{\sigma_{i_p}^2})
\end{aligned} \qquad (4)$$

Each Gaussian validity function has two sets of adjustable parameters: centers (c_{ij}) and standard deviations (σ_{ij}), generally there are M×P parameters of the nonlinear hidden layer. Optimization or learning methods are used to adjust the two sets of parameters, the rule consequent parameters of the locally linear models

(w_{ij}) and the rule premise parameters of validity functions (c_{ij} and σ_{ij}). Global optimization of linear consequent parameters is simply obtained by least squares technique. The global parameter vector contains M×(p+1) elements:

$$w = \begin{bmatrix} w_{1p} & w_{1p} & \cdots & w_{1p} & w_{21} & w_{22} & \cdots & w_{M0} & \cdots & w_{Mp} \end{bmatrix} \tag{5}$$

And the associated regression matrix X for N measured data samples is:

$$\underline{X} = \begin{bmatrix} \underline{X}_1 & \underline{X}_2 & \cdots & \underline{X}_M \end{bmatrix} \tag{6}$$

where

$$\underline{X}_i = \begin{bmatrix} \emptyset_i(\underline{u}(1)) & u_1(1)\emptyset_i(\underline{u}(1)) & \cdots & u_p(1)\emptyset_i(\underline{u}(1)) \\ \emptyset_i(\underline{u}(2)) & u_1(2)\emptyset_i(\underline{u}(2)) & \cdots & u_p(2)\emptyset_i(\underline{u}(2)) \\ \vdots & \vdots & & \vdots \\ \emptyset_i(\underline{u}(N)) & u_1(N)\emptyset_i(\underline{u}(N)) & \cdots & u_p(N)\emptyset_i(\underline{u}(N)) \end{bmatrix} \tag{7}$$

Therefore, as we know least square estimation and optimal parameters is obtained as follow:

$$\underline{\hat{y}} = \underline{X}.\underline{\hat{w}} \tag{8}$$

where

$$\underline{\hat{w}} = (\underline{X}^T\underline{X})^{-1}\underline{X}^T\underline{y} \tag{9}$$

The LoLiMoT algorithm consists of an outer loop in which the rule premise structure is determined and a nested inner loop in which the rule consequent parameters are optimized by local estimation. This loop can be summarized as follows:

1. Start with an initial model: Construct the validity functions for the initial input space partitioning and estimate the LLM parameters.
2. Find worst LLM: Calculate a local loss function for each of the LLMs.
3. Check all divisions: For the worst LLM the following steps are carried out:
    3.1. Construction of the multi-dimensional membership functions for both generated hyper rectangles; Construction of all validity functions.
    3.2. Estimation of the rule consequent parameters for newly generated LLMs.
    3.3. Calculations of the loss function for the current overall model.
4. Find best division: The best division in step 3 is selected. The number of LLMs is incremented.
5. Test for convergence: If the termination criterion is met then stop, else go to step 2.

Figure 1 illustrates the operation of the LoLiMoT algorithm in the first four iterations for a two-dimensional input space.

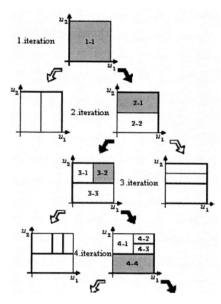

Fig. 1. First four iterations of LoLiMoT training algorithm

2.3 Wavelet Analysis

Fourier analysis consists of breaking up a signal into sine waves of various frequencies. Similarly, the wavelet analysis is the breaking up of a signal into shifted and scaled versions of the original (or mother) wavelet [4].

Wavelets allow a time series to be viewed in multiple resolutions. Each resolution reflects a different frequency. The wavelet technique takes averages and differences of a signal, breaking the signal down into spectrum. Each step of the wavelet transform produces two sets of values: a set of averages and a set of differences (the differences are referred to as wavelet coefficients). Each step produces a set of averages and coefficients that is half the size of the input data. The averages then become the input for the next step.

The discrete wavelet transform: Calculating wavelet coefficients at every possible scale is very time and space consuming. If we choose scales and positions based on powers of two (so-called dyadic scales and positions), the nour-analysis will be much more efficient, and just as accurate. We obtain such an analysis from the discrete wavelet transform (DWT). An efficient way to implement this scheme using filters was developed by Mallat (1998). This very practical filtering algorithm yields a fast wavelet transform. A signal is broken into two compo- nents, an approximation and a detail. If we repeat the procedure on the approximations, we obtain multiple DWT.

There are a wide variety of popular wavelet algorithms, including Daubechies wavelets [5], Mexican Hat wavelets and Morlet wavelets [6]. These wavelet algorithms have the advantage of better resolution for smoothly changing time series. But they have the disadvantage of being more expensive to calculate than the Haar [7] wavelets. The higher resolution provided by these wavelets is not worth the cost for financial time series, which are characterized by jagged transitions.

Mother Haar wavelet is defined as follows:

$$\varphi(t) = \begin{cases} 1 & 0 \le t < 1/2, \\ -1 & 1/2 \le t < 1, \\ 0 & \text{otherwise} \end{cases} \tag{10}$$

Where the scaling function is:

$$\emptyset(t) = \begin{cases} 1 & 0 \le t < 1 \\ 0 & \text{otherwise} \end{cases} \tag{11}$$

3 The Proposed Method

One of the most popular learning algorithms on the LLNF models is local Linear Model Tree (LoLiMoT) which is an incremental learning algorithm to adjusting the antecedent and consequent parameters of such models. LLNF models with this incremental learning algorithm have been used for modeling and control of several complex systems and the efficiency of this approach in description of complex phenomena is very good. However, LoLiMoT algorithm has some limitations in its structure. First of all, this is an incremental learning algorithm and it is not possible to return to some former state during learning phase. In other words, if a local linear region be formed there is not possible to go back and delete that region. Furthermore, in this algorithm splitting is always done in a way that the sub-regions be equal halves. The existance of these two limitations caused the motivation of modifying this algorithm. So if we make the system easier, we can partly overcome these limitations, whereas we can use the good performance of this algorithm in some applications such as prediction. In this paper, we use the wavelet analysis to reform this algorithm.

As we know, Wavelet analysis allows the use of long time intervals where more precise low-frequency information is needed, and shorter regions where we want high-frequency information. Thus by using the wavelet analysis we can break up the input signal where use these simple signals to LoLiMoT algorithm. Due to the using of wavelet analysis and LoLiMoT algorithm we called this method "WLoLiMoT". We will see that this proposed method has improved the efficiency of this algorithm using multi resolution wavelet.

The prediction method based on wavelet analysis and Local Linear Model Tree can be implemented as follows:

1) Decomposition of time series with Multi resolution analysis (MRA): obtain given time series $Q(t)$ by the wavelet decomposition of level j and approximation c_j and detail sections d_i (i=1,2,...,j). The main signal $Q(t)$ can be reconstructed by c_j and d_i :

$$Q(t) = c_j + \sum_j d_j \qquad (12)$$

2) Construction of prediction model using LoLiMoT: construct prediction model for the wavelet transformation approximation section and detail sections using LoLiMoT.

3) Reconstruction of prediction process: predicted values (approximation and detail sections) of time series Q(t) is obtained using local neurofuzzy predictor. Suppose that $\hat{c}_j$ and $\hat{d}_i$ (i=1,...,j) represent predicted values of approximation and detail sections respectively. Reconstruction of each section can be used as the final prediction result.

$$\hat{Q}(t) = \hat{c}_j + \sum_j \hat{d}_j \qquad (13a)$$

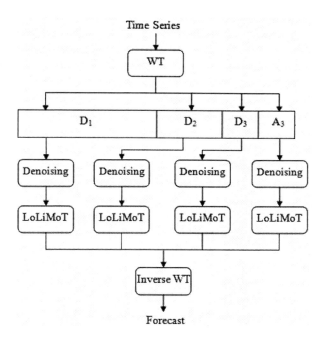

Fig. 2. The diagram of proposed method

4 Case Studies

Sun spots represent magnetic variable conditions in visible surface of the sun. They are very strong magnetic, relatively algid and therefore dim regions. The surface temperature when occurrence of a sun spot is about 3800 Kelvin degree against 5800 Kelvin degree normal temperature in the photosphere and its magnetic field is about 2000 to 4000 G against only a few G normal field intensity in the photosphere. The number of sun spots index R, based on Wolf criteria and definition of the number of Zurich sun spots is obtained by

$$R=K(10g+f) \tag{13b}$$

in which g is the number of group spots, f the number of single spots and K a constant depends on sensitivity and parameters of observatory and equipments.

Great deals of researches have been being done in studying the number of sun spots effect on different factors of human life. Scientists discuss about long term effect of this observation on the status of climate. This index is very good criteria of solar activity which is source of many space weather phenomena.

The data set is the number of sun spots which is publicly available from the World Data Center for Sunspot Index [8]. The first 2989 numbers of this data set and the remaining is used as training and testing data respectively.

At first, we have decomposed the number of sun spots time series to approximation section and detail sections for training and testing data. Here the Haar wavelet is used at three levels. Then, we construct the prediction model based on LoLiMoT for approximation section and detail sections shown in figure 2.

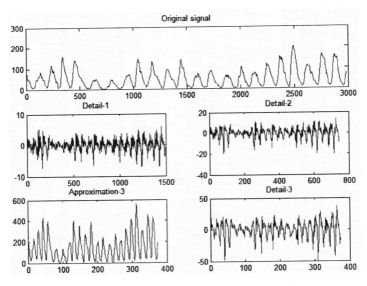

Fig. 5. The decomposition of training data time series

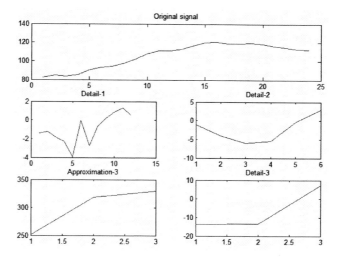

Fig. 6. The decomposition of test data time series

The figures from analysis of time series are shown in figure 5 and figure 6.

To evaluate the prediction efficiency is used the efficiency index NMSE (Normalized Mean Square Error), in which x_j, $\hat{x}_j$, $\bar{x}$ represent actual value, predicted value and average value respectively [9].

$$\text{NMSE} = \frac{\sum_j (x_j - \hat{x}_j)^2}{\sum_j (x_j - \bar{x})^2} \tag{14}$$

The figures derived from applying the proposed prediction method on the training and testing data are shown in below figures.

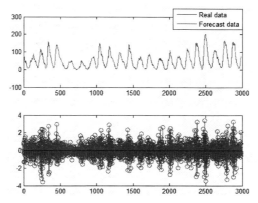

Fig. 7. Top: real value and predicted value of the training data using the proposed method, Bottom: prediction error

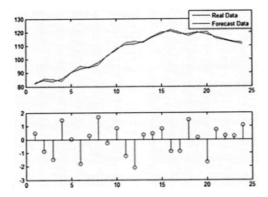

Fig. 8. Top: real value and predicted value of the test data using the proposed method, Bottom: prediction error

MSE for evaluating the accuracy of the proposed method is shown below

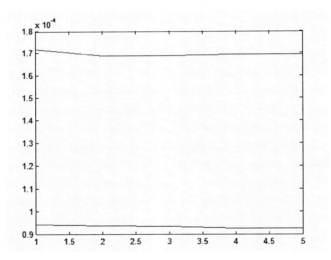

Fig. 9. MSE error for training and test data

The value of the error criterion resulted from the proposed model is compared with other methods in Table 1. As it is evident from Table 1 the error criterion using the proposed model is smaller than other methods.

Table 1. NMSE error for some methods

Method	NMSE
LoLiMoT [9]	0.046
RBF [9]	0.032
McNish-Linkon [10]	0.08
Sello [11]	0.34
WLoLiMoT	0.016

5 Conclusions

Many methods for identifying nonlinear systems by local linear model have been introduced; one of the most important methods used in these fields is LoLiMoT, which is an incremental learning algorithm. Wavelet analysis make capable to using the prolonged interval where more accurate low-frequency information is needed and shorter areas where we need high –frequency information. In this paper, the proposed method for predicting time series using MRA and LoLiMoT has higher accuracy, and also it doesn't increase response time. In fact, this method with decomposing the time series into its components breaks the complex system into simpler levels and then uses the prediction algorithm based on LoLiMoT. The proposed method can also be used to prediction of space weather phenomena, the power load forecasting, predicting equipment failure and secure predicting for power system.

References

1. Takagi, T., Sugeno, M.: Fuzzy Identification of Systems and its Applications to Modeling and Control. IEEE Trans. System, Man, Cybernetics SMC-15, 116–132 (1985)
2. Nelles, O.: Local linear model tree for on-line identification of time invariant nonlinear dynamic systems. In: Proc. International Conference on Artificial Neural Networks (ICANN), Bochum, Germany, pp. 115–120 (1996)
3. Nelles, O.: Nonlinear system identification. Springer, Berlin (2001)
4. Chui, C.K.: An Introduction to Wavelets. Academic Press, New York (1992)
5. Daubechies, I.: Ten Lectures on Wavelets. SIAM, Philadelphia (1992)
6. Mallat, S.: A Wavelet Tour of Signal Processing. Academic Press, New York (1998)
7. Haar, A.: Zur Theorie der orthogonalen Funktionensysteme. German, Mathematische Annalen 69(3), 331–371 (1910)
8. http://sidc.oma.be/index.php3
9. Gholipour, A., Araabi, B.N., Lucas, C.: Prediction chaotic time series using neural and neuro-fuzzy models: A comparison study. Neural Processing Letters 24, 217–239 (2006)
10. McNish, A.G., Lincoln, J.V.: Prediction of Sunspot Numbers. Transactions American Geophysical Union 30, 673 (1949)
11. Sello, S.: Solar cycle forecasting: a nonlinear dynamic approach. Astronomy and Astrophysics 377(1), 312–320 (2001)

Author Index

Abtahi, Amir 185
Ahmadi, H. 243
Ahzi, S. 3, 13
Araabi, Babak N. 345
Arani, Elahe 345

Baumgartner, D. 3
Berlik, Stefan 293
Boroomand, F. 301
Brück, Rainer 267

Cochez, M. 47
Czerner, Peter 321

Dadalau, Alexandru 79
Dumont, G. 243

Fereidunian, A. 301
Ferriol, M. 47
Fielding, Michael 215
Friederich, B. 47

George, D. 13
Gracio, J. 13

Hafla, Alexander 79
Hanoun, Samer 215
Hayashi, Yuji 331
Heinrich, Michael 109
Hessami, A.G. 227
Hildebrand, Lars 313
Hockauf, Matthias 93

Jamalabadi, H.R. 301

Kaufmann, J. 133
Klärner, Matthias 109
Kroll, Lothar 23, 67, 109, 133

Laachachi, A. 47
Lampke, Thomas 121
Lesani, H. 301
Lucas, C. 301
Lucas, Caro 203, 345

Mäder, Thomas 143, 155
Müller, Tobias 143
Montino, Ralf 321
Moore, M. 227
Mottahedi, Mahdi 79
Murai, Koji 331

Nahavandi, Saeid 215
Nendel, Sebastian 23
Nendel, Wolfgang 109
Nestler, Daisy 23, 59, 93, 121, 143, 155
Nibennaoune, Z. 13
Niroumand, Amir M. 171

Odenwald, S. 133

Paessler, E. 133
Ploch, Thomas 313
Podlesak, Harry 93

Rahman, R. Abdel 3
Remond, Y. 3, 13
Roder, Kristina 59
Ruch, D. 13, 47

Saadat, Mozafar 37
Saif, Mehrdad 171
Sassani, F. 243
Siebeck, Steve 93
Steger, Heike 23, 121

Tafreshi, R. 243
Toniazzo, V. 47
Tröltzsch, Jürgen 23, 67, 109

Verl, Alexander 79

Wagner, Swetlana 93
Walther, Marco 109
Wielage, Bernhard 23, 59, 93, 121, 143, 155
Wu, I-Chin 279

Zamani, M.A. 301
Zilouchian, Ali 185